T0210706

Lecture Notes in Artificial Intelligence 9513

Subseries of Lecture Notes in Computer Science

More information about this series at http://www.springer.com/series/1244

Luis Almeida · Jianmin Ji
Gerald Steinbauer · Sean Luke (Eds.)

RoboCup 2015:
Robot World Cup XIX

 Springer

Editors
Luis Almeida
University of Porto
Porto
Portugal

Gerald Steinbauer
Graz University of Technology
Graz
Austria

Jianmin Ji
University of Science and Technology
Hefei
China

Sean Luke
George Mason University
Fairfax, VA
USA

ISSN 0302-9743   ISSN 1611-3349   (electronic)
Lecture Notes in Artificial Intelligence
ISBN 978-3-319-29338-7   ISBN 978-3-319-29339-4   (eBook)
DOI 10.1007/978-3-319-29339-4

Library of Congress Control Number: 2015960427

LNCS Sublibrary: SL7 – Artificial Intelligence

Printed on acid-free paper

This Springer imprint is published by SpringerNature
The registered company is Springer International Publishing AG Switzerland

Preface

These are the proceedings of the 19th Annual RoboCup International Symposium, held on July 23, 2015, at the Anhui International Conference and Exhibition Center, in Hefei, China.

The Symposium is annually colocated with and associated with RoboCup. RoboCup is a venue to promote robotics and artificial intelligence by offering difficult challenge problems in robotics, simulation, and related areas. The Symposium augments this by providing a forum for the presentation of scientific research related to many aspects of robotics, artificial intelligence, computer vision, multiagent systems, simulation, and other areas, with a special focus on the various RoboCup competitions and demonstrations. This range of research is highly interdisciplinary, and is directly applied to nontrivial and cutting-edge tasks in robotics. The goal of the Symposium is to showcase important and vital work in these fields. This year RoboCup had 349 teams from 47 countries and regions in 12 leagues. Overall some 3000 participants and volunteers were involved, with over 34,000 spectators.

At the 2015 Symposium there were two tracks: a general track and a special development track meant to showcase open software and hardware contributions to the RoboCup community. The Symposium received a total of 39 submissions (31 in the general track, eight in the development track), which were reviewed by the 69 members of the Program Committee. All submissions received a minimum of three and an average of four reviews. In all, 20 papers were accepted: seven for oral presentation, thirteen for poster presentation. Additionally, the proceedings include papers detailing the efforts of champion teams from eleven different RoboCup competitions.

Associated with the Symposium were three other workshops and forums: the RoboCup 2015 Industry Forum, the Workshop on Educational Robotics (WEROB 2015), and the Demo Challenge and Workshop on Benchmarking Service Robots. The proceedings include one paper from the Benchmarking Workshop, entitled "Synthetical Benchmarking of Service Robots: A First Effort on Domestic Mobile Platforms", by Min Cheng, Xiaoping Chen, Keke Tang, Feng Wu, Andras Kupcsik, Luca Iocchi, Yingfeng Chen, and David Hsu.

The Symposium had four invited speakers: three keynotes and an invited presentation from AAAI-15. The presenters were:

- Wen Gao is Professor in the School of Electrical Engineering and Computer Science at Peking University. His presentation was entitled "A Smart Vision System for Robot Navigation".
- Yuki Nakagawa is founder and CEO of RT Corporation, which develops robotic systems. Her presentation was entitled "Force Management for Articulated Robots".
- Daniel Lee is Professor and director of the GRASP Lab in the Department of Electrical and Systems Engineering, at the University of Pennsylvania. His presentation was entitled "Machine Learning for Robots: Perception, Planning and Motor Control".

- The invited AAAI-15 presentation, aiming at strengthening ties between AAAI and the RoboCup Federation, was "From Goals to Behaviors: Automating the Search for Goal-Achieving Finite State Machines", by Siddharth Srivastava, and it was based on the AAAI-15 paper "Tractability of Planning with Loops", by Siddharth Srivastava, Shlomo Zilberstein, Abhishek Gupta, Pieter Abbeel, and Stuart Russell.

This year the Symposium offered two awards, one for Best Science Paper and one for Best Engineering Paper:

- The Best Science Paper award was shared by two papers. The first was "Language-Based Sensing Descriptors for Robot Object Grounding", by Guglielmo Gemignani, Manuela Veloso, and Daniele Nardi. The second was "Interactive Learning of Continuous Actions from Corrective Advice Communicated by Humans", by Carlos Celemin and Javier Ruiz-del-Solar.
- The Best Engineering Paper award went to "Evaluation of the RoboCup Logistics League and Derived Criteria for Future Competitions", by Tim Niemueller, Sebastian Reuter, Alexander Ferrein, Sabina Jeschke, and Gerhard Lakemeyer.

We would like to thank the Program Committee members, participants, and other reviewers for their efforts in making the Symposium a reality. Finally, we thank the general chair of RoboCup 2015, Xiaoping Chen, and the associate chairs, Wan-Mi Chen, Fei Liu, Feng Wu, and Shi Li, for their hard work and leadership. We hope to see you all at RoboCup 2016.

December 2015

Luis Almeida
Jianmin Ji
Gerald Steinbauer
Sean Luke

Organization

Symposium Co-chairs

Luis Almeida   IT/University of Porto, Portugal
Jianmin Ji   University of Science and Technology of China
Gerald Steinbauer   Graz University of Technology, Austria
Sean Luke   George Mason University, USA

Program Committee

Levent Akin   Bogazici University, Turkey
Hidehisa Akiyama   Fukuoka University, Japan
Jacky Baltes   University of Manitoba, Canada
Bikramjit Banerjee   University of Southern Mississippi, USA
Sven Behnke   University of Bonn, Germany
Reinaldo A.C. Bianchi   Centro Universitario da FEI, Brazil
Joydeep Biswas   Carnegie Mellon University, USA
Ansgar Bredenfeld   Dr. Bredenfeld UG, Germany
Esther Colombini   Aeronautics Institute of Technology, Brazil
Anna Helena Reali Costa   University of Sao Paulo, Brazil
Bernardo Cunha   University of Aveiro, Portugal
Klaus Dorer   University of Applied Sciences Offenburg, Germany
Christian Dornhege   University of Freiburg, Germany
Amy Eguchi   Bloomfield College, USA
Luis F. Lupian   La Salle University, USA
Alessandro Farinelli   Verona University, Italy
Alexander Ferrein   FH Aachen University of Applied Sciences, Germany
Maria Gini   University of Minnesota, USA
Fredrik Heintz   Linköping University, Sweden
Glen Henshaw   Naval Research Laboratory, USA
Koen Hindriks   Delft University of Technology, The Netherlands
Dirk Holz   University of Bonn, Germany
Luca Iocchi   Sapienza University of Rome, Itay
Nobuhiro Ito   Aichi Institute of Technology, Japan
Gerhard Kraetzschmar   Bonn-Rhein-Sieg University of Applied Sciences,
                                Germany
Gerhard Lakemayer   RWTH Aachen University, Germany
Nuno Lau   University of Aveiro, Portugal
Pedro Lima   University of Lisbon, Portugal
Fei Liu   Second Military Medical University, China

Daniel Lofaro	George Mason University, USA
Eric Matson	Purdue University, USA
Tekin Meriçli	Bogazici University, Turkey
Çetin Meriçli	Carnegie Mellon University, USA
Elena Messina	National Institute of Standards and Technology, USA
Zhao Mingguo	Tsinghua University, China
Eduardo Morales	National Institute of Astrophysics, Optics and Electronics, Mexico
Daniele Nardi	Sapienza University of Rome, Italy
Tatsushi Nishi	Osaka University, Tokyo
Itsuki Noda	National Institute of Advanced Industrial Science and Technology, Japan
Daniel Polani	University of Hertfordshire, UK
Luis Paulo Reis	University of Minho, Portugal
A. Fernando Ribeiro	University of Minho, Portugal
Thomas Röfer	German Research Centre for Artificial Intelligence
Subramanian Ramamoorthy	University of Edinburgh, UK
Raul Rojas	Free University of Berlin, Germany
Javier Ruiz-Del-Solar	University of Chile
Erol Sahin	Middle East Technical University, Turkey
Sanem Sariel	Istanbul Technical University, Turkey
Jesus Savage	National Autonomous University of Mexico
Paul Scerri	Carnegie Mellon University, USA
Raymond Sheh	Curtin University, Australia
Saeed Shiry	Amirkabir University of Technology, Iran
Alexandre Simões	São Paulo State University, Brazil
Elizabeth Sklar	University of Liverpool, UK
Mohan Sridharan	The University of Auckland, New Zealand
Peter Stone	University of Texas at Austin, USA
Luis Enrique Sucar	National Institute of Astrophysics, Optics and Electronics, Mexico
Komei Sugiura	National Institute of Information and Communications Technology, Japan
Yasutake Takahashi	University of Fukui, Japan
Yasinori Takemura	Nishinippon Institute of Technology, Japan
Flavio Tonidandel	Centro Universitario da FEI, Brazil
Markus Vincze	Vienna University of Technology, Austria
Arnoud Visser	University of Amsterdam, The Netherlands
Ubbo Visser	University of Miami, USA
Oskar von Stryk	Technische Universität Darmstadt, Germany
Jingchuan Wang	Shanghai Jiao Tong University, China
Alfredo Weitzenfeld	University of South Florida, USA
Franz Wotawa	Graz University of Technology, Austria
Feng Wu	University of Science and Technology of China

Additional Reviewers

Mehmet Akar	Bogazici University, Turkey
Philipp Allgeuer	University of Bonn, Germany
Okan Aşık	Bogazici University, Turkey
Kai Chen	University of Science and Technology of China
Eric Chown	Bowdoin College, USA
Hafez Farazi	University of Bonn, Germany
Reinhard Gerndt	Ostfalia Hochschule für angewandte Wissenschaften, Germany
Binnur Görer	Bogazici University, Turkey
Andreas Hertle	University of Freiburg, Germany
Nico Hochgeschwender	Bonn-Rhein-Sieg University of Applied Sciences, Germany
Victor Hofstede	University of Amsterdam, The Netherlands
Dirk Holz	University of Bonn, Germany
Ulrich Karras	Festo Didactic, Germany
Andras Kupcsik	National University of Singapore
Zhe Liu	Shanghai Jiao Tong University, China
Marcell Missura	University of Bonn, Germany
Walter Nowak	Locomotec, Germany
Soroush Sadeghnejad	Sharif University of Technology, Iran
Stefan Schiffer	RWTH Aachen University, Germany
Marco A.C. Simões	Bahia State University, Brazil
Zhixuan Wei	Shanghai Jiao Tong University, China

RoboCup 2015 Organizing Committee

General Chair

Xiaoping Chen   University of Science and Technology of China

Public Affairs Associate Chair

Wanmi Chen   Shanghai University, China

Site Setup Associate Chair

Fei Liu   Second Military Medical University, China

Industrial Activity Associate Chair

Feng Wu   University of Science and Technology of China

Youngster Affairs Associate Chair

Shi Li   Beijing Institute of Automation, China Academy of Sciences

Local Chair, Small Size League

Ren Wei   Zhejiang University, China

Local Chair, Middle Size League

Junhao Xiao   National University of Defense Technology, China

Local Chair, Humanoid League

Minguoa Zhao   Tsinghua University, China

Local Chair, Standard Platform League

Qijun Chen   Tongji University, China

2D Simulation League

Xiao Li, Local Chair   University of Science and Technology of China

Local Chair, 3D Simulation League

Baofu Fang   Hefei University of Technology, China

Local Chair, Rescue Robot League

Yingqiu Xu   Southeast University, China

Local Chair, Rescue Simulation League

Fu Jian   Central South University, China

Local Chair, @Home League

Yingfeng Chen   University of Science and Technology of China

Local Chair, @Work League

Junlin Xiong   University of Science and Technology of China

RoboCup 2015 International Advisory Committee

Itsuki Noda   National Institute of Advanced Industrial Science
                                and Technology, Japan
Chiangjiu Zhou   Singapore Polytechnic
A. Fernando Ribeiro   University of Minho, Portugal
Claude Sammut   The University of New South Wales, Australia

Contents

Poster Presentations

Development Track

Benchmarking Workshop

Best Paper Award for its Scientific Contribution

Language-Based Sensing Descriptors for Robot Object Grounding

Guglielmo Gemignani[1]([✉]), Manuela Veloso[2], and Daniele Nardi[1]

[1] Department of Computer, Control, and Management Engineering
"Antonio Ruberti", Sapienza University of Rome, Rome, Italy
{gemignani,nardi}@dis.uniroma1.it
[2] Computer Science Department, Carnegie Mellon University,
5000 Forbes Avenue, Pittsburgh, PA 15213, USA
veloso@cmu.edu

Abstract. In this work, we consider an autonomous robot that is required to understand commands given by a human through natural language. Specifically, we assume that this robot is provided with an internal representation of the environment. However, such a representation is unknown to the user. In this context, we address the problem of allowing a human to understand the robot internal representation through dialog. To this end, we introduce the concept of *sensing descriptors*. Such representations are used by the robot to recognize unknown object properties in the given commands and warn the user about them. Additionally, we show how these properties can be learned over time by leveraging past interactions in order to enhance the grounding capabilities of the robot.

Keywords: Sensing descriptors · Human-robot interaction · Natural language processing

1 Introduction

One of the main goals of RoboCup@Home is to develop an assistant and companion for humans in domestic settings. The idea is to allow robots to naturally interact with non-expert users in these environments. However, when first interacting with an unknown robot, users may be able to imagine its capabilities, while not knowing how to instruct it. For example, when seeing a manipulator in front of multiple blocks, a user might assume that the robot is able to manipulate them, while being unaware of the commands understood. To this end, several approaches have been proposed to enable untrained users to interact with robots through either constrained or unconstrained natural language.

In this paper, we consider the scenario in which a human needs to instruct an autonomous robot through a natural language interface. We assume that this robot is provided with a specific internal representation of the environment

G. Gemignani—contributed to this work while visiting Carnegie Mellon University.

L. Almeida et al. (Eds.): RoboCup 2015, LNAI 9513, pp. 3–15, 2015.
DOI: 10.1007/978-3-319-29339-4_1

that is unknown to the user. For example, a robot might be able to understand colors but not orderings. Also, it may be able to recognize shapes but may not be able to resolve spatial referring expressions. In this scenario, we address the problem of allowing a robot to recognize what object properties can or cannot be grounded with its current sensing capabilities. Moreover, we address the problem of learning new object attributes by exploiting past interactions with the user. While addressing these problems, our goal is to enable an untrained user to understand, through the interaction with the system, which object properties the robot can understand. These interactions can then be used to enhance the grounding capabilities of our robots. Note that in this paper, we will use the term grounding to refer to the concept of "physical symbol grounding" as defined by Vogt [1].

To this end, we contribute a novel approach that enables the robot to recognize unknown objects properties contained in the received commands and warn the user about them. We note that the majority of the techniques proposed in literature make the implicit assumption that if a robot can semantically parse an utterance, then it will be able to ground it. We believe that this assumption may not always hold, since while a robot may be able to correctly parse a sentence and extract its semantics, it may not be able to ground it due to a missing sensing capability. Hence, we internally represent sensing capabilities through *sensing descriptors* and use them to recognize unknown object properties. At this point, the robot can notify the user and request an alternative command. In addition, the robot can learn new object properties by leveraging these interactions with the user. After learning, the robot is able to execute the natural language commands, as in Fig. 1. Our contribution has been used to instruct several robots, including a Baxter manipulator able to perform complex manipulation tasks. In this paper, we describe all the components of our approach along with in depth illustrative examples with the Baxter manipulator robot.

Commands
- pick up the cubic block
- grab the yellow block
- touch the second block
- point at the left block
- take the narrow block

Fig. 1. Baxter manipulator robot used in our experiments and examples of commands that our approach is able to successfully execute.

In the remainder of the paper, we first present an overview of related work, focusing on past research on natural language processing applied to robotic systems. Next, we provide an overview of our natural language approach describing all of our contributions thoroughly. Then, we present an application of the approach to the case of a Baxter manipulator. This setting is then used to quantitatively evaluate the proposed approach. Finally, we conclude with a discussion of our contribution and remarks on future work.

2 Related Work

Our research topic is mostly related to the literature on natural language human-robot interaction. Initial studies on natural language understanding can be traced back to SHRDLU [2], a system able to process natural language instructions to perform actions in a virtual environment. Inspired by this system, multiple researchers extended SHRDLU's capabilities into real-world scenarios, soon starting to tackle related problems, including natural language on robotics systems.

Research has applied speech-based approaches to deploy robotic systems in a wide variety of environments. For example, these approaches have been used in manipulators [7–9], aerial vehicles [10], and wheeled platforms [11,12]. Moreover, several prototypes have been developed for social robots carrying out specialized tasks, such as attending as a waiter [13], as a receptionist [15] or as a bartender [14]. Some of these specialized tasks target industrial goals, such as assembly [16], or moving objects [17]. Dialog has also been used to teach robots how to accomplish a given task, such as giving a tour [18], delivering objects [19], or manipulating them [20]. Finally, other related works have combined speech-based approaches with other types of interactions [21,22]. Specifically, in the former work the authors have developed a theory of mind for the interacting user, built upon perspective taking, multi-modal communication, and a symbol grounding capability. Instead, in the latter case, the authors present a multi-modal approach for building on-line a semantic map of the environment.

More recently, several domain-specific systems that allow users to instruct robots through natural language have been presented in literature. For example, Kollar et al. [3] and MacMahon et al. [4] present different methods for following natural language route instructions by decoupling the semantic parsing problem from the grounding problem. In these works, the input sentences are first translated to intermediate representations, which are then grounded into the available knowledge base. Instead, Chernova et al. [5] show how to enable natural language human-robot interaction in a scenario of collaborative human-robot tasks, by data-mining past interactions between humans. Dzifcak et al. [6] address the problem of translating natural language instructions into goal descriptions and actions by exploiting $\lambda-$calculus. However, these approaches are not able to incrementally enhance their natural language understanding from the continuous interaction with the user.

Such a problem has been faced by Kollar et al. [23]. By exploiting the dialog with the user, in this work the authors present a probabilistic approach able to learn referring expressions for robot primitives and physical locations in a map. Our approach is inspired to this latter work. However, we make an additional step forward, assuming the user to be unaware of the capabilities and the internal representation of the robot. With this assumption, we propose an approach for allowing a robot to recognize unknown object properties contained in the received commands and warn the user about them. With this approach, on one hand the user is able to understand over time what a robot can and cannot ground. On the other hand, the robot can leverage past interactions to learn new object properties. The next section describes how our approach can achieve these goals.

3   Approach

In this section, first we motivate and introduce the concept of *sensing descriptors*. Next, we present our approach for human-robot natural language interaction based on such a concept. Finally, we show how the system can leverage previous interactions with users to learn previously unknown referring expressions for the objects perceived.

3.1   Sensing Descriptors

Usually, when dealing with robots and natural language user commands, a standard processing chain is adopted to decouple the semantic parsing problem from the grounding problem [3,4,19,23]. First, the natural language utterances are converted into text through an automatic speech recognition (ASR) system. Next, the text is converted into a specific representation that captures the semantic meaning of the uttered command. This conversion is carried out either through grammars or probabilistic approaches. The obtained representation is then "contextualized" in the operational environment through a grounding process. The final result is an executable function and a set of parameters passed as input.

In general, during this process each natural language command is grounded through a combination of sensing actions and queries to a given knowledge base. However, this approach does not take into account the sensing capabilities of the robot. In fact, we note that approaches proposed in literature often assume that if the robot can semantically parse an utterance, then it will be able to ground it. However, a robot may be able to correctly parse a sentence and extract its semantics without being able to ground the command due to a missing sensing capability. Hence, we propose to explicitly represent in the knowledge base these capabilities and use them to recognize parts of the commands that could only be grounded through a sensing ability not available to the robot. To this end, we introduce the concept of *sensing descriptor*.

Each sensing operation carried out by a robot can be defined as a function that takes as input a particular type of sensed data and outputs a value expressed in the internal representation of the robot. This value will be an instance of a sensing descriptor. Formally, a sensing operation can be defined as:

$$f_{sensing} : D \rightarrow SD$$

were D is the particular type of data sensed and SD is a specific sensing descriptor. As an example, let's consider the operation of sensing the color of a particular object. The input will be the RGB values of the pixels sensed by a camera. The output will be one or more instances of the sensing descriptor *color* (e.g., $[255, 0, 0]$ or *red* depending on the internal representation of the robot). These sensing descriptors can be used to check if the utterances received from a user can be grounded with the current capabilities of a robot. We perform this check as an intermediate step between the semantic parsing and the grounding process, as explained in the next section.

3.2 Human-Robot Natural Language Interaction

Figure 2 shows an overview of our processing chain. Specifically, this processing approach is divided in four consecutive steps. First, speech is converted into text using a free-form speech-to-text engine. Text from speech is confirmed by the user. Thus, without loss of generality, the input of the system is established as natural language text.

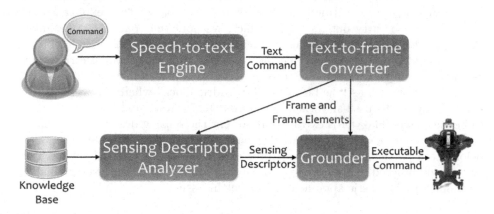

Fig. 2. Overview of our natural language processing chain. Instead of directly grounding the frames extracted from the commands, we perform an additional step that analyzes the sensing descriptors included in the frame elements.

Next, the text is converted into a specific representation characterizing the semantics of the sentence. This step is performed through the aid of specific grammars that drive the recognition process by attaching a proper semantic output to each grammar rule. The output has the form of a *semantic frame* representing a "situation" in the world (typically an action) inspired by the notion defined in the *Frame Semantics linguistic* theory [24]. The meaning of each frame can be enriched by semantic arguments, called *frame elements*, that are part of the input sentence. The output of the recognition process is then converted to a parse tree that contains syntactic and semantic information.

This information is used to instantiate a frame, similarly to [25]. As an example, the command "pick up the red block" will be mapped to the GETTING frame. The sub-phrase "the red block" will instead represent the specific frame element THEME, which represents the target of the GETTING action.

At this point, instead of directly grounding the frames in the internal representation of the robot we explicitly represent each sensing descriptor that can be recognized and grounded by the robot, also defining the range of values that it can assume. Formally, in our knowledge base we represent every sensing descriptor SD_i that can be handled by a robot, also representing all its possible known instances $sd_j \in SD_i$. We use these sensing descriptors to check if the obtained frame elements can be grounded with the current sensing capabilities of the robot. Hence, we define *sensing descriptor extractor* a function ψ able to extract from each frame element all the contained instances of sensing descriptors. Formally, if we define FE the frame element type, the *sensing descriptor extractor* can be specified as:

$$\psi : FE \rightarrow \{SD_1, SD_2, ...SD_n\}$$

where SD_i is a specific sensing descriptor extracted from the given frame element.

There are many possible ways to implement this function. In our approach, the sensing descriptor extractor is represented as a parser that exploit grammatical rules to carry out its task. In fact, we note that particular grammar elements are associated to referring expressions that require sensing capabilities to be grounded. Hence, for our specific case, we propose an heuristic rule that selects all the adjectives found in the frame elements. This rule is used to handle element frames such as "the big red and cylindric block" where the word "and" may or may not be used and where the words "big", "red", and "cylindric" need to be extracted. The words extracted represent the sensing descriptor instances that will be checked in the knowledge base. If all the instances are found to belong to a particular sensing descriptor expressed in the knowledge base, the system will proceed to ground the command, otherwise we either leverage dialog or adopt a probabilistic approach to resolve this issue.

3.3   Handling Unknown Sensing Descriptors

When an instance of a sensing descriptor is not found in the knowledge base two different scenarios may occur:

- The referring expression belongs to an unknown sensor descriptor and it has never been used by a user;
- The referring expression belongs to a sensor descriptor not available to the robot but it has been previously used to refer to a particular object.

In the first case, the robot asks the user to provide an alternative referring expression to the object, while keeping track of all the referring expressions used in the different interactions. These expressions are in fact the unknown sensing descriptors found in the frame elements that are not represented in the knowledge base of the robot. Since there is a limited amount of sensing properties that can be expressed, eventually the user will refer to the object in a way that the robot can understand, enabling the robot to associate all the previously used referring expression to the grounding found. To this end, we explicitly represent this association in the knowledge base by using the binary logic predicate *sd_grounding(X, Y)*. In this predicate, *X* represents the unknown instance of a sensing descriptor, while *Y* represents the grounding found through the multiple interactions with the user.

For example, let us consider a robot only able to recognize colors. Additionally, let us assume that a user needs to refer to a cylindric red object. At a first interaction a user might refer to the object as "the cylindric block". When warned by the robot that the term "cylindric" can not be understood, the user will provide a command with an alternative referring expression. Eventually, the user will refer to the object as "the red block", enabling the robot to correctly ground the expression and assert *sd_grounding(cylindric, block_1)* in his knowledge base. Figure 3 shows an example of a dialog between a manipulator robot and a user that our system is able to understand and the information that the robot is able to extract and store in the knowledge base.

User: pick up the cylindric block.
Robot: I do not understand "cylindric".
Are you referring to "blue"?
User: No.
Robot: Ok, please rephrase the command.
User: pick up the red block.

he_sd_assoc(cylindric, red)

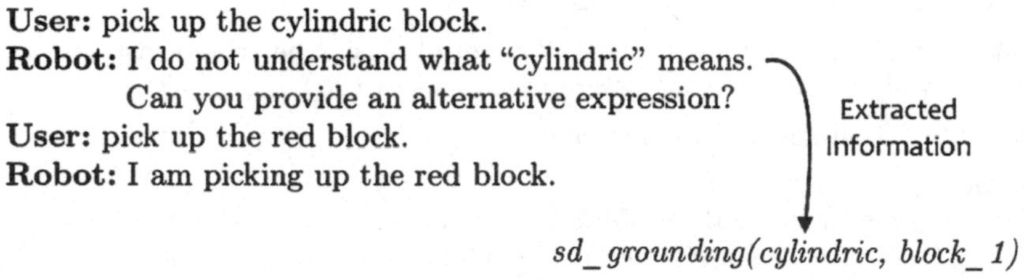

User: pick up the cylindric block.
Robot: I do not understand what "cylindric" means.
Can you provide an alternative expression?
User: pick up the red block.
Robot: I am picking up the red block.

Extracted
Information

sd_ grounding(cylindric, block_ 1)

Fig. 3. Example of a dialog between the robot and a user that our system is able to understand and the information extracted and stored in the knowledge base.

To each association instance in the knowledge base, a number is also attached to keep track of how many times the referring expression has been used to refer to a particular object. This counter is needed to handle the alternative scenario that may occur. In this second scenario, a referring expression belonging to an unknown sensor descriptor has been previously used to refer to a particular object. In this case, we adopt a probabilistic approach to ground the expressions. Specifically, if we define KB as the knowledge base available to the robot, R the referring expression being analyzed and G the possible groundings for it, we can obtain the most probable grounding by selecting the one that maximizes Bayes rule:

$$p(G|R; KB) = \frac{p(R|G; KB) \cdot p(G; KB)}{\sum_R p(R|G; KB) \cdot p(G; KB)}.$$

Here, the prior over groundings $p(G; KB)$ is computed by looking at the counts of each element of G in the knowledge base. The other term $p(R|G; KB)$ is instead obtained by counting the number of times a particular referring expression has been used to refer to a particular grounding, and dividing by the overall number of referring expressions used for the same grounding. Formally, if we define *count* the function that returns the number of times that a particular association has been encountered, we can compute $p(R|G; KB)$ as:

$$p(R|G; KB) = \frac{count(association(R, G))}{\sum_R count(association(R, G))}.$$

After having grounded the expressions, we allow the user to give a feedback to the robot to update the counter attached to each association instance. Algorithm 1 reports the overall natural language processing approach. Specifically, the algorithm takes as inputs the natural language command expressed as text and a specific knowledge base. The command is first analyzed to obtain its representation in terms of frames and frame elements (line 3). Next, the sensing descriptor instances are extracted from each frame elements through the sensing descriptor extractor ψ (line 5). Once extracted, the instances are checked against the

---

Algorithm 1. Ground Command

Input: Text command C, knowledge base KB

Data: Frame f, set of frame elements FE, set of sensing descriptor instances SD, set of unknown sensing descriptor instances USD

Output: Executable action function Φ

1 **begin**
2   // Extract frames and set of frame elements
3     $f, FE \leftarrow$ extractFramesAndFrameElements(C)
4   // Extract the set of sensing descriptor instances
5     $SD \leftarrow \psi(FE)$
6   // Select unknown sensing descriptor instances
7     $USD \leftarrow$ selectUnknownInstances(SD, KB)
8   **if** $USD \neq \{\}$ **then**
9   // Exploit Dialog and Previous Experience to ground command
10       $\Phi \leftarrow$ handleUnknownSensingDescriptors(USD, SD, KB, f, FE)
11   **else**
12   // Otherwise normally ground command
13       $\Phi \leftarrow$ ground(f, FE)
14   **return** Φ
15 **end**

available knowledge base to find any that cannot be grounded with the current sensing capabilities of the robot (line 7). If an unknown instance is found, the robot exploits dialog and the previous knowledge acquired to assign a grounding to the referring expressions (line 10). Otherwise, the command is grounded into the knowledge base available to the robot to obtain the final executable function (line 13).

4   Experimental Evaluation

In this section we describe in detail how the presented approach has been deployed on a Baxter manipulator robot able to manipulate a set of blocks placed in front of it. This setting has been used to quantitatively evaluate our proposed approach. Since the evaluation space of the experiment was large and generating results with humans was extremely time consuming, the experiments were conducted by using a simulator faithful to the chosen setting[1]. A representative sample of the scenarios described in the paper was successfully run on the manipulator interacting with humans, achieving results that are consistent with those reported in the following sections.

4.1   Setup

Baxter has two 7 degree of freedom arms, cameras on both arms, and a mounted Microsoft Kinect. Baxter has been programmed to perform the actions touch, grab, move, point to, and push. These primitives are used to manipulate a set of blocks located on a table in front of the robot. The manipulated blocks have different shapes and colors. Additionally, each block has a unique id, associated with a specific QR code. Given this setting, we considered the sensing descriptors shown in Table 1. Specifically, five different blocks were considered:

– A short, wide, triangular, blue block;
– A short, narrow, cubic, brown block;
– A short, wide, bridge-shaped, yellow block;
– A tall, narrow, rectangular, green block;
– A tall, narrow, cylindric, red block.

Additionally, these blocks were associated with the number one through five, respectively. Figure 1 shows the described scenario.

Before accepting commands, the robot was allowed to analyze the scene in order to accumulate knowledge about the operational environment. This knowledge was stored in the form of logic predicates in a knowledge base. The spoken commands given to the robot were converted into text through a free-form ASR[2]. For this particular scenario, a dedicated grammar was developed to convert the natural language commands to the previously described frame representation.

---

[1] https://github.com/RethinkRobotics/sdk-docs/wiki/Baxter-simulator.
[2] The Google free-form ASR has been used.

Table 1. Sensing descriptors considered in the chosen scenario and possible values.

Sensing descriptors	Possible values
color	{blue, brown, yellow, green, red, orange, purple}
shape	{triangular, cubic, bridge-shaped, rectangular, cylindric}
block id	{first, second, ..., fifth}
height	{short, tall}
width	{narrow, wide}
spatial location	{left, center, right}

To extract the sensing descriptors from the frame elements, a POS Tagger[3] was used to grammatically analyze the words in the command. Particularly, we adopted the heuristic of extracting the adjectives related to target objects, considering them instances of a specific sensing descriptor. With this approach we were able to allow users to understand how to instruct the robot while interacting with it.

4.2 Approach Evaluation

In order to show the effectiveness of our algorithm, we compared our approach with an algorithm commonly used in literature. Specifically, the chosen two-step approach first converts the received commands to frames exploiting grammars. Then, it directly grounds the commands without exploiting any information about sensing descriptors. When the algorithm receives a command that can be grounded to multiple targets (e.g., "touch the narrow block" in this scenario), it selects a random target between the possible ones.

The two approaches have been tested by first generating all the possible commands that can be given to the robot in this setting. Figure 1 shows some example commands generated. Next, 50 commands were randomly chosen and incrementally given in input to the robot. When the robot wasn't able to understand an object attribute, the property was changed with another one not yet used. This process was repeated until the robot understood the command. Such an operation has been carried out for both approaches and averaged for 100 times by varying the number of sensing descriptors known by the robot. For each run we measured the cumulative number of interactions needed to execute all the 50 commands. Figure 4 shows the results obtained in the experiment.

From the graph, it can be noticed that on average our algorithm required significantly less interactions to ground the randomly chosen commands. Moreover, it is worth noticing the effects of the different available levels of information on the two approaches. In fact, when the two robots were capable of understanding and grounding most of the used sensing descriptors, the two approaches had a

---

[3] We exploited the Stanford POS Tagger to extract the sensing descriptor instances from the frame elements.

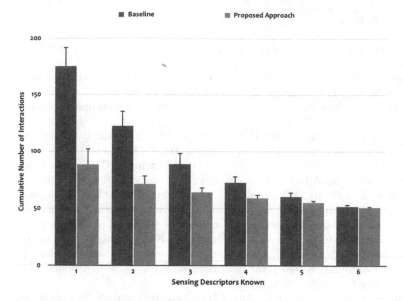

Fig. 4. Results for the experiment performed on both processing chains averaged for 100 times by varying the number of sensing descriptors known by the robot.

comparable result. Instead, when a lower amount of information was available, our approach greatly outperformed the other one, leading to a decrease in interactions needed to understand the command, up to approximately 50 % in the chosen scenario.

5   Conclusion

In this paper, we considered an autonomous robot provided with an internal representation of the environment, unknown to a user interacting with it through natural language. In this setting, we addressed the problem of allowing humans to understand the internal representation of the robot through dialog. Moreover, we enabled our robot to learn previously unknown object properties leveraging the past interactions with the user. We successfully deployed our approach on a Baxter manipulator robot able to carry out tasks assigned by several users through natural language. Specifically, our experiments report in-detail the performance of our algorithm in this scenario, suggesting an improvement in the grounding effectiveness compared to another commonly used approach.

As a future work, we are studying extensions of the proposed approach. In fact, as a long term goal, we would like to generalize the approach allowing our robots to not only recognize unknown object properties but also every unknown concept contained in the received commands.

References

1. Vogt, P.: The physical symbol grounding problem. Cogn. Syst. Res. **3**(3), 429–457 (2002)
2. Winograd, T.: Procedures as a representation for data in a computer program for understanding natural language. Technical report (1971)
3. Kollar, T., Tellex, S., Roy, D., Roy N.: Toward understanding natural language directions. In: HRI (2010)
4. MacMahon, M., Stankiewicz, B., Kuipers, B.: Walk the talk: connecting language, knowledge, and action in route instructions. In: AAAI (2006)
5. Chernova, S., Orkin, J., Breazeal, C.: Crowdsourcing HRI through online multiplayer games. In: AAAI Fall Symposium on Dialog with Robots (2010)
6. Dzifcak, J., Scheutz, M., Baral, C., Schermerhorn, P.: What to do and how to do it: translating natural language directives into temporal and dynamic logic representation for goal management and action execution. In: ICRA (2009)
7. Zuo, X., Iwahashi, N., Taguchi, R., Funakoshi, K., Nakano, M., Matsuda, S., Sugiura, K., Oka, N.: Detecting robot-directed speech by situated understanding in object manipulation tasks. In: International Symposium of Robots and Human Interactive Communication (2010)
8. Spangenberg, M., Henrich, D.: Towards an intuitive interface for instructing robots handling tasks based on verbalized physical effects. In: The 23rd IEEE International Symposium on Robot and Human Interactive Communication (2014)
9. Connell, J.H.: Extensible grounding of speech for robot instruction. In: Robots that Talk and Listen (2014)
10. Kollar, T., Tellex, S., Roy, N.: A discriminative model for understanding natural language route directions. In: AAAI Fall Symposium on Dialog with Robots (2010)
11. Kruijff, G., Zender, H., Jensfelt, P., Christensen, H.: Situated dialogue and spatial organization: what, where.. and why. Int. J. Adv. Robot. Syst. **4**(2), 125–138 (2007)
12. Bastianelli, E., Bloisi, D.D., Capobianco, R., Cossu, F., Gemignani, G., Iocchi, L., Nardi, D.: On-line semantic mapping. In: ICAR (2013)
13. Bannat, A., Blume, J., Geiger, J.T., Rehrl, T., Wallhoff, F., Mayer, C., Radig, B., Sosnowski, S., Kühnlenz, K.: A multimodal human-robot-dialog applying emotional feedbacks. In: Ge, S.S., Li, H., Cabibihan, J.-J., Tan, Y.K. (eds.) ICSR 2010. LNCS, vol. 6414, pp. 1–10. Springer, Heidelberg (2010)
14. Stiefelhagen, R., Ekenel, H., Fugen, C., Gieselmann, P., Holzapfel, H., Kraft, F., Nickel, K., Voit, M., Waibel, A.: Enabling multimodal human-robot interaction for the karlsruhe humanoid robot. IEEE Trans. Rob. **23**(4), 840–851 (2007)
15. Nisimura, R., Uchida, T., Lee, A., Saruwatari, H., Shikano, K., Matsumoto, Y.: Aska: receptionist robot with speech dialogue system. In: International Conference on Intelligent Robots and Systems (2002)
16. Foster, M.E., Giuliani, M., Isard, A., Matheson, C., Oberlander, J., Knoll, A.: Evaluating description and reference strategies in a cooperative human-robot dialogue system. In: International Joint Conference on Artificial Intelligence (2009)
17. Tellex, S., Kollar, T., Dickerson, S., Walter, M.R., Banerjee, A.G., Teller, S.J., Roy, N.: Understanding natural language commands for robotic navigation and mobile manipulation. In: AAAI (2011)
18. Rybski, P., Yoon, K., Stolarz, J., Veloso, M.: Interactive robot task training through dialog and demonstration. In: HRI (2007)
19. Gemignani, G., Bastianelli, E., Nardi, D.: Teaching robots parametrized executable plans through spoken interaction. In: AAMAS (2015)

20. Gemignani, G., Klee, S.D., Nardi, D., Veloso, M.: On task recognition and generalization in long-term robot teaching. In: AAMAS (2015)
21. Lemaignan, S., Alami, R.: Talking to my robot: from knowledge grounding to dialogue processing. In: Human-Robot Interaction (2013)
22. Bastianelli, E., Bloisi, D.D., Capobianco, R., Cossu, F., Gemignani, G., Iocchi, L., and Nardi, D.: On-line semantic mapping. In: International Conference on Advanced Robotics (2013)
23. Kollar, T., Perera, V., Nardi, D., Veloso, M.: Learning environmental knowledge from task-based human-robot dialog. In: ICRA (2013)
24. Fillmore, C.J.: Frames and the semantics of understanding. Quad. di Semantica 6(2), 222–254 (1985)
25. Thomas, B.J., Jenkins, O.C.: Roboframenet: verb-centric semantics for actions in robot middleware. In: ICRA (2012)

Interactive Learning of Continuous Actions from Corrective Advice Communicated by Humans

Carlos Celemin$^{(\boxtimes)}$ and Javier Ruiz-del-Solar

Department of Electrical Engineering and Advanced Mining Technology Center,
Universidad de Chile, Santiago, Chile
{carlos.celemin,jruizd}@ing.uchile.cl

Abstract. An interactive learning framework that allows non-expert humans to shape a policy through corrective advice, using a binary signal in the action domain of the robot/agent, is proposed. One of the most innovative features of COACH (COrrective Advice Communicated by Humans), the proposed framework, is a mechanism for adaptively adjusting the amount of human feedback that a given action receives, taking into consideration past feedback. The performance of COACH is compared with the one of TAMER (Teaching an Agent Manually via Evaluative Reinforcement), ACTAMER (Actor-Critic TAMER), and an autonomous agent trained using SARSA(λ) in two reinforcement learning problems: ball dribbling and Cart-Pole balancing. COACH outperforms the other learning frameworks in the reported experiments. In addition, results show that COACH is able to transfer successfully human knowledge to agents with continuous actions, being a complementary approach to TAMER, which is appropriate for teaching in discrete action domains.

Keywords: Robot learning · Interactive learning · Human teachers · Human feedback in action domains · Ball dribbling · Robot soccer

1 Introduction

The use of computational/machine learning techniques such as Reinforcement Learning (RL) allows robots, and agents in general, to address complex decision-making tasks. However, one of the main limitations of the use of learning approaches in real-world problems is the large number of learning trials required to learn complex behaviors. This can make prohibitive the use of many learning approaches in problems such as autonomous driving, x-copter flight control, or soccer robotics, where the implementation of learning trials with real robots, in the real world, may have a high cost.

This drawback can be addressed by using human feedback during learning, i.e., the learning process can be assisted by a human teacher who supervises the agent-environment interaction. There are two main schemes for using human feedback in order to modify the policy of a learning agent: a first one in which the trainer indicates to the agent what to do, i.e., human feedback in the actions domain [1–6]; and a second one in which the trainer evaluates the actions through rewards and punishments, i.e., human feedback in the evaluative domain [7–14].

© Springer International Publishing Switzerland 2015
L. Almeida et al. (Eds.): RoboCup 2015, LNAI 9513, pp. 16–27, 2015.
DOI: 10.1007/978-3-319-29339-4_2

In this context, the main goal of this article is to propose COACH (COrrective Advice Communicated by Humans), a new interactive learning framework that borrows elements from both schemes. COACH is inspired by the Shaping paradigm [16], which allows to interactively train an agent through signals of positive and negative reinforcement. But, as in the *Advice Operators* paradigm [6], the human feedback indicates the agent how the action has to be modified (increased or decreased).

Thus, COACH is based on the TAMER framework [15], but instead of using human feedback for evaluating the results of the action as TAMER does, human feedback is given in the action domains as in the Advice Operators paradigm [6], but without using an off-line and data-driven based supervised learning process. COACH manages the human feedback and the interactive update of the policy in a similar way that TAMER. One of the most innovative features of COACH is a mechanism for adaptively adjusting the amount of human feedback that a given action receives, taking into consideration past feedback. An additional feature of COACH is the possibility of providing human-feedback using either a keyboard or hand-gesture interface during training.

The proposed learning framework is validated and compared with TAMER, ACTAMER [17], and a classical SARSA(λ) method, in two problems: (i) ball dribbling with humanoid robots [18], and (ii) the very well-known Cart-Pole problem [19]. COACH outperforms all other learning frameworks in the reported experiments.

The paper is organized as follows. In Sect. 2 the COACH learning framework is presented, and in Sect. 3 the hand-gesture visual interface is described. The experimental validation and conclusions are given in Sects. 4 and 5, respectively.

2   The COACH Learning Framework

The COACH learning framework uses human binary feedback as a correction in the action domain, in order to update the current policy for the state wherein that action was executed. The trainer has to provide its feedback immediately after the execution of the action to be learnt.

COACH lets the trainer to shape the policy of an agent through occasional feedback. The method updates a policy model based on a supervised learning strategy supported by four main modules: *Human Feedback Modeling*, which characterizes the sequence of pieces of human advice, and determines how much feedback must be added to the executed action; *Policy Supervised Learner* and *Human Feedback Supervised Learner*, which updates the parameters of the policy model and the human feedback model, respectively; and *Credit Assigner*, which handles the time delay of human feedback (see Fig. 1).

In COACH, when the *Policy* module observes the state vector $\vec{s}$, it executes a continuous action $P(\vec{s})$ according to the policy model $P : S \rightarrow \mathbb{R}$ (this is a difference regarding the TAMER modeling, which bases the policy on a human trainer's reinforcement function from the state-action space $H_{TAMER} : S \times A \rightarrow \mathbb{R}$). Then, the human trainer observes the effect of the action in the environment, and gives an advice h. The signal h is the binary feedback (+1 or -1), which states how the current executed action has to be modified for that state $\vec{s}$ (increase or decrease its value, respectively).

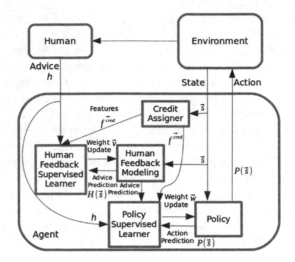

Fig. 1. Block diagram of the COACH learning framework.

The state $\vec{s}$, the executed action $P(\vec{s})$, and the human feedback h, are taken by the learning modules for updating the parameters of P (weights vector $\vec{w}$). Then, in the next time step, P has a new parameters set. When the trainer does not provide any feedback signal, h is taken as zero. The trainer is only allowed to give a binary correction, because COACH works under the assumption that a person cannot estimate the exact value of an appropriate correction, human just provide a trend of the modification (e.g. more/less force, velocity, energy, etc.).

Policy: The policy model P can be implemented using any function approximator. COACH uses a linear model of Gaussian features as TAMER and continuous SARSA (in general the Radial Basis Functions (RBF) features are the most used for function approximation in RL [19, 20]). The expression for $P(\vec{s})$ is the inner product between the $\vec{w}$ vector and the features vector $\vec{f}$ mapped by the Gaussian Kernels from the state space described by the state vector $\vec{s}$:

$$P(\vec{s}) = \vec{w}^T \cdot \vec{f} \tag{1}$$

The $\vec{w}$ vector is updated using a gradient descend approach as:

$$\Delta w_l = \alpha \cdot error \cdot \frac{\partial P(\vec{s})}{\partial w_l} = \alpha \cdot error \cdot f_l \tag{2}$$

$$error = h \cdot e \tag{3}$$

with e a constant error magnitude, h the sign of the error, given by he human feedback signal, and l the weight's/feature's index.

Human Feedback Modeling and Supervised Learner: The trainer intentions, observed in the binary feedback signal, can be considered a source of information that not only provides the sign (direction) of the corrections, but also its magnitude. Hence, in the COACH framework a model of the human feedback $H : S \rightarrow \mathbb{R}$ is built, which characterizes the human feedback signal over each region of the state space. As in the case of the policy model P, a linear parameterization of RBF features is used for representing the human feedback model H. Therefore, two Supervised Learner modules are required in the framework, one for P and one for H (see Fig. 1).

In the proposed modeling, sequences of feedback signals with constant sign over a specific state $(\vec{s_a})$, would mean the trainer suggest a large change of the magnitude of the associated action $P(\vec{s_a})$. On the other hand, sequences of alternating values of the sign in the human feedback would mean that the trainer is trying to provide a finer change around a given set point. Thus, using the information of H for computing an adaptive learning rate is appropriate for avoiding the dilemma of setting the magnitude of the error e either large (it does not let to perform fine adjustments) or small (it does not let to carry out large corrections quickly).

The model of the human feedback H, is built using the same features vector $\vec{f}$ of P, and a weight vector $\vec{v}$ as:

$$H(\vec{s}) = \vec{v}^T \cdot \vec{f} \tag{4}$$

Both P and H models map the same state space, and are based on the same kind of function approximator. Also, their respective *Supervised Learners* use the human feedback signal for updating the parameters. However, the H model is updated using the prediction error based on the difference of h and $H(\vec{s})$. Therefore, using a gradient descend approach, the weights associated to the H model are updated as:

$$\Delta v_l = \beta \cdot (h - H(\vec{s})) \cdot \frac{\partial H(\vec{s})}{\partial v_l} = \beta \cdot \left(h - \vec{v}^T \cdot \vec{f} \right) \cdot f_l \tag{5}$$

with β the learning rate and l the weight's/feature's index.

Then, the adaptive learning rate of the policy model learning process is computed as:

$$\alpha(\vec{s}) = |H(\vec{s})| + bias = \left| \vec{v}^T \cdot \vec{f} \right| + bias \tag{6}$$

where *bias* is the default value of the learning rate.

The magnitude of $|H(\vec{s})|$ is close to 1 when most of the last human feedback signals for a specific state $\vec{s_a}$ have the same value (either +1 or −1). On the contrary, alternating values of the feedback signal decrease the magnitude of $|H(\vec{s})|$. Hence, $\alpha(\vec{s})$ is set to a large value when feedback signals of constant value are received, and $\alpha(\vec{s})$ is set to a smaller value when feedback signals of alternating value are received.

Finally, the weights associated to the P model are now updated using $\alpha(\vec{s})$ as:

$$\Delta w_l = \alpha(\vec{s}) \cdot error \cdot \frac{\partial P(\vec{s})}{\partial w_l} = \alpha(\vec{s}) \cdot error \cdot f_l \tag{7}$$

with l the weight's/feature's index.

Credit Assigner: The corrective advice has to be given after the agent executes each action. But in decision-making problems of high frequency, a human trainer is not able to assess the effect of each action at each time step, this produce a delay between the action execution and the human response. The Credit Assigner proposed in TAMER, approaches this problem by associating the feedback not only to the last state-action pair, but to a past window of pairs. Each state-action pair is weighted with the corresponding probability that characterizes the human delay. COACH uses the TAMER's credit assigner; hence this module computes a new features vector $\overrightarrow{f^{cred}}$ for replacing the original vector $\vec{f}$ in the H and P update process.

Complete Algorithm: The H model is used for supporting the *Policy Supervised Learner* module, which updates the P model. The *Credit Assigner* module takes the states vector and computes credit assignments based on the history of past states. The Supervised Learners modules do not read directly the state vector $\vec{s}$, but instead take the features vector provided by the *Credit Assigner* module. This features vector is the average weighted sum of the past features.

Algorithm 1: Learning from Corrective Advice of a human trainer, using an adaptive learning rate for the policy update (complete framework).

1:      $e \leftarrow constant$ // error magnitude
2:      $\beta \leftarrow constant$ // learning rate
3:      $bias \leftarrow constant$ // offset for α
4:      **while** *true* **do**
5:          $\vec{s} \leftarrow getStateVec()$
6:          $\vec{f_1} \leftarrow getFeatures(\vec{s})$
7:          $P(\vec{s}) \leftarrow \vec{w}^T \cdot \vec{f_1}$
8:          $TakeAction(P(\vec{s}))$
9:          wait for next time step
10:         $h \leftarrow gethumanCorrectiveAdvice()$
11:         **for** $1 \leq t \leq T$
12:             $c_t \leftarrow assignCredit(t)$
13:             $\overrightarrow{f^{cred}} = \overrightarrow{f^{cred}} + \left(c_t \cdot \vec{f_t}\right)$
14:         **end for**
15:         **if** $h =! 0$
16:             $H(\vec{s}) = \vec{v}^T \cdot \overrightarrow{f^{cred}}$
17:             $\Delta v_l = \beta \cdot \left(h - H(\vec{s})\right) \cdot f_l^{cred}; l = 1,..,N_{feat}$
                $v_l = v_l + \Delta v_l$
18:             $\alpha(\vec{s}) = \mid H(\vec{s}) \mid + bias = \mid \vec{v}^T \cdot \overrightarrow{f^{cred}} \mid + bias$
19:             $error \leftarrow h \cdot e$
20:             $\Delta w_l = \alpha(\vec{s}) \cdot error \cdot f_l^{cred}; l = 1,..,N_{feat}$
                $w_l = w_l + \Delta w_l$
21:         **end if**
22   **end while**

Algorithm 1 presents the whole learning process. First, the error magnitude e and the learning rate β for the weights update and the *bias* of (6) are defined (lines 1–3). The loop between lines 4–22 occurs once per time step. The agent observes the new state $\vec{s}$ (line 5), and it computes the basis functions/features of the policy model $P(\vec{s})$ (line 6). Thus, the agent takes $\vec{s}$ from the state space and maps it into the feature-space for obtaining the features vector $\vec{f}$. $P(\vec{s})$ is obtained as the inner product between $\vec{f}$ and the weight's vector $\vec{w}$ (line 7). Afterwards, $P(\vec{s})$ is executed (line 8).

After action execution (lines 8–9), the human trainer sends a feedback signal h (line 10). Then the credit assigner computes the vector $\overrightarrow{f^{cred}}$ (lines 11–14). For the credit assignment, the index t is varied according with the amount of past time steps T that compound the history window (line 11); c_t obtains the credit associated with the tth prior time step (line 12), which depends on the probability density function used for modeling the human delay. $\overrightarrow{f_t}$ is the features vector computed in line 6, t time steps ago, which is weighted by the respective c_t and cumulated in $\overrightarrow{f^{cred}}$ (line 13).

Afterwards, the human feedback model H (lines 16–17) is updated, and the value of the variable learning rate is calculated (line 18). Finally, the policy model P is updated (line 19–20): first the prediction *error* of P, whose magnitude/sign is given by e/h is computed (line 19), and then, the weights of $P(\vec{s})$ are updated (line 20).

3 Human-Machine Interface

Given that human feedback is a key component of the proposed learning framework, a new hand-gesture interface that allows providing feedback to the agent, is proposed for validating and contrasting the learning frameworks' performances through this interface and a keyboard based interface. The interface allows detecting 5 gestures (positive correction, negative correction, a neutral gesture used when users do not need to provide feedback, a reward, and a punishment) in dynamic setups (variable illumination, non-uniform backgrounds). It uses background subtraction to detect regions of interest (ROI), i.e. hand candidates, Kalman filtering for tracking the hand candidates, and Local Binary Patterns (LBP) and SVM for the final detection of hand-gestures (see Fig. 2). These three functionalities are described in the following paragraphs:

– *Detection of Regions of Interest (ROI):* Movement blobs are first detected using background subtraction. Then, adjacent blobs are merged and filtered using morphological filters, and the largest blob is selected as a hand candidate and feed to the tracking system.

In parallel, a second process applies background subtraction to color edges: First, a binary edge image is computed, and then, color information is incorporated into the edges. Afterwards, background subtraction and area filtering is applied in the edge's domain. Finally, the output of the area-filtering module is intersected with the color edges, in order to manage occlusions properly (see Fig. 2(b)). The output is a blob with the detected moving, color edges.

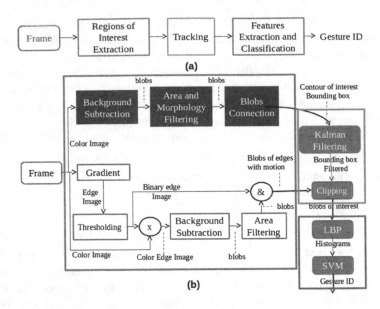

Fig. 2. Hand Gesture Recognition system. (a) General scheme, (b) detailed scheme.

- *Tracking:* The parameters of the bounding box given by the prior module are used as observations by a Kalman filter, which estimates the final hand candidates, based on the fusion of the current ROI information and the prior ones. Afterwards, the blob with the moving edges is intersected with the Kalman-filtered bounding box.
- *Features Extraction and Classification:* The image window given by the *Tracking* module is analyzed in order to classify the captured gesture. Histograms of LBP features are computed inside the image window. This feature vector feeds five SVM classifiers, one trained for each gesture, where the gestures are detected.

The described system is able to detect gestures with a mean detection rate of 91 %. A more detailed description of this system is given in [21].

4 Experiments and Results

The performance of the COACH framework is validated and compared with the ones of TAMER [15], ACTAMER [17] (an Actor Critic approach based on TAMER) which are pure interactive machine learning algorithms like COACH, and an autonomous agent trained using SARSA(λ). Two learning problems are used for the comparisons: ball dribbling with Humanoid robots [18] and the Cart-Pole [19]. In the first problem, 10 subjects interacted with each interactive learning framework (COACH, TAMER and ACTAMER) only through a keyboard interface, while in the second problem 15 subjects trained an agent using each framework, first with the keyboard interface, and then with the proposed hand-gesture interface. In this second experiment, the goal is to evaluate the use of these two alternative interfaces.

For each problem to be solved, the subjects interacted with each learning framework four times: the first and second trials are used for practicing (human operator training), while the other two for the interactive training of the agent. In the case of the SARSA(λ) method, 50 training runs were executed.

Although, only the SARSA(λ) agent uses an environmental reward function during learning, the cumulative environmental reward and the average environmental reward per episode were computed and used as performance metrics.

4.1   Ball Dribbling with Humanoid Robots

In the context of humanoid robotics soccer, ball dribbling is a relevant problem in which a robot walks to a target as fast as possible by keeping the ball possession. Keeping the ball possession means having the ball close to the robot's feet while walking (see a description of this behavior in [18]). In a recent work this problem has been modeled as two simpler tasks: ball pushing, and alignment to the ball and target. The first task reduces the problem to dribbling on a straight line (one dimension), and it is tackled using autonomous RL [18]. An episode is completed when the robot goes across the complete soccer field with the ball. For dribbling the ball in a straight line the robot estimates the distance to the ball ρ, and decides its speed. Therefore this problem has a very small state space but with high level of uncertainty, due to the fact that the feet motion is not observed by the decision-making system of the robot.

This work proposes a modification of the reward function used in [18], consisting on incorporating a parameter that defines a security robot-ball distance ρ_{max} that must not be exceeded:

$$r = \begin{cases} 100 + v_x, & \rho \leq \rho_{max} \\ -100 - (\rho - \rho_{max}) + v_x, & \rho > \rho_{max} \end{cases} \tag{8}$$

This reward function re defines the ball-dribbling task: the goal is to walk as fast as possible, without exceeding the distance ρ_{max}.

The robot speed v_x is set between 0 and 100 mm/s. For the algorithms with discretization of the action space (SARSA and TAMER) ten different magnitudes of speed were defined; the state space was divided uniformly in thirty features between 0 and 3 m, and the ρ_{max} parameter was set to 300 mm.

In this problem the gesture recognition interface was not used due to its computational burden. The obtained results are shown in Fig. 3. As it can be observed in this

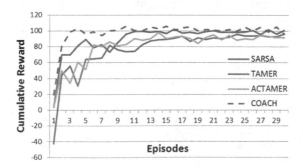

Fig. 3. Average cumulative reward for the ball dribbling problem.

figure, the three interactive learning algorithms converged faster than the non-interactive one (SARSA), and COACH achieves the fastest convergence (in three episodes), and the highest average performance. The second best convergence was obtained by TAMER.

4.2   Cart-Pole

This well known problem in RL literature is an episodic task with the goal of keeping balanced a pole on top of a cart. The actions are the forces applied by the cart; the state space has four dimensions defined by the cart's position and velocity, and the pole's angle and angular velocity [19]. An episode is finished (a failure occurs) when the pole falls to a given angle regarding the vertical axis, or if the cart exceeds the bounds of the scene. In our modeling, the continuous, 4-dimensional state space is approximated and divided uniformly using 256 Gaussian RBF features as in [17]. This problem has been approached by TAMER and ACTAMER in their original papers. All learning frameworks were tested under the same conditions.

In the experiments with the hand gesture recognition system, with COACH are used: neutral, positive, and negative correction gestures; for TAMER and ACTAMER are used: neutral, reward and punishment gestures. In Fig. 4 are shown the obtained learning curves of this problem trained with both the keyboard and the hand-gesture interfaces. The experiments with interactive agents were finished in the episode 150 with a maximum of allowed time steps of 5,000, considered here as the optimum performance. The TAMER and ACTAMER algorithms reach the lowest performances, although these algorithms achieve faster learning than SARSA in the early episodes (by the episode 20). The autonomous SARSA agent converges by the episode 500 with higher final performance than the ones achieved by TAMER and ACTAMER. COACH outperforms the other algorithms since the first episode, and it achieves a performance four and three times higher than the one of SARSA's (the second highest performance) with the keyboard and the hand-gesture interfaces, respectively. In addition COACH achieves the fastest convergence among all the agents; the users only need almost 25 episodes for teaching the policies.

The use of the keyboard interface allowed obtaining better results. Despite the hand-gesture interface is a more natural way of communication for human-machine interaction, the keyboard interface was easier for most of the users because it allowed them to alternate the signals given to the agent at a higher speed.

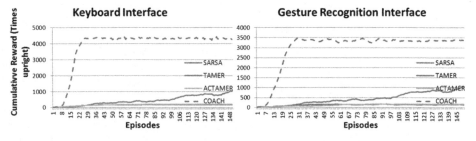

Fig. 4. Average cumulative reward for the cart-pole problem.

Table 1 summarizes the obtained results. It can be observed that COACH allows obtaining a faster convergence than all other methods, and also a higher performance. In the case of the Cart-Pole problem, the final performance obtained by COACH is much higher than the ones obtained by TAMER or ACTAMER, which is very low even compared to the one obtained when using SARSA.

Table 1. Algorithms comparison. EC: Number of episodes for convergence. AP: Average of final performance of the last 5 episodes.

	Ball dribbling		Cart-pole (keyboard)		Cart-pole (hand-gestures)	
	EC	AP	EC	AP	EC	AP
TAMER	11	96.2	89	188	40	150
ACTAMER	11	93.8	68	196	79	166
COACH	3	104.8	25	4,317	22	3,383
SARSA	21	91.6	461	1,125	461	1,125

It is interesting to note that in [17] it was possible to attain higher performance for the Cart-Pole problem only when using hybrid methods that combine the use of type-TAMER algorithms and autonomous RL. In a first stage, those hybrid methods learn suboptimal policies from humans. Then the acquired knowledge is transferred for supporting agents with autonomous RL processes. In this work, similar high performances were obtained with faster convergence, directly by COACH agents, which are based on a pure interactive machine learning approach that uses the feedback in a more intuitive sense.

5 Conclusions

Humans assisting agents in the process of shaping a policy is a strategy very useful in problems where the environment-agent interaction could be observed by the human senses. Interactive learning algorithms have the ability to allow the transfer of knowledge from humans to machines in this kind of problems. COACH lets to a human trainer with no expertise, to shape a policy with a binary signal. COACH sets the feedback in the domain of the actions space as in Learning from Demonstration approaches, which is more natural and intuitive than the evaluative feedback domain. Due to the above and the obtained results, it is stated that COACH is an easy-to-use framework for teaching an agent that have to perform continuous actions.

Moreover, COACH models the human feedback which allows extracting more information from the human trained. Thus allows advancing on understanding how humans teach autonomous agents.

With the reported experiments we have validated the hypothesis that interactive learning frameworks for shaping the policies of learning agents may be the solution in applications where a human is able to observe the agent-environment interaction and to advice it. This is especially important because in many applications the a priori definition of the reward function is not easy to obtain for an autonomous RL agent.

Results showed that human participation in learning process is a powerful strategy that succeeds and outperforms the classical autonomous RL, depending of the kind of feedback given by humans and the problem's nature. Although TAMER algorithms have more general applications (discrete and also continuous action problems), COACH is a complementary approach to TAMER, while TAMER is appropriate for teaching in discrete domains, COACH is more suitable for continuous domain applications.

It is important to remark, that the experimental procedure allowed us to see that due to people are not experts with the tasks being solved and also no experts in teaching agents, sometimes they do mistakes giving accidentally incorrect signals (e.g. punishing actions they think are the correct one, or increasing a force they do not plan to change). Then, turning back to the prior policy is a challenge easier to achieve with COACH compared to TAMER, because with TAMER the users change the policy over a specific state, punishing the action, but they cannot control or forecast which is the new action executed by the policy in that state, whereas with COACH, the human has an insight about the changes in the policy after feeding back. The aforementioned fact has impact in the time reduction obtained in the learning process with COACH.

The difficult level of teaching a task to a machine is due to two important facts: the problem's nature, and the abilities of humans using the human-machine interface for teaching the task. In this work the difficulty associated to the interface was analyzed: a keyboard and a hand-gesture interface were used as input devices. The hand-gesture interface is more intuitive for users, but it constraints them, because when using hand-gestures it is not easy to provide different orders (i.e. different gestures) in short periods of time, as the case when using a keyboard. Nevertheless, results show that COACH allowed users to reach optimal policies too, in this last case. This leads to the conclusion that COACH is robust and reliable to be used in difficult problems where either the nature of the agent-environment interaction is complex or where the human-machine interface is complex to use.

Acknowledgements. This work was partially funded by FONDECYT Project 1130153, CONICYT-PCHA/Doctorado Nacional/2015-21151488, and the Department of Electrical Engineering, University of Chile.

References

1. Argall, B.D., Chernova, S., Veloso, M., Browning, B.: A survey of robot learning from demonstration. Rob. Auton. Syst. **57**(5), 469–483 (2009)
2. Breazeal, C., Scassellati, B.: Robots that imitate humans. Trends Cogn. Sci. **6**(11), 481–487 (2002)
3. Abbeel, P., Ng, A.Y.: Apprenticeship learning via inverse reinforcement learning. In: Proceedings of the Twenty-First International Conference on Machine Learning. ACM (2004)
4. Meriçli, C., Veloso, M., Akin, H.L.: Complementary humanoid behavior shaping using corrective demonstration. In: 10th IEEE-RAS International Conference on Humanoid Robots (Humanoids), pp. 334–339. IEEE (2010)

5. Meriçli, Ç., Veloso, M., Akin, H.L.: Task refinement for autonomous robots using complementary corrective human feedback. Int. J. Adv. Rob. Syst. **8**(2), 68 (2011)
6. Argall, B.D., Browning, B., Veloso, M.: Learning robot motion control with demonstration and advice-operators. In: IEEE/RSJ International Conference on Intelligent Robots and Systems IROS 2008, pp. 399–404. IEEE (2008)
7. Mitsunaga, N., Smith, C., Kanda, T., Ishiguro, H., Hagita, N.: Adapting robot behavior for human–robot interaction. IEEE Trans. Rob. **24**(4), 911–916 (2008)
8. Tenorio-Gonzalez, A.C., Villaseñor-Pineda, L., Morales, E.F.: Dynamic reward shaping: training a robot by voice. In: Kuri-Morales, A., Simari, G.R. (eds.) IBERAMIA 2010. LNCS, vol. 6433, pp. 483–492. Springer, Heidelberg (2010)
9. León, A., Morales, E.F., Altamirano, L., Ruiz, J.R.: Teaching a robot to perform task through imitation and on-line feedback. In: San Martin, C., Kim, S.-W. (eds.) CIARP 2011. LNCS, vol. 7042, pp. 549–556. Springer, Heidelberg (2011)
10. Suay, H.B., Chernova, S.: Effect of human guidance and state space size on interactive reinforcement learning. In: RO-MAN 2011, pp. 1–6. IEEE (2011)
11. Pilarski, P.M., Dawson, M.R., Degris, T., Fahimi, F., Carey, J.P., Sutton, R.S. Online human training of a myoelectric prosthesis controller via actor-critic reinforcement learning. In: IEEE International Conference on Rehabilitation Robotics (ICORR), pp. 1–7. IEEE (2011)
12. Yanik, P.M., Manganelli, J., Merino, J., Threatt, A.L., Brooks, J.O., Green, K.E., Walker, I.D.: A gesture learning interface for simulated robot path shaping with a human teacher. IEEE Trans. Hum.-Mach. Syst. **44**, 41–54 (2014)
13. Thomaz, A.L., Hoffman, G., Breazeal, C.: Reinforcement learning with human teachers: understanding how people want to teach robots. In: The 15th IEEE International Symposium on Robot and Human Interactive Communication, ROMAN 2006, pp. 352–357. IEEE (2006)
14. Thomaz, A.L., Breazeal, C.: Asymmetric interpretations of positive and negative human feedback for a social learning agent. In: The 16th IEEE International Symposium on Robot and Human Interactive Communication, RO-MAN 2007, pp. 720–725. IEEE (2007)
15. Knox, W.B., Stone, P.: TAMER: training an agent manually via evaluative reinforcement. In: 7th IEEE International Conference on Development and Learning, ICDL 2008, pp. 292–297. IEEE (2008)
16. Knox, W.B., Stone, P.: Interactively shaping agents via human reinforcement: the TAMER framework. In: Proceedings of the Fifth International Conference on Knowledge Capture, pp. 9–16. ACM (2009)
17. Vien, N.A., Ertel, W., Chung, T.C.: Learning via human feedback in continuous state and action spaces. Appl. Intell. **39**(2), 267–278 (2013)
18. Leottau, L., Ruiz-del-Solar, J., Celemin, C.: Ball dribbling for humanoid biped robots: a reinforcement learning and fuzzy control approach. In: Bianchi, R.A., Akin, H., Ramamoorthy, S., Sugiura, K. (eds.) RoboCup 2014. LNCS, vol. 8992, pp. 549–561. Springer, Heidelberg (2015)
19. Sutton, R.S., Barto, A.G.: Reinforcement Learning: an Introduction. MIT Press, Cambridge (1998)
20. Busoniu, L., Babuska, R., De Schutter, B., Ernst, D.: Reinforcement Learning and Dynamic Programming Using Function Approximators. CRC Press, Boca Raton (2010)
21. Celemin, C.: A hand-gesture interface for interactive learning. Internal report, Advanced Mining Technology Center, Universidad de Chile (2014). (in Spanish)

Best Paper Award for its Engineering Contribution

Evaluation of the RoboCup Logistics League and Derived Criteria for Future Competitions

Tim Niemueller[1]([⊠]), Sebastian Reuter[2], Alexander Ferrein[3], Sabina Jeschke[2], and Gerhard Lakemeyer[1]

[1] Knowledge-Based Systems Group, RWTH Aachen University, Aachen, Germany
{niemueller,gerhard}@kbsg.rwth-aachen.de
[2] Institute Cluster IMA/ZLW & IfU, RWTH Aachen University, Aachen, Germany
{Sebastian.Reuter,Sabina.Jeschke}@ima-zlw-ifu.rwth-aachen.de
[3] MASCOR Institute, Aachen University of Applied Sciences, Aachen, Germany
ferrein@fh-aachen.de

Abstract. In the RoboCup Logistics League (RCLL), games are governed by a semi-autonomous referee box. It also records tremendous amounts of data about state changes of the game or communication with the robots. In this paper, we analyze the data of the 2014 competition by means of Key Performance Indicators (KPI). KPIs are used in industrial environments to evaluate the performance of production systems. Applying adapted KPIs to the RCLL provides interesting insights about the strategies of the robot teams. When aiming for more realistic industrial properties with a 24/7 production, where teams perform shifts (without intermediate environment reset), KPIs could be a means to score the game. This could be tried first in a simulation sub-league.

1 Introduction

Benchmarking of autonomous mobile robots and industrial scenarios alike are difficult due to many dynamic factors. The scenarios might be too diverse to compare or the environment is not observable (enough). This makes it problematic to evaluate such domains objectively. The *RoboCup Logistics League* (RCLL) is a medium complex domain inspired by actual challenges in industrial applications – in particular that of intra-logistics in a smart factory environment, that is, moving goods in a factory among a number of machines for processing. When developing the league, it was ensured that the domain remained partially observable – enough, so that one could autonomously judge the game.

In an industrial setting, companies strive to improve in terms of *Key Performance Indicators (KPI)*. KPIs are, for example, the time required to move a part through its production process along several machines, or how many products are currently worked on (work in progress) at a time.

Our goal is to *make KPI applicable in the RCLL* in a meaningful way. As a first step, we have analyzed games of the RCLL competition in 2014 focusing on the two top performing teams Carologistics and BBUnits. We provide an evaluation in terms of KPIs mapped to the RCLL game. This is possible, because the

© Springer International Publishing Switzerland 2015
L. Almeida et al. (Eds.): RoboCup 2015, LNAI 9513, pp. 31–43, 2015.
DOI: 10.1007/978-3-319-29339-4_3

referee box, a program that controls and monitors the game, also records relevant data like game state changes and robot communication. The KPIs adapted for the RCLL provide the *performance metrics* by which we can analyze this data. Based on this analysis we give possible explanations on the differences in performance seen from the two teams. The information gained also allows for improving the RCLL as a testbed for industrial applications.

Additionally, on the road to a more realistic industrial setting it is conceivable to aim for a 24/7 production where teams take over shifts without an intermediate environment reset. That would allow for better judging of system robustness and flexibility of the task-level coordination of a team. However, this requires new metrics to score the game, which the adapted KPIs might provide. The RCLL simulation [1] might be a suitable basis to try this in a reasonable way.

In the following Sect. 2, we introduce the RCLL in more detail. In Sect. 3, we give an overview of related work regarding robotic competitions and benchmarks. KPIs and their adaptation to the RCLL is presented in Sect. 4, before applying them for analyzing the RCLL 2014 finale in Sect. 5. We conclude in Sect. 6.

2   RoboCup Logistics League

RoboCup [2] is an international initiative to foster research in the field of robotics and artificial intelligence. The basic idea of RoboCup is to set a common testbed for comparing research results in the robotics field. RoboCup is particularly well-known for its various soccer leagues. In the past few years, application-oriented leagues received increasing attention. In 2012, the new industry-oriented RoboCup Logistics League (RCLL, previously LLSF), was founded to tackle the problem of production logistics. Groups of up to three robots have to plan, execute, and optimize the material flow in a smart factory scenario and deliver products according to dynamic orders. Therefore, the challenge consists of creating and adjusting a production schedule and coordinate the group of robots. In the following, we describe the rules of 2014, that we used for our evaluation.

Fig. 1. Carologistics (three Robotino 2 with laptops on top) and BavarianBendingUnits (two larger Robotino 3) during the RCLL finals at RoboCup 2014 (Color figure online).

The RCLL competition takes place on a field of $11.2\,\text{m} \times 5.6\,\text{m}$ (Fig. 1). Two teams are playing at the same time competing for points, (travel) space and time. Each team has an exclusive input storage (blue areas) and delivery zone (green area in Fig. 1). Machines are represented by RFID readers with signal lights on top indicating the machine state. At the beginning all pucks (representing the products) have the raw material state, are in the input storage, and can be refined (through several stages) to final products using the production machines. These machines are assigned a type randomly at the start of a match which determines what inputs are required and what output will be produced, and how long this conversion will take [1]. Finished products must then be taken to the active gate in the delivery zone. The game is controlled by the referee box (refbox), a software component which instructs and monitors the game [3]. It posts orders dynamically that state the product type (required final puck state), how many items are requested, and a time window when the order must be delivered. Pucks are identified by a unique ID stored on an RFID tag to maintain the puck's virtual state. After the game is started, no manual interference is allowed, robots receive instructions only from the refbox. Teams receive points for producing complex products, delivering ordered products, and recycling. The RCLL is also very interesting from a planning and scheduling point of view [4].

2.1   The Referee Box

Overseeing the game requires tracking of more than 40 pucks and their respective states, watching machine areas of 24 machines to detect pucks that are moved out of bounds, checking for the completion of production steps along the production chain awarding points and keeping a score. This can easily overwhelm a human referee and make the competition hard to understand for the audience. Therefore, we introduced a (semi-) autonomous referee box (refbox) in 2013. It controls and monitors all machines on the field, tracks the score, and provides

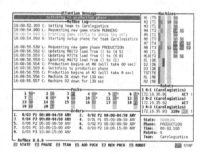

Fig. 2. The Referee Box UI

information for visualization to the audience. The interface for the human referees (e.g., to start or pause the game) is shown in Fig. 2. The refbox communicates with all robots on the field. Some core aspects are listed in the following.

Control. The refbox must oversee the game implementing the rules defined in the rule book[1]. For this very purpose it uses the rule-based system CLIPS [5]. This part is responsible for awarding points if the robots accomplished a (partial) task.

Communication. It must communicate with the robots on the field to provide information, send orders, and receive reports.

---

[1] The current rules can be found at http://www.robocup-logistics.org/rules.

Representation. A textual or graphical application is required to visualize the current state of the game and to receive command input from the human referees.

Interfacing. The referee box needs to communicate with the programmable logic controller (PLC) which is used to set the light signals and read the RFID sensors.

Data Recording. The refbox records each and any message received or sent over the network, all state changes of the internal fact base that is used to control the game, and comprehensive game reports. This is crucial for this work.

3   Related Work: Competitions and Benchmarks

Competitions and benchmarking through competitions have become very popular for many research fields from the AI planning and scheduling community (e.g. [6,7]) leading ultimately to the development of PDDL and its extensions over SAT solvers [8] to game-based benchmark for learning algorithms [9] and robotics research. Since its beginnings in the 90's (see [10]), a large number of robotics competitions were launched in all fields of robot applications from autonomous driving (e.g. DARPA Grand Challenges, http://www.darpa/mil/grandchallenge) to disaster response (for instance, European Land Robot Trial, http://www.elrob.org) to landmine disposal (e.g., Minesweeper, http://www.landminefree.org). The motivations for running a competition are manifold. There are aspects to promote or compare research output and approaches. For exchanging ideas and experiences, symposia or user-group meetings are often organized together with a competition to foster the open exchange of solutions and ideas. Additionally, competitions are very motivating and can, in particular, activate students to be part of a competition team.

Among the established robotics competitions, the RoboCup competition [2] is a very successful example. While one of the frequently mentioned motivations of RoboCup is to compare approaches that work well in practice, the comparison of different approaches is nonetheless difficult. A reason is, in part, that robots systems are highly integrated and it is, in general, not possible to exchange software modules or test functionalities in isolation easily. In [11], the authors argue that competition challenges should lead to better algorithms and systems by a continual development process. Anderson et al. [12] critically review the contributions of a number of competitions. Proper benchmarks are not simply given and defined by performing a robotic competition. The organizers of a competition have to define determining factors in order to develop a robotic competition into a benchmark. Many competitions work toward this goal. Under the roof of the RoboCup Federation, in particular, the RoboCup Rescue [13] and RoboCup@Home [14] competitions have to be mentioned. In the RoboCup Rescue competition, for instance, benchmarks for assessing the quality of generated environment maps are established (see e.g. [15]). In RoboCup@Home, the rules change from year to year and an innovative scoring system helps to define a benchmark for fully integrated domestic service robots. Other approaches focus more on certain components such as motion algorithms [16]. The recent RoCKIn project (http://rockinrobotchallenge.eu/) aims at setting up a robot competition that increases the scientific an technological knowledge [17,18].

In the next section, we will define the key performance indicators for production systems. These performance indicators can be used in order to judge the performance of a team. Analyzing the data recorded by the referee box using KPIs, the RCLL could indeed define a logistic benchmark in the future.

4   Key Performance Indicators

The traditional goal of production systems (in the sense of systems producing goods, not rule-based production systems) is to maximize production output while minimizing production costs. In the context of increasing market competition, product delivery times and reliability gain importance as buying criteria alongside price and quality of the product [19]. High delivery reliability and short delivery times of products demand for short throughput times of all required intermediate parts and high schedule reliability of all sub-processes within the logistic system [20]. The demand for short throughput time (time span for an order to be created) and high schedule reliability (extent to which planned orders are finished in time) conflicts with the minimization of costs which calls for a high utilization of production resources [21]. Furthermore, the minimization of throughput time and the maximization of output rate contradict each other: As a maximization of output depends on a high level of work in progress (WIP, production orders that are processed in parallel), short throughput times can only be achieved by a low level of WIP [20].

For example, a high utilization of production entities implies a high level of WIP to prevent shortages within the material flow. But it will also slow down the throughput time, because it requires a lot of transport resources. Hence, high machine utilization and short throughput times cannot be achieved together [22].

This conflict among the objectives *logistic performance* and *logistic costs* is called the scheduling dilemma of logistics [22]. Figure 3 shows Key Performance Indicators as measures for *logistic performance* and *logistic costs* [21]. KPIs are used in industry to make the efficiency of logistic systems assessable.

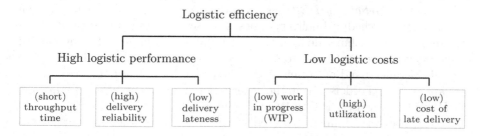

Fig. 3. Key Performance Indicator (KPI) within Production Logistics

The logistic performance can be described by the measures throughput time, delivery reliability and delivery lateness of orders. The *throughput time TTP* for an operation is defined as start of the order processing ($T_{\text{operation start}}$) till the end of the order processing ($T_{\text{operation end}}$) [21]: $\text{TTP} = T_{\text{operation end}} - T_{\text{operation start}}$. An exemplary throughput of a product of type P2 is shown in Fig. 4. The production of a product P2 is consists of a manufacture of a intermediate product S1 and S2. The critical path – the minimal throughput time of a product P2 – is formed by the throughput time of the intermediate product S2 and the final assembly. The manufacturing of an intermediate product S2 consists of two operations on the machines T1 and T2 as well as the time span needed for transportation of the intermediate products (S1 and S2) and the waiting times.

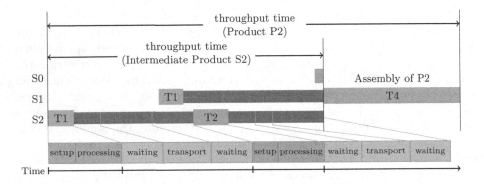

Fig. 4. Throughput Time Components

The *delivery lateness DEL* is a measure for the deviation of the actual ($T_{\text{actual delivery date}}$) and the planned delivery date ($T_{\text{planned delivery date}}$) [21]: $\text{DEL} = T_{\text{actual delivery date}} - T_{\text{planned delivery date}}$. As the actual delivery of an order can be before and after the specified delivery date, a positive lateness describes an order that was delivered too late and a negative lateness describes an order that was delivered too early. The lateness of an order has a negative impact on the overall delivery reliability of the production system.

The *delivery reliability DERE* is an indicator to measure if a production systems sticks to scheduled delivery times. It describes the percentage of orders that are delivered within a defined delivery reliability tolerance. The number of in-time deliveries refers to all production orders that are completed within the specified tolerance band of permissible delivery lateness. The number of orders (NO) are all posted orders within the observation period. The *delivery reliability DERE* can be expressed as [21]: $\text{DERE} = \text{DEL}/\text{NO} * 100\%$.

The *logistic costs* influence the effectiveness of a logistics system just as the logistic performance does. As the logistic cost increase, the product price increases and decreases the customers willingness to buy the product. Measure for the logistic costs are work in progress, utilization and cost of late delivery.

The *work in progress WIP* describes the amount of orders that are started within a production system but are not yet completed. It can be calculated by subtracting the system output from the system input. For discretization, the period of observation can be split into equidistant time slots such as standard hours. Thus, the development of the WIP can be tracked.

The *utilization U* describes the ratio of idle and working time of production resources such as production machines or transportation entities. In terms of a production machine the utilization describes the amount of time the machine is processing an item ($T_{\text{operation}}$) in relation to the duration of a reference period ($T_{\text{duration of reference period}}$) [21]: $U = \sum T_{\text{operations}}/T_{\text{duration of reference period}} * 100\%$.

The *cost of late delivery COLD* are expenses due to a delay of an order delivery. Cost could be due to increased cost for express shipment or default penalties, or the cost of late delivery can be expressed as a lack of customer trust.

4.1   KPIs Applied in the RoboCup Logistics League

The RCLL aims to simulate a realistic, yet simplified, production environment. With the given resources (stationary production machines and mobile robots for transportation) the teams have to maximize the production output with respect to a certain set of products. In this section, we map KPIs to the RCLL.

The *throughput time TTP* in our scenario is defined as the time from the insertion of the first input product (of any accepted type) for a machine until all required inputs have been provided and the processing has been completed. For example, in Fig. 5 in the second line for M2, the Busy-Blocked-Busy cycle is the TTP for a production of 84 s. The *delivery lateness DEL* is directly applicable given that orders have a delivery time window stating a latest time for delivery. The *delivery reliability DERE* can by calculated by dividing the number of delivered products by the total number of products ordered. In the RCLL, *work in progress (WIP)* can be interpreted as machines currently being blocked for an order. This contains machines blocked for the production of intermediate as well as final products, i.e. by the green and orange blocks in Fig. 5. The *utilization U* of a machine is calculated by dividing the actual busy time by the overall game time, i.e. in Fig. 5 all bright green areas. The *cost of late delivery COLD* are expressed in the scoring scheme of the RCLL. A delivery in the requested time window is awarded with 10 points, while a late delivery only scores 1 point, setting COLD to 9 points. Furthermore the RCLL punishes over-production by also reducing the score from 10 to 1 point.

The teams have to balance logistic performance and logistic costs. On the one hand, the teams aim to maximize the logistic performance by short throughput times and low delivery lateness leading to a high output rate and high delivery reliability. On the other hand, the teams have to take care of high WIP which is a prerequisite for high resource utilization, but has a negative effect on throughput time and delivery reliability.

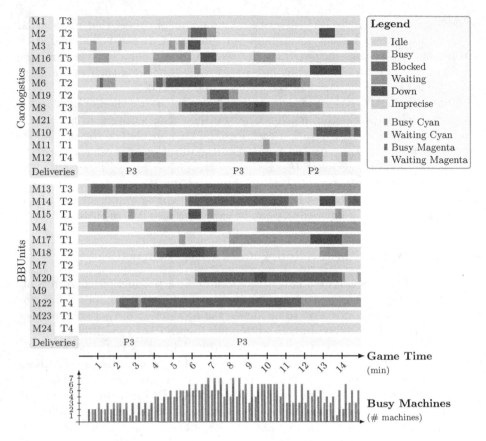

Fig. 5. Machine states over the course of the final game at RoboCup 2014. The lower graph shows the occupied machines per 20 s time block (Color figure online).

5   Analysis of the RCLL 2013 and 2014

For the presented data analysis we have used recordings of the RoboCup competition 2014. The data comprises about 75 GB of refbox data of communication, the state changes of the internal knowledge-based system, text logs, and comprehensive reports of all games played. The data is organized using MongoDB which provides fast and efficient access [23].

The basic analysis was performed using aggregation and map-reduce features of the database as well as retrieval and analysis scripts written in Python. While we have records for all games of the competitions, for brevity we focus on the two top performers in 2014, the Carologistics and BBUnits teams.

5.1   Exemplary Application of KPI to the RoboCup 2014 Finale

We will exemplary apply the KPIs for the RoboCup 2014 final of the RCLL between the Carologistics (cyan) and the BBUnits (magenta) which ended with a score of 165 to 124[2]. We base our analysis on Figs. 5 and 6 for this game.

Figure 5 shows the machines (M1–M24) grouped per team above the time axis. Each row expresses the machine's state over the course of the game. Gray means it is currently unused (idle). Green means that it is actively processing (busy) or blocked while waiting for the next input to be fed to the machine. After a work order has been completed, the machine is waiting for the product to be picked up (orange). The machine can be down for maintenance for a limited time (dark red). Sometimes the machine is used imprecisely, that is, the product is not placed properly under the RFID device. The row 'Deliveries' shows products that are delivered at a specific time. Below the time axis, Fig. 5 shows the busy machines over time. Each entry consists of a cyan and magenta column and represents a 20 s period. The height of each column shows the number of machines that are producing in that period (bright team color) or waiting for the product to be retrieved (dark team color).

Figure 6 shows the orders grouped by teams, which product type was requested and how many were requested. In each row, the colored box denotes the delivery time window in which the product has to be delivered. If the box is green, the order was fulfilled (partial fulfillment means that a smaller number of products was delivered than requested), if it is red no product was delivered in time. The red circles mark the time of delivery. Both teams were able to fulfill the second and third P3 order (partially), but only cyan managed to deliver a P2 product.

The *throughput time TTP* of an order within a machine is denoted by the green (light green and dark green) boxes in Fig. 5. Cyan generally retrieves (partially) finished work orders faster, while magenta often leaves machines blocked for considerable time (dark green areas). The *delivery lateness DEL* can be best seen in Fig. 6. The delivered orders (green boxes) would result in a DEL of zero, while unfulfilled orders would result the maximum DEL of the full game time (in seconds). In some games, orders were delivered after the delivery time window and therefore would get a smaller positive DEL. In the given game, the *delivery reliability DERE* of cyan is 50 %, that of magenta is 33 %. The *work in progress WIP* machines are shown as busy machines in Fig. 5. As we can see, the WIP was generally equal or higher for magenta, having more machines in use. Looking at the machine states however, most of this time is blocked time in which a machine waits for the next input. If combining with the machines waiting for removal of the finished product, magenta has more machines unusable for new productions on average. The typical *machine utilization U* is currently low in the RCLL due to an emphasis on the logistics aspect that causes long travel times. In the finals, the overall utilization of all machines was about 2.3 % by cyan and 1.8 % by magenta (thus cyan has utilized the machines more than 25 % better).

---

[2] Video of the final is available at https://youtu.be/_iesqH6bNsY.

If there had been a late delivery (which did occur in other games), the *cost of late delivery COLD* would be severe (9 points). What we do see is that some orders were missed completely (resulting in a maximum DEL of losing the full 10 delivery points). In particular, no team managed to fulfill the P1 order. Only cyan managed to complete the work order at all (cf. Fig. 5, row for T3 machine).

KPI Discussion. It seems that especially the lower throughput time TTP of cyan contributed to their success. Machines can be used again much faster. For example, only this made it possible to match the delivery time of the P2 order that magenta missed due to very long blocking times. Considering the waiting times makes this even more severe. The cyan team followed the strategy to store products finished before a matching order was received. This meant that the involved machines could be used again much faster (the waiting times of cyan are much shorter). Even with more work in progress machines of magenta, the cyan team used more T1 machines (3 instead of 2). Magenta even left M17 with a finished puck untouched for half of the game. A contributing factor here could be that magenta lost a robot during the game due to a software problem. It seems that the other robots could not recover the state of M17 (instead they later produced at another T1 machine). Concluding it seems that the cyan strategy focused more on low TTP and high throughput, while magenta's strategy was to maximize the overall machine usage and the WIP.

While BBUnits lost a robot in this game, similar statements can be made about a play-offs game between the teams a day earlier that ended 158 to 122 for the Carologistics where both teams had all robots running continuously.[3]

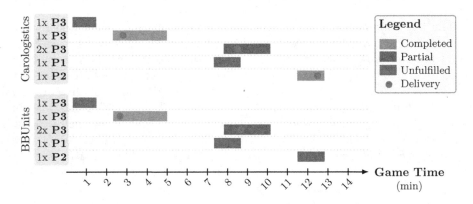

Fig. 6. Adherence to delivery schedule (finals RoboCup 2014). Each row represents an order for the indicated team on the left. The blocks denote their respective delivery windows in the game time represented on the Y-axis. Green boxes mean (partially) completed orders, red unfulfilled ones. Red dots indicate the time of delivery (Color figure online).

---

[3] Video of the play-offs game is available at https://youtu.be/77V-7LzMBY8.

5.2 Overall Tournament Evaluation

Analyzing the data of all games at RoboCup2014 (within the Round-Robin Phase, Playoffs and Finals) in terms of machine state graphs (Fig. 5) and adherence to delivery schedules (Fig. 6) as well as using KPIs as statistical queries yields insights for the development of the competition as a whole.

A key insight is that the current dynamic order scheme parameterization is unsuitable for the given resources (robots and machines). Even the best teams at most delivered 3 of 6 ordered products in any game. This seems to be, in particular, because the order time windows are too short. Especially with the modified game in 2015 with vastly more product variants this must be taken into account, since opportunistic production is virtually impossible.

A possible solution would be to considerably increase the time of a game. This would give the robot teams more time to work on the orders and we could gather more data to valuate the KPIs for a game. It also increases demands for system robustness, a crucial factor for industrial applications. The increased time could be tried first in a simulation league. Work is currently underway to create a common and open simulation for the RCLL based on [1] by the Carologistics and BBUnits team[4], which could provide the basis for the project.

6 Conclusion

In recent years we have developed the RCLL as a domain of medium complexity towards being a testbed for industry-inspired robotic applications. The domain is partially observable by the referee box which allows to record detailed data about the course of the game. This data combined with Key Performance Indicators known from industrial environments allow for analyzing games objectively. We can also use this analysis combined with statistical evaluation to optimize the competition to be more balanced and to improve it as a testbed for industrial robotic applications in smart factory environments.

In an example analysis of the finals in 2014 we have determined some factors based on KPI that may explain the outcome of the game, i.e. that the winning team Carologistics' strategy was focused on short throughput times rather than a high number of machines busy at the same time as the competitor BBUnits did. While we have seen that the order schedule should be tuned to better fit the given resources for more interesting games, teams also need to investigate better scheduling strategies that allow to use the given resources more effectively. KPIs can be one aspect of determining the utility in this regard.

To aim for a more realistic scenario, it is conceivable to develop the RCLL towards a long-time evaluation in the sense of a *24/7 robot competition*. Each team gets assigned a shift in which it has to realize the material flow in the production system without a reset of the environment. Within this scenario a more complex grading scheme is needed as the state of the production system is changing in terms of the amount of work in progress, blocked machines and

---

[4] The project is available at https://github.com/robocup-logistics.

orders that are currently selected for production. The introduced KPIs are a possible approach to adapt the *grading scheme* to this scenario. It will also require that the teams take different initial states into account and that they provide accurate information to the refbox during a handover to the next team. Especially the development of a *simulation league* can help to facilitate this in a shorter time frame. It would allow teams to adapt more gently. Work in this direction is on-going as described in Sect. 5.2.

More information, the recorded data as well as the evaluation scripts are available at http://www.fawkesrobotics.org/p/llsf2014-eval.

References

1. Zwilling, F., Lakemeyer, G., Niemueller, T.: Simulation for the RoboCup logistics league with real-world environment agency and multi-level abstraction. In: Bianchi, R.A.C., Akin, H.L., Ramamoorthy, S., Sugiura, K. (eds.) RoboCup 2014. LNCS, vol. 8992, pp. 220–232. Springer, Heidelberg (2015)
2. Kitano, H., Asada, M., Kuniyoshi, Y., Noda, I., Osawa, E.: Robocup: the robot world cup initiative. In: 1st International Conference on Autonomous Agents (1997)
3. Niemueller, T., Lakemeyer, G., Ferrein, A., Reuter, S., Ewert, D., Jeschke, S., Pensky, D., Karras, U.: Proposal for advancements to the LLSF in 2014 and beyond. In: ICAR - 1st Workshop on Developments in RoboCup Leagues (2013)
4. Niemueller, T., Lakemeyer, G., Ferrein, A.: The RoboCup logistics league as a benchmark for planning in robotics. In: WS on Planning and Robotics (PlanRob) at International Conference on Automated Planning and Scheduling (ICAPS) (2015, to appear)
5. Wygant, R.M.: CLIPS: a powerful development and delivery expert system tool. Comput. Ind. Eng. **17**, 1–4 (1989)
6. Long, D., Kautz, H., Selman, B., Bonet, B., Geffner, H., Koehler, J., Brenner, M., Hoffmann, J., Rittinger, F., Anderson, C.R., et al.: The AIPS-98 planning competition. AI Mag. **21**(2), 13–33 (2000)
7. Howe, A.E., Dahlman, E.: A critical assessment of benchmark comparison in planning. J. Artif. Intell. Res. **17**(1), 1–33 (2002)
8. Balint, A., Belov, A., Järvisalo, M., Sinz, C.: Sat challenge 2012 random sat track: description of benchmark generation. In: Proceedings of SAT CHALLENGE 2012, p. 72 (2012)
9. Karakovskiy, S., Togelius, J.: The mario AI benchmark and competitions. IEEE Trans. Comput. Intell. AI Games **4**(1), 55–67 (2012)
10. Balch, T.R., Yanco, H.A.: Ten years of the AAAI mobile robot competition and exhibition. AI Mag. **23**(1), 13–22 (2002)
11. Anderson, M., Jenkins, O., Osentoski, S.: Recasting robotics challenges as experiments [competitions]. IEEE Robot Autom. Mag. **18**(2), 10–11 (2011)
12. Anderson, J., Baltes, J., Cheng, C.T.: Robotics competitions as benchmarks for AI research. Knowl. Eng. Rev. **26**(01), 11–17 (2011)
13. Kitano, H., Tadokoro, S.: Robocup rescue: a grand challenge for multiagent and intelligent systems. AI Mag. **22**(1), 11–17 (2001)
14. Wisspeintner, T., Van Der Zant, T., Iocchi, L., Schiffer, S.: Robocup@home: scientific competition and benchmarking for domestic service robots. Interact. Stud. **10**(3), 392–426 (2009)

15. Madhavan, R., Lakaemper, R., Kalmár-Nagy, T.: Benchmarking and standard-ization of intelligent robotic systems. In: International Conference on Advanced Robotics (ICAR) (2009)
16. Calisi, D., Iocchi, L., Nardi, D.: A unified benchmark framework for autonomous mobile robots and vehicles motion algorithms (movema benchmarks). In: WS on Experimental Methodology and Benchmarking in Robotics Research at RSS (2008)
17. Ahmad, A., Awaad, I., Amigoni, F., Berghofer, J., Bischoff, R., Bonarini, A., Dwiputra, R., Fontana, G., Hegger, F., Hochgeschwender, N., et al.: Specification of general features of scenarios and robots for benchmarking through competitions. RoCKIn Deliverable D 1 (2013)
18. Amigoni, F., Bonarini, A., Fontana, G., Matteucci, M., Schiaffonati, V.: Bench-marking through competitions. In: European Robotics Forum-Workshop on Robot Competitions: Benchmarking, Technology Transfer, and Education (2013)
19. Nyhuis, P., Wiendahl, H.P.: Fundamentals of Production Logistics: Theory Tools and Applications. Springer Science & Business Media, Heidelberg (2008)
20. Wriggers, F., Busse, T., Nyhuis, P.: Modeling and deriving strategic logistic mea-sures. In: IEEE International Conference on Industrial Engineering and Engineer-ing Management (2007)
21. Lödding, H.: Handbook of Manufacturing Control: Fundamentals, Description, Configuration. Springer Science & Business Media, Heidelberg (2012)
22. Nyhuis, P., Wiendahl, H.P.: Logistic production operating curves-basic model of the theory of logistic operating curves. CIRP Annals-Manufact. Tech. $55(1)$, 441–444 (2006)
23. Niemueller, T., Lakemeyer, G., Srinivasa, S.S.: A generic robot database and its application in fault analysis and performance evaluation. In: International Confer-ence on Intelligent Robots and Systems (IROS) (2012)

Champions Papers

The Carologistics Approach to Cope with the Increased Complexity and New Challenges of the RoboCup Logistics League 2015

Tim Niemueller[1]([⊠]), Sebastian Reuter[2], Daniel Ewert[2], Alexander Ferrein[3],
Sabina Jeschke[2], and Gerhard Lakemeyer[1]

[1] Knowledge-based Systems Group, RWTH Aachen University, Aachen, Germany
niemueller@kbsg.rwth-aachen.de
[2] Institute Cluster IMA/ZLW & IfU, RWTH Aachen University, Aachen, Germany
[3] Mobile Autonomous Systems and Cognitive Robotics Institute,
FH Aachen, Aachen, Germany

Abstract. The RoboCup Logistics League (RCLL) has seen major rule changes increasing the complexity, e.g. by raising the number of product variants from 3 to almost 250, and introducing new challenges like the handling of physical processing machines. We describe various aspects of our system that allowed to improve the performance in 2015 and our efforts to advance the league as a whole.

1 Introduction

In 2015 the RoboCup Logistics League (RCLL) has changed considerably by introducing new machines on the field that require more elaborate handling, and by increasing the complexity through increasing the number of possible products from 3 to almost 250. The Carologistics team was able to adapt best to the new circumstances. This was possible, because from the beginning of the team in 2012, flexible and robust solutions were chosen and developed. Members of three partner institutions bring their individual strengths to tackle the various aspects of the RCLL: designing hardware modifications, developing functional software components, system integration, and high-level control of a group of mobile robots. Only the effective combination of these approaches explains the team's overall performance.

Our team has participated in RoboCup 2012–2015 and the RoboCup German Open (GO) 2013–2015. We were able to win the GO 2014 and 2015 (cf. Fig. 1) as well as the RoboCup 2014 and 2015 in particular demonstrating flexible task coordination, and robust collision avoidance and self-localization. We have publicly released our software stack used in 2014 in particular including our high-level reasoning components for all stages of the game[1] [1].

---

[1] Software stack available at http://www.fawkesrobotics.org/p/llsf2014-release.

© Springer International Publishing Switzerland 2015
L. Almeida et al. (Eds.): RoboCup 2015, LNAI 9513, pp. 47–59, 2015.
DOI: 10.1007/978-3-319-29339-4_4

In the following we will describe some of the challenges originating from the new game play in 2015. In Sect. 2 we give an overview of the Carologistics platform and the changes that were necessary to adapt to the new game play. Some parts have been used during the German Open 2015, but several components were extended and improved afterwards. We continue highlighting our behavior components in Sect. 4 and our continued involvement for advancing the RCLL as a whole in Sect. 5 before concluding in Sect. 6.

Fig. 1. Teams carologistics (robots with additional laptop) and solidus (pink parts) during the RCLL finals at RoboCup 2015, Hefei, China, that ended 46 to 16 (Color figure online).

1.1   Game Play Changes 2015

The goal is to maintain and optimize the material flow in a simplified Smart Factory scenario. Two competing groups of up to three robots each use a set of exclusive machines spread over a common playing field to produce and deliver products (cf. [2–4]).

In 2015, the RCLL has changed considerably by introducing actual processing machines based on the Modular Production System (MPS) platform by Festo Didactic [3] as shown in Figs. 2 and 4. For more details on the new game play we refer to [5]. The new machines require to equip the robot with a gripper for product handling, adaptation to the general game play due to vastly extended production schedules, and suggest switching to the Robotino 3 platform, which is larger and supports a higher payload.

Fig. 2.   Carologistics   robotino approaching a ring station MPS.

In the *exploration phase* the robots are given zones on the playing field in which they are supposed to look for machines, and – if one is found – identify and report them. In 2015, the machines can be freely positioned within the zones, in particular at any randomly chosen orientation. Additionally, the referee box can no longer provide ground truth information about the poses of machines in the production phase, making a successful exploration a mandatory requirement. Additionally the exploration procedure per machine became more complex. The robots must first identify whether there is a machine in a zone or not, generally requiring multiple positions per zone to be checked. In our system, we sweep the zone with the laser scanner looking for machine edges and a camera looking for markers to make a quick decision whether a zone is empty or not. If a machine is identified, robots must align on the output side of a machine (frequently requiring to go around the machine which costs time) to identify the marker and light signal with high confidence.

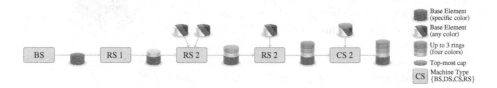

Fig. 3. Refinement steps for the production of a highest complexity product in the RCLL 2015 (legend on the right).

In the *production phase*, work orders are much more diverse in 2015 increasing the number of product variants from 3 to about 250. An example production chain is shown in Fig. 3. This in turn requires that the high-level reasoning component needs to make more decisions dynamically and opportunistic production is virtually impossible. It also requires to coordinate the robots more closely to achieve delivery in the desired time windows. Another challenge is that points are

Fig. 4. Play-offs game at the German open 2015

only awarded on successful delivery, even the points for intermediate processing steps. This requires the robots to be much more robust and the handling to be nearly perfect since harm done by failures in the production later along the chain is much more severe than before.

2   The Carologistics Platform

The standard robot platform of this league is the
Robotino by Festo Didactic [6]. The Robotino is
developed for research and education and features
omni-directional locomotion, a gyroscope and webcam,
infrared distance sensors, and bumpers. The teams may
equip the robot with additional sensors and computation devices as well as a gripper device for product
handling.

2.1   Hardware System

The robot system currently in use is based on the
Robotino 3. The modified Robotino used by the Carologistics RoboCup team is shown in Fig. 5 and features three additional webcams and a Sick laser range
finder. The webcam on the top is used to recognize
the signal lights, the one above the number to identify machine markers, and the one below the gripper
is used experimentally to recognize the conveyor belt.

Fig. 5.   Carologistics
robotino 2015

We have recently upgraded to the Sick TiM571 laser scanner used for collision
avoidance and self-localization. It has a scanning range of 25 m at a resolution
of 1/3 degrees. An additional laptop increases the computation power.

Several parts were custom-made for our robot platform. Most notably, a
custom-made gripper based on Festo fin-ray fingers and 3D-printed parts is used
for product handling. The gripper is able to adjust for lateral and height offsets.

2.2   Architecture and Middleware

The software system of the Carologistics robots combines two different mid-
dlewares, Fawkes [7] and ROS [8]. This allows us to use software components
from both systems. The overall system, however, is integrated using Fawkes.
Adapter plugins connect the systems, for example to use ROS' 3D visualization capabilities. The overall software structure is inspired by the three-layer
architecture paradigm [9]. It consists of a deliberative layer for high-level reasoning, a reactive execution layer for breaking down high-level commands and
monitoring their execution, and a feedback control layer for hardware access and
functional components. The lowest layer is described in Sect. 3. The upper two
layers are detailed in Sect. 4. The communication between single components –
implemented as *plugins* – is realized by a hybrid blackboard and messaging approach [7]. This allows for information exchange between arbitrary components.
Information is written to or read from *interfaces*, each carrying certain information, e.g. sensor data or motor control, but also more abstract information
like the position of an object. The information flow is somewhat restricted – by
design – in so far as only one component can write to an interface. Reading,

however, is possible for an arbitrary number of components. This approach has proven to avoid race conditions when for example different components try to instruct another component at the same time. The principle is that the interface is used by a component to provide state information. Instructions and commands are sent as messages. Then, multiple conflicting commands can be detected or they can be executed in sequence or in parallel, depending on the nature of the commands.

3   Advances to Functional Software Components

A plethora of different software components is required for a multi-robot system. Here, we discuss some components and advances of particular relevance to the game as played in 2015.

3.1   Basic Components

The lowest layer in our architecture which contains functional modules and hardware drivers. All functional components are implemented in Fawkes. Drivers have been implemented based on publicly available protocol documentation, e.g. for our laser range finders or webcams. For this year, we have extended the driver of our laser range finder for the Sick TiM571 model. To access the Robotino base platform hardware we make use of a minimal subset of OpenRobotino, a software system provided by the manufacturer. For this year, we have upgraded to using the version 2 API in order to use the Robotino 3 platform. Localization is based on Adaptive Monte Carlo Localization which was ported from ROS and then extended. For locomotion, we integrated the collision avoidance module [10] which is also used by the AllemaniACs[2] RoboCup@Home robot.

3.2   Light Signal Vision

A computer vision component for robust detection of the light signal state on the field has been developed specifically for this domain. For 2015, we have improved the detection component to limit the search within the image by means of the detected position of the machine, which is recognized through a marker and with the laser range finder. This provides us with a higher robustness towards ambiguous backgrounds, for example colored T-shirts in the audience. Even if the machine cannot be detected, the vision features graceful degradation by using

Fig. 6. Vision-based light-signal detection during production (post-processed for legibility).

a geometric search heuristic to identify the signal, losing some of the robustness towards the mentioned disturbances (Fig. 6).

---

[2] See the AllemaniACs website at http://robocup.rwth-aachen.de.

3.3   Conveyor Belt Detection

The conveyor belts are rather narrow compared to the products thus require a precise
handling. The tolerable error margin is in
the range of about 3 mm. The marker on a
machine allows to determine the lateral offset from the gripper to the conveyor belt. It
gives a 3D pose of the marker with respect to
the camera and thus the robot. However, this
requires a precise calibration of the conveyor
belt with respect to the marker. While ide-

Fig. 7. Vision-based conveyor belt
detection (training images).

ally this would be the same for each machine, in practice there is an offset which
would need to be calibrated per station. Combined with the inherent noise in the
marker detection this approach requires filtering and longer accumulation times.
The second approach is to detect a line using the laser scanner. This yields more
precise information which we hope to further improve with the higher resolution
of the Sick TiM571.

The third approach we are investigating is using a dedicated vision component to detect the conveyor belt in an image. So far, we have implemented a
detection method based on OpenCV's cascade classifier [11,12]. First, videos are
recorded that are split into images from which positive and negative samples are
extracted. These samples are then fed into a training procedure that extracts
local binary features from which the actual classifiers are built. This training
requires tremendous amounts of computing power and is generally done offline.
The detection, however, is swift and allows for on-line real-time detection of the
conveyor belt as shown in Fig. 7. We are currently evaluating the results and
considering further methods to improve on these results.

4   High-Level Decision Making and Task Coordination

The behavior generating components are separated into three
layers, as depicted in Fig. 8:
the low-level processing for perception and actuation, a mid-
level reactive layer, and a
high-level reasoning layer. The
layers are combined following
an adapted hybrid deliberative-
reactive coordination paradigm.

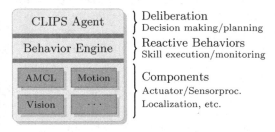

Fig. 8. Behavior layer separation

The robot group needs to
cooperate on its tasks, that is, the robots communicate information about their
current intentions, acquire exclusive control over resources like machines, and
share their beliefs about the current state of the environment. Currently, we
employ a distributed, local-scope, and incremental reasoning approach [5]. This

means that each robot determines only its own action (local scope) to perform next (incremental) and coordinates with the others through communication (distributed), as opposed to a central instance which plans globally for all robots at the same time or for multi-step plans.

In the following we describe the reactive and deliberative layers of the behavior components. For computational and energy efficiency, the behavior components need also to coordinate activation of the lower level components.

4.1   Lua-Based Behavior Engine

In previous work we have developed the Lua-based Behavior Engine (BE) [13]. It serves as the reactive layer to interface between the low- and high-level systems. The BE is based on hybrid state machines (HSM). They can be depicted as a directed graph with nodes representing states for action execution, and/or monitoring of actuation, perception, and internal state. Edges denote jump conditions implemented as Boolean functions. For the active state of a state machine, all outgoing conditions are evaluated, typically at about 15 Hz. If a condition fires, the active state is changed to the target node of the edge. A table of variables holds information like the world model, for example storing numeric values for object positions. It remedies typical problems of state machines like fast growing number of states or variable data passing from one state to another. Skills are implemented using the light-weight, extensible scripting language Lua.

4.2   Incremental Reasoning Agent

The problem at hand with its intertwined world model updating and execution naturally lends itself to a representation as a fact base with update rules for triggering behavior for certain beliefs. We have chosen the CLIPS rules engine [14], because using incremental reasoning the robot can take the next best action at any point in time whenever the robot is idle. This avoids costly re-planning (as with approaches using classical planners) and it allows us to cope with incomplete knowledge about the world. Additionally, it is computationally inexpensive. More details about the general agent design and the CLIPS engine are in [15].

The agent for 2015 is based on the previous years [15]. One major improvement is the introduction of a task concept. Until the 2014 version, the executive part of the agent consisted of separate rules for each action to perform. This resulted in a considerable amount of rules often repeating similar conditions. While this is not a problem in terms of performance, it does make maintaining the agent code more involved. For 2015, we have introduced the concept of tasks. Tasks group several steps necessary to perform a certain behavior, for example producing a low complexity product consisting only of a base element and a cap, into a single entity. Then, generic code can execute the steps of a task. We retained the flexibility and extensibility of our approach by allowing to add execution rules for steps which need special treatment or parametrization, and monitoring rules to react to disturbances during execution, for example a product being dropped during transport.

Additionally, we are progressing on making the world model synchronization generic. So far, we have had an explicit world model consisting of a specified set of elements. Now, we are aiming for a more generic world model consisting of key-value pairs. This will allow for simpler updates to the world model as modifications do not require changing an explicit world model schema anymore.

4.3 Multi-robot Simulation in Gazebo

The character of the RCLL game emphasizes research and application of methods for efficient planning, scheduling, and reasoning on the optimal work order of production processes handled by a group of robots. An aspect that distinctly separates this league from others is that the environment itself acts as an agent by posting orders and controlling the machines' reactions. This is what we call *environment agency*. Naturally, dynamic scenarios for autonomous mobile robots are complex challenges in general, and in particular if multiple competing agents are involved. In the RCLL, the large playing costs are prohibitive for teams to set up a complete scenario for testing, let alone to have two teams of robots. Members of related communities like planning and reasoning might not want to deal with the full software and system complexity, yet they welcome relevant scenarios to test and present their research. Therefore, we have created an *open simulation environment* [15,16] based on Gazebo.[3]

After RoboCup 2014 the BBUnits team joined the effort and we cooperatively extended the simulation for the new game as shown in Fig. 9. The simulation is developed publicly[4] and we hope for more teams to join the effort. With the new game, the need for a simulation is even more pressing as the cost for the field has drastically increased. Additionally, we envision that long-term games and future changes to the game can be tried in the simulation before implementing it in the real world [17].

Fig. 9. Simulation of the RCLL 2015 with MPS stations.

5   League Advancements and Continued Involvement

We have been active members of the Technical, Organizational, and Executive Committees and proposed various ground-breaking changes for the league like merging the two playing fields or using physical processing machines in 2015 [2,3]. Additionally we introduced and currently maintain the autonomous referee box for the competition and develop an open simulation environment.

---

[3] More information, media, the software itself, and documentation are available at http://www.fawkesrobotics.org/projects/llsf-sim/.

[4] The project is hosted at https://github.com/robocup-logistics.

5.1   RCLL Referee Box and MPS Stations

The Carologistics team has developed the autonomous referee box (refbox) for the RCLL which was deployed since 2013 [2]. It strives for full autonomy, i.e. it tracks and monitors machine states, creates (randomized) game scenarios, handles communication with the robots, and interacts with a human referee.

In 2015, the refbox required updates for the modified game involving MPS stations. A noteworthy difference is that in the current game, the products can no longer be tracked as they do not have an RFID chip anymore. There are plans to remedy this later, for example using bar codes or RFID chips again on products. This would in particular allow to grant production points not only on delivery, but once a production step has been performed. The new MPS stations require programs that provide sensor data to and execute commands from the refbox. For example, all stations require a setup step to be performed, for example to instruct the particular ring color a ring station should mount in the next step. Festo Didactic sponsored an internship for a student who performed the development embedded in the Carologistics team. It involved developing a program for each of the four station types to reliably carry out the production steps and implementing a robust communication to the stations. As Wifi is used for communication which is inherently unreliable in particular during RoboCup events, the refbox and stations must be robust towards temporary connection losses. The basic prototype was used successfully at the GO2015 and we are continuing the development to further stabilize and improve the game.

5.2   League Evaluation

We have made an effort [17] to analyze past games based on data recorded during games by the referee box and propose new additional criteria for game evaluation based on Key Performance Indicators (KPI) used in industrial contexts to evaluate factory performance. For example, one such KPI is the throughput time TTP which describes the time required at a specific station to complete processing and retrieve the processed workpiece. We have analyzed several games and will focus now on the finals of RoboCup 2014 between the Carologistics (cyan) and BBUnits (magenta) which ended with a score of 165 to 124.[5] It seems that especially the lower throughput time TTP of cyan contributed to the Carologistics' success. Machines can be used again much faster. This is shown by the green times in Fig. 10, which is effective utilization of the machines. The BBUnits team has much longer waiting times (orange) that resulted for one in machines to be blocked for other uses and for another meant that the production chain for a product was not advanced quickly. For more details we refer to [17]. For a generalization to mobile industrial robotics scenarios cf. [18].

We imagine that KPIs can play a role in the future of the league and could be tested in simulated tournaments before. The Technical Committee has added a technical challenge to the rules 2015 which involves performing such a simulated game in order to foster adoption and participation in the simulation efforts.

---

[5] Video of the final is available at https://youtu.be/_iesqH6bNsY.

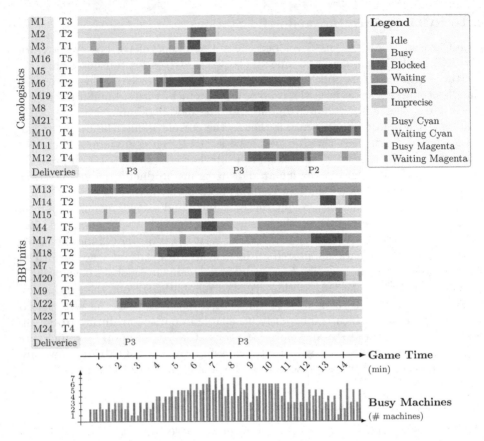

Fig. 10. Machine states over the course of the final game at RoboCup 2014. The lower graph shows the occupied machines per 20 s time block (Color figure online).

5.3   Public Release of Full Software Stack

Over the past eight years, we have developed the *Fawkes Robot Software Framework* [7] as a robust foundation to deal with the challenges of robotics applications in general, and in the context of RoboCup in particular. It has been developed and used in the Middle-Size [19] and Standard Platform [20] soccer leagues, the RoboCup@Home [21,22] service robot league, and now in the *RoboCup Logistics League* [15] as also shown in Fig. 11.

The Carologistics are the first team in the RCLL to publicly release their software stack. Teams in other leagues have made similar releases before. What makes ours unique is that it provides a complete and *ready-to-run package with the full software* (and some additions and fixes) that we used in the competition in 2014. This in particular *includes* the complete *task-level executive* component, that is the strategic decision making and behavior generating software. This component was typically held back or only released in small parts in previous

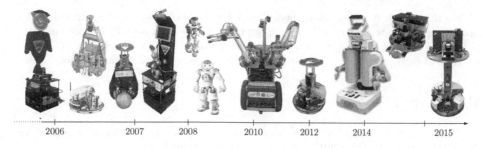

| 2006 | 2007 | 2008 | 2010 | 2012 | 2014 | 2015 |

Fig. 11. Robots running (parts of) Fawkes which were or are used for the development of the framework and its components.

software releases by other teams (for any league). We are currently preparing a similar release of the software stack 2015.

5.4 Agent-Based Programming for High-Level Control

In Sect. 4.2, we have discussed our incremental high-level reasoning component. Furthermore, we have used the RCLL scenario as an evaluation testbed for the agent-based programming and reasoning system YAGI [23]. YAGI (Yet Another Golog Interpreter) is an approach to develop an alternative implementation to the well-known logic-based agent control language GOLOG. It combines imperative programming with reasoning to formulate the behaviors of the robot.

6   Conclusion

In 2015, we have in particular adapted to the new game. We upgraded to the Robotino 3 platform, developed new custom hardware additions like a gripper, and adapted and extended the behavior and functional components. We have also continued our contributions to the league as a whole through active participation in the league's committees, publishing papers about the RCLL and proposing new performance criteria. The development of the simulation we initiated has been transferred to a public project where other teams have joined the effort. Most notably, however, have we released the complete software stack including all components and configurations as a ready-to-run package.

The website of the Carologistics RoboCup Team with further information and media can be found at http://www.carologistics.org.

Acknowledgments. The team members in 2015 are Daniel Ewert, Alexander Ferrein, Mostafa Gomaa, Sabina Jeschke, Gerhard Lakemeyer, Nicolas Limpert, Matthias Löbach, Randolph Maaßen, David Masternak, Victor Mataré, Tobias Neumann, Tim Niemueller, Sebastian Reuter, Johannes Rothe, and Frederik Zwilling.

We gratefully acknowledge the financial support of RWTH Aachen University and FH Aachen University of Applied Sciences.

We thank Sick AG and Adolf Hast GmbH & Co. KG for sponsoring our efforts by providing hardware and manufacturing support. F. Zwilling and T. Niemueller were supported by the German National Science Foundation (DFG) research unit *FOR 1513* on Hybrid Reasoning for Intelligent Systems (http://www.hybrid-reasoning.org).

References

1. Niemueller, T., Reuter, S., Ferrein, A.: Fawkes for the robocup logistics league. In: RoboCup Symposium 2015 - Development Track (2015)
2. Niemueller, T., Ewert, D., Reuter, S., Ferrein, A., Jeschke, S., Lakemeyer, G.: RoboCup logistics league sponsored by festo: a competitive factory automation benchmark. In: RoboCup Symposium 2013 (2013)
3. Niemueller, T., Lakemeyer, G., Ferrein, A., Reuter, S., Ewert, D., Jeschke, S., Pensky, D., Karras, U.: Proposal for advancements to the LLSF in 2014 and beyond. In: ICAR - 1st Workshop on Developments in RoboCup Leagues (2013)
4. Niemueller, T., Lakemeyer, G., Ferrein, A.: Incremental task-level reasoning in a competitive factory automation scenario. In: Proceedings of AAAI Spring Symposium 2013 - Designing Intelligent Robots: Reintegrating AI (2013)
5. Niemueller, T., Lakemeyer, G., Ferrein, A.: The robocup logistics league as a benchmark for planning in robotics. In: 25th International Conference on Automated Planning and Scheduling (ICAPS) - Workshop on Planning in Robotics (2015)
6. Karras, U., Pensky, D., Rojas, O.: Mobile robotics in education and research of logistics. In: IROS 2011 - Workshop on Metrics and Methodologies for Autonomous Robot Teams in Logistics (2011)
7. Beck, D., Lakemeyer, G., Ferrein, A., Niemueller, T.: Design principles of the component-based robot software framework fawkes. In: Ando, N., Balakirsky, S., Hemker, T., Reggiani, M., von Stryk, O. (eds.) SIMPAR 2010. LNCS, vol. 6472, pp. 300–311. Springer, Heidelberg (2010)
8. Quigley, M., Conley, K., Gerkey, B.P., Faust, J., Foote, T., Leibs, J., Wheeler, R., Ng, A.Y.: ROS: an open-source robot operating system. In: ICRA Workshop on Open Source Software (2009)
9. Gat, E.: Three-layer architectures. In: Kortenkamp, D., Bonasso, R.P., Murphy, R. (eds.) Artificial Intelligence and Mobile Robots, pp. 195–210. MIT Press, Cambridge (1998)
10. Lakemeyer, G., Beck, D., Ferrein, A., Schiffer, S., Jacobs, S.: Robust collision avoidance in unknown domestic environments. In: Baltes, J., Lagoudakis, M.G., Naruse, T., Ghidary, S.S. (eds.) RoboCup 2009. LNCS, vol. 5949, pp. 116–127. Springer, Heidelberg (2010)
11. Viola, P., Jones, M.: Rapid object detection using a boosted cascade of simple features. In: IEEE Conference on Computer Vision and Pattern Recognition (CVPR) (2001)
12. Lienhart, R., Maydt, J.: An extended set of haar-like features for rapid object detection. In: Proceedings of International Conference on Image Processing (2002)
13. Niemüller, T., Lakemeyer, G., Ferrein, A.: A lua-based behavior engine for controlling the humanoid robot nao. In: Baltes, J., Lagoudakis, M.G., Naruse, T., Ghidary, S.S. (eds.) RoboCup 2009. LNCS, vol. 5949, pp. 240–251. Springer, Heidelberg (2010)
14. Wygant, R.M.: CLIPS: a powerful development and delivery expert system tool. Comput. Ind. Eng. **17**(1–4), 546–549 (1989)

15. Niemueller, T., Ferrein, A., Lakemeyer, G., Reuter, S., Jeschke, S., Ewert, D.: Decisive factors for the success of the carologistics robocup team in the robocup logistics league 2014. In: Bianchi, R.A.C., Akin, H.L., Ramamoorthy, S., Sugiura, K. (eds.) RoboCup 2014. LNCS, vol. 8992, pp. 155–167. Springer, Heidelberg (2015)
16. Niemueller, T., Zwilling, F., Lakemeyer, G.: Simulation for the robocup logistics league with real-world environment agency and multi-level abstraction. In: Bianchi, R.A.C., Akin, H.L., Ramamoorthy, S., Sugiura, K. (eds.) RoboCup 2014. LNCS, vol. 8992, pp. 220–232. Springer, Heidelberg (2015)
17. Niemueller, T., Reuter, S., Ferrein, A., Jeschke, S., Lakemeyer, G.: Evaluation of the robocup logistics league and derived criteria for future competitions. In: RoboCup Symposium 2015 - Development Track (2015)
18. Niemueller, T., Lakemeyer, G., Reuter, S., Jeschke, S., Ferrein, A.: Benchmarking of cyber-physical systems in industrial robotics. In: Song, H., Rawat, D.B., Jeschke, S., Brecher, C., (eds.) Cyber-Physical Systems: Foundations, Principles and Applications. Elsevier, Amsterdam (2016) (in press)
19. Beck, D., Niemueller, T.: AllemaniACs 2009 Team Description. Technical report, Knowledge-based Systems Group, RWTH Aachen University (2009)
20. Ferrein, A., Steinbauer, G., McPhillips, G., Niemueller, T., Potgieter, A.: Team ZaDeAt 2009 - Team Report. Technical report, RWTH Aachen University, Graz University of Technology, and University of Cape Town (2009)
21. Schiffer, S., Lakemeyer, G.: AllemaniACs Team Description RoboCup@Home. Knowledge-based Systems Group, RWTH Aachen University, TDP (2011)
22. Ferrein, A., Niemueller, T., Schiffer, S., and G.L.: Lessons learnt from developing the embodied AI platform caesar for domestic service robotics. In: AAAI Spring Symposium 2013 - Designing Intelligent Robots: Reintegrating AI (2013)
23. Ferrein, A., Maier, C., Mühlbacher, C., Niemueller, T., Steinbauer, G., Vassos, S.: Controlling logistics robots with the action-based language YAGI. In: IROS Workshop on Task Planning for Intelligent Robots in Service and Manufacturing (2015)

KeJia-LC: A Low-Cost Mobile Robot Platform — Champion of Demo Challenge on Benchmarking Service Robots at RoboCup 2015

Yingfeng Chen$^{(\boxtimes)}$, Feng Wu, Ningyang Wang, Keke Tang, Min Cheng, and Xiaoping Chen

Multi-Agent Systems Laboratory, Department of Computer Science and Technology, University of Science and Technology of China, Hefei 230027, China
{chyf,wny257,kktang,ustccm}@mail.ustc.edu.cn,
{wufeng02,xpchen}@ustc.edu.cn
http://www.wrighteagle.org

Abstract. In this paper, we present the system design and the key techniques of our mobile robot platform called *KeJia-LC*, who won the first place in the demo challenge on Benchmarkinng Service Robots in RoboCup 2015. Given the fact that *KeJia-LC* is a low-cost version of our KeJia robot without shoulder and arm, several new technical demands comparing to RoboCup@Home are highlighted for better understanding of our system. With the elaborate design of hardware and the reasonable selection of sensors, our robot platform has the features of low cost, wide generality and good extensibility. Moreover, we integrate several functional softwares (such as 2D&3D mapping, localization and navigation) following the competition rules, which are critical to the performance of our robot. The effectiveness and robustness of our robot system has been proven in the competition.

Keywords: Low-cost mobile robot platform · Benchmarking service robots · RoboCup competition

1 Introduction

Competitions have been broadly used as an effective venue for promoting scientific and technological progress in robotic research. Many robotic competitions are held around the world every year, such as RoboCup competition, DARPA Robotics Challenge, Robot Cleaning Competition [1]. Particularly for the domestic service robots area, RoboCup@Home has been widely considered as the most influential testbed. It's nearly ten years since the first RoboCup@Home competition was held in 2006. Now we can clearly see the great advance that has been manifested by the competition in various robotic research including navigation, speech recognition, human-robot interaction and object recognition.

© Springer International Publishing Switzerland 2015
L. Almeida et al. (Eds.): RoboCup 2015, LNAI 9513, pp. 60–71, 2015.
DOI: 10.1007/978-3-319-29339-4_5

Most recently, a survey article [8] has been published for comprehensively evaluating this progress over the past decade.

Although the RoboCup@Home competition indeed boosts the research on domestic service robots, it still has limitation in the aspect of scientific significance [1]. To address this, the RoCKIn project [10] has been proposed and aims to design and initiate a scientific robotic competition using the global Motion Capture System (MoCap), which can provide the ground truth of robot's performance. The demo challenge on Benchmarking Service Robots (BSR) shares the same motivation to the RoCKIn. More importantly, it also proposes quantitative measurements for robotic research especially for low-cost platform.

The BSR workshop & demo challenge is a sub-event of RoboCup 2015, and successfully attracted ten teams to join in the competition lasting for three days. During the competition, our *KeJia-LC* robot demonstrated the abilities of 3D navigation, dynamic environment perception, accurate localization, etc. In the remainder of this paper, we will detail our robot hardware and software design that were developed for this competition.

2   Background of the Demo Challenge on BSR

The BSR competition has two main intentions that have been reflected clearly in the rules. Firstly, it attempts to accelerate the emergence of a standard mobile robot platform that should be low-cost, universal and extensible. Only a low-cost platform, at least not expensive, could possibly adapt to the affordability of different teams. The cost of the robots (shown in Fig. 1) in the RoboCup@Home league is usually higher than the acceptable level of some teams who are eager to participate. The expensive "threshold price" has been an issue hindering the further expansion of the RoboCup@Home league, and it is the reason why the cost of the robots in this competition is limited to a certain amount. Secondly, the BSR competition itself also intends to be more scientific, and even a benchmarking tool which could be treated as an objective performance evaluation of robot systems. As it is known, RoboCup@Home [12,13] is more focused on assessing the performance of integrated robot system through executing several high-level tasks, sometimes the teams just know the results that they fail in tests, but it's difficult for them to trace the causes and reproduce the runtime scenarios. The BSR competition utilizes a optical MoCap to obtain the ground truth of robot's trajectories, which provide a possible method to evaluate robots' performance using specific quantitative measurements.

At present stage, the rules of BSR competition mainly involve the precision of odometry, localization and navigation, the perception of environment and the reaction to dynamic change.

2.1   Rules and Skills

According to the rules[1], the whole BSR competition is divided into two stages and a final. In Stage I, the teams are required to complete several simple tasks

---

[1] http://www.robocup2015.org/upload/Benchmarking/Rules_draft_1.2release.pdf.

with odometer sensor under different load conditions, any other auxiliary sensors are prohibitive. The score of Stage I depends on the distance between the expected motion distance/angle and the actual measured values. In Stage II, robots are required to reach several waypoints in correct order under different load conditions. Between different waypoints, robots need to avoid different types of obstacles, such as hollow chairs, small objects and moving people. Perhaps, the most interesting obstacle is the double-arch door (shown in the Fig. 2), there are two slider beams in such a door and the height is adjusted randomly in the competition, hence the robot must decide whether and which channel it's able to pass through. The score of Stage II depends on the number of the waypoints robots reaches successfully and the precision of the robots' stopping positions. The form of the Final Stage is same with Stage II, but the degree of difficulty increases.

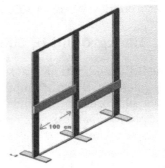

Fig. 1. Robots in RoboCup@Home 2015   **Fig. 2.** Double-arch door

Besides the mentioned rules, the robot must meet the cost restriction. Detailly, the hardware cost or the market retail price of the basic mobile platform should be less than 10,000 RMB (about 1,600 USD). Meanwhile, the price of any extended sensors should be not more than 50 % of the basic mobile robot platform. In addition, we also need to consider the load capacity of robot.

2.2 Different Requirements and Techniques with RoboCup@Home

Our team WrightEagle@Home have taken part in six consecutive RoboCup@Home competitions since 2010, all our related techniques are developed with the requirements of RoboCup@Home. In order to adapt to the new rules of BSR competition, we need to find out the unexplored technological demands.

Cost Restriction. The cost restriction has a great influence on the hardware design, we have to try out a set of hardware equipments which satisfy both the constrains of cost and performance. The situation of RoboCup@Home league is quite different, the teams are allowed to select any kind of equipment and sensors. In most case, a common laser scanner is much more expensive than the whole robot in the BSR competition.

Precision of Navigation. The BSR competition is more concerned about the precision of robot navigation, which directly determines the test score. Unlike in the RoboCup@Home competition, there is no specific restriction of reaching points, just within the tolerable range of where they are supposed to be.

Load Capacity. The load capacity of robot is seldom considered especially in domestic environment. In BSR competition, the mobile robot platform is expected at least to bear the weight of 20 kg (nearly the weight of a bag of rice), this regulation could be thought of increasing the versatility under different occasions.

Perception of 3D Environment. With the rapid development of sensor technology, many new 3D sensors (e.g., Microsoft Kinect, Xtion Pro Live) are emerging, which provide reliable perception of 3D environment. To perceiving obstacles in the BSR competition, the robots are strongly recommended to equip with additional 3D sensors rather than exclusive planar lasers. In fact, it's also hard to perceive the dynamic changes with cheap laser scanner.

Uneven Floor. In the latest rulebook of RoboCup@Home [11], it notes that minor unevenness such as carpet, transitions in floor covering between different areas, and minor gaps are expected. But in setting of real competition area, these thorny things are avoided to the greatest extent. In the BSR competition, the robot are explicitly expected to step over unevenness with the maximal height of 3 cm.

3   Hardware Design of *KeJia-LC*

The *KeJia-LC* robot platform aims to be a low-cost, general and extensible robot chassis. As shown in Fig. 3, the shape of the chassis is a square box with size of $50 * 50 * 30$ cm, which is suitable for passing across narrow passages and avoiding obstacles in common domestic environment. Under the metal shell of chassis, some metal strips constitute the framework of the platform, this structure gives the chassis good loading capacity (even the weight of an adult). In front of the chassis, there exists a groove with depth of 10 cm, it could endure heavy objects because the bottom of the groove is supported on the baseplate. The whole chassis is driven by differential structure, it consists two concentric wheels and a passive omni-directional wheel, which endows *KeJia-LC* platform with a good maneuverability and stability. On the chassis, an adjustable support bracket is fixed, where the two sensors (e.g., Kinect and laser scanner) are mounted.

In the hardware design of *KeJia-LC*, we have taken many issues into account, and some special concerns are listed as follow:

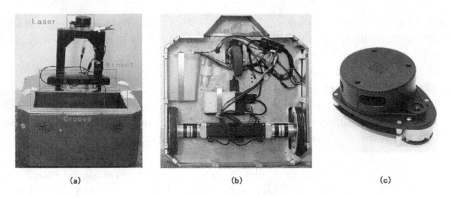

Fig. 3. (a) The front of the chassis (b) The bottom of the chassis (c) The RPLIDAR laser scanner

(1) *Selection Principles of Motor Drive:* The motor drive system mainly includes three parts: electric motor, motor reducer and motor controller, each of them has influence on the performance of the mobile platform and the cost. The mobile platform is expected to have a powerful motivator, therefore we choose a homemade high-quality DC motor with considerable volume. As to motor reducer, we adopt the planetary reducer rather than the harmonic reducer for these reasons: (a) The price of harmonic reducer commonly is much more expensive than planetary, which will greatly increase the budget. (b) The platform may suffer violent vibration when moving on uneven floor, which will cause serious damage to harmonic reducer but less to planetary. The chosen motor controller has rich control interfaces, moreover its price is reasonable. In general, the cost of the whole motor drive system is about 3,000 RMB.

(2) *Selection Principles of Sensors:* The extended sensors of *KeJia-LC* robot platform include a kinect and a laser scanner presently. The kinect has been widely used for 3D environment perception with many robots, and lots of related techniques have been developed, which are quite useful for our competition work. To seek for a low-cost, usable laser scanner is a quite hard job, fortunately, we found a product named RPLIDAR (shown in Fig. 3), it performs 360 degree scanning with the detection range of 6 m. With the usage of RPLIDAR[2], we implemented our mapping, localization and navigation modules.

4   Software Design

4.1   Odometry Parameters Calibration

According to the rules of Stage I, only the robot has accurate odometry without external sensors could get a high score. To get a precise odometry of our robot, we

---

[2] http://www.robopeak.com/blog/?p=587.

need to determine which parameters should be tuned accurately. As illustrated in Fig. 4, the movement of the robot platform is composed by rolling of the two wheels. The trajectory of the robot in a short time is usually considered as an arc (Fig. 5), thus, the calculation formulas are derived as:

$$
\begin{cases}
\Delta\theta = (l_2 - l_1)/d \\
\Delta x = sin\Delta\theta * (l_1 + l_2)/2\Delta\theta \\
\Delta y = (1 - cos\Delta\theta) * (l_1 + l_2)/2\Delta\theta \\
x' = x - sin\theta * \Delta y + cos\theta * \Delta x \\
y' = y + sin\theta * \Delta x + cos\theta * \Delta y \\
\theta' = \theta + \Delta\theta
\end{cases}
\tag{1}
$$

where (x, y, θ) is the odometry data at the last time, $(\Delta x, \Delta y, \Delta\theta)$ is the tiny movement in the robot's coordinate, the (x', y', θ') is the new calculated odometry data. Thereinto, the l_1, l_2 get from:

$$
l_i = \Delta E_i * 2\pi r/R \qquad (i = 1, 2)
\tag{2}
$$

In Eq. 2, the ΔE_i is the increased position values of encoder i in one update cycle and the R means the reduction ratio. It's obvious that the wheels distance d and the wheels radius r are the key parameters to the odometry precision.

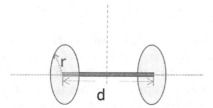

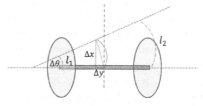

Fig. 4. The structural model of differential wheels

Fig. 5. The motion model of differential wheels

In order to simplify the tuning operation, we divide the process into two steps. In the first step, we choose to obtain the true value of the wheels radius r, because the odometry is only determined by r when moving straightly. With the great deal of feedback data collected by the MoCap, it's natural to the regard parameter tuning as a data fitting problem. We simply assume that the trajectories of robot straight movement are ideal lines, so just the displacements between the start positions and the stop positions need to be recorded by MoCap (ignoring the orientation). After the parameter r is determined, we command the robot to spin and collect the orientation measurements, then wheels distance d also can be computed with the same method.

4.2 Mapping, Localization and Navigation

Mapping Technique. The basic requirement of participants in Stage II is to navigate in the competition field autonomously. Before this, a presentation of the competition area should be built, hence the mapping technique is introduced. The particle filter based simultaneous localization and mapping (SLAM) [4,5] methods have been proven to be effective and pragmatic in robotic applications, and many mature implementations are available online. These mapping techniques work well in our case in general, and some changes are made in accordance with the practical situation. The range of the RPLIDAR laser is limited (6 m), which is often let the robot to be "blind" in open area. Owing to the accuracy of the well-tuned odometry, even if the laser data provide no sufficient information to ascertain robot pose, our robot still could keep a confident estimation of its position for certain long distance.

Besides the 2D grid map, we also created the 3D map of the environment with Kinect. RGB-D mapping have been a hot research subject recently [2,6], many algorithms are proposed. Here, we don't focus on making new progress on RGB-D mapping domain, but try to apply the existing approaches to our competition work. The mapping method from RTAB-MAP[3] [9] is employed to align the point cloud frame to global coordinate, the outcome is the point cloud map of environment. The drawbacks of the point cloud map are that the sensor noise and dynamic objects can't be handled directly and that it is not convenient to integrate with navigation module. After obtaining the point cloud map, we convert it to the OctoMap proposed by [14], which use a tree-based structure to offer maximum flexibility and updatability.

These two maps are created using different techniques and in the unified coordinate system, and in normal conditions, they are matched well with each other. Both maps describe the primary objects that rarely change in the environment, and the local map of navigation will be copied from the static maps and updated with the sensor data, which will be detailed in the next section.

Localization and Navigation. The localization is the prerequisites of accurate navigation. Our localization module is based on adaptive Monte Carlo localization method [3], which uses self-adapted particle filter to track the robot pose against a known 2d map. To acquire more precise pose, we add a scan match post-process with all the particles, which could eliminate the sampling error. The localization precision also depends on the resolution of the map, the resolution is usually set to 2–5 cm for common indoor environment. In consideration of the measurement accuracy of the cheap laser (1 % error within the range of 6 m), resolution lower than 2 cm is unnecessary for our application.

The major challenge of navigation module is to deal with the dynamic obstacles that can't be perceived only by planar laser. With the 3D map of the environment built previously, we project the spatial obstacles onto a plane with regard to the height of the robot [7]. Similar to the previous researches, our navigation

---

[3] http://introlab.github.io/rtabmap/.

module is combined with global and local planners. The local planner is operated on a moving local map centered with robot, at the very start the local map is copied from global static map, then updated with the coming of laser data and kinect data. With all these techniques, our robot is capable of avoiding all kinds of dynamic obstacles and even pass through narrow "tunnels" – not to mention the particular double-arch doors in the BSR competition.

4.3   Coordinate Transformation Between MoCap and Maps

The purpose of Stage II is testing the precision of the robot's stopping positions, which should be only affected by localization and navigation modules. But in fact, different teams have their own methods to build the maps of the competition area, it's impractical to force all maps to be in a same coordinates with MoCap. Thus, 10 coordinate values (in the MoCap coordinate) of landmarks are given to the teams (shown in the Fig. 6), all the landmarks are apt to discern in their own maps.

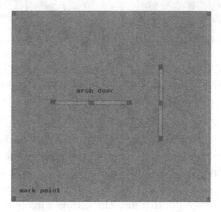

Fig. 6. The red points are the 10 selected landmarks (Color figure online)

Once the coordinate values are provided, the transformational relation between MoCap and robot's map can be uniquely identified by at least three pairs of matched points. In the real case, the mapping errors are not linearly distributed within the realistic scenarios. So, we should take full advantage of the 10 pairs of points to decrease the errors in the conversion.

As described above, we give a mathematical formalization of this problem. Let $P_i = [x_i, y_i, 1]$ represent the i point in the MoCap coordinate, the $P_i' = [x_i', y_i', 1]$ indicates the i point in map coordinate, and our aim is to get a transfer matrix T that could minimize the Euclidean distance:

$$\hat{T} = \arg\min \sum_i (P_i \cdot T - P_i')^2 \tag{3}$$

Given Eq. 3, the conversion error is suppressed fewer than 7 cm on average.

5   Competition Results

5.1   Competition Results

The competition field is shown as Fig. 7, the size of the area is about 7*7 m, covered by 12 distributed optical cameras of MoCap and 4 video cameras at each corner. A triple-marks set (Fig. 8) is installed on the robot, afterward, the robot will be regard as a rigid body attached with the set.

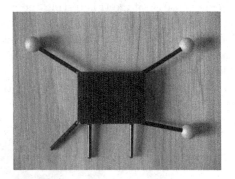

Fig. 7. The competition field of BSR

Fig. 8. The triple-marks set used for tracking robot's pose

The Stage I. In the Stage I, we chose all three weight of loadings (10 kg, 20 kg and optional 40 kg) and the ultimate results are listed in the Table 1[4]. The linear motion error is very small and neglectable with the well-tuned parameter (i.e., the wheels radius r). As to the rotate motion error, it's much more intractable. Under the different loading conditions, the same wheels distance parameter d results in diverse errors, which may be caused by the unbalanced weight distribution. Solving this problem, we calculates a set of parameters in different loading conditions, and in the competition, we chosen a proper parameter d from the list according to the actual loading weight.

The Stage II and Final. Before Stage II, we controlled the robot to go around the competition to establish the 2D and 3D maps. To overcome the limited range of RPLIDAR, we made the robot to move close to the wall and obstacles. The outcomes of the mapping process are shown in Fig. 10, the resolution of both types of maps is 5 cm (Fig. 9).

Our advantage in Stage II and the Final is quite significant: our *KeJia-LC* robot platform could avoid the objects flexibly and distinguish the state of the double-arch door quickly (Fig. 11). Besides, the hardware superiority played an

---

[4] The results of all teams can be seen from http://result.robocup2015.org/show/item?id=77.

Table 1. Score statistic of our team

Weight (kg)	Expected		Actual		Error rate
	Distance (m)	Angle (°)	Distance (m)	Angle (°)	
0	4	720	3.998	716.24	0.00333
20	4	720	3.999	721.22	0.00112
30	4	720	4.001	723.95	0.00339
40	4	720	4.003	721.65	0.00167

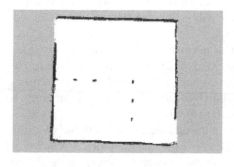

Fig. 9. The 2D map of the competition field

Fig. 10. The 3D map of the competition field

important rule, the ability of stepping over unevenness brought us extra bonus, making a big lead against all the other teams.

KeJia-LC robot reached all specified waypoints successfully and passed the arch-door twice smoothly, where the average error is about 8 cm, a little higher than the resolution of the maps. On the other hand, a frequently occurring flaw is that the robot has difficulty in making a fine-tuning when already near to the

Fig. 11. The robot is judging the state of the door

Fig. 12. The robot is passing through the opened door

waypoint, sometimes the robot rotated slowly for a long time to find a better stopping position (Fig. 12).

6   Conclusion and Future Work

In this paper, we present our *KeJia-LC* mobile robot platform, which is specially designed for the demo challenge on BSR in RoboCup 2015. In order to achieve high score in competition, related modules (including mapping, localization and navigation) were developed and integrated according to the rules. Relying on the stable and splendid performance of our robot in competition, out team won the first place with a big advantage.

Although the contradiction between the performance and cost is inevitable, while it's still possible to make a trade off. Our *KeJia-LC* robot is a try in this direction, the restriction of single sensor can be compensated with the data fusion from multiple low-cost sensors. How to adapt the mature robotic algorithms to new constrains is rarely explored and worth studying.

Inspired by the odometry parameters calibration, in the future, we could make a further step on the problem of parameter calibration, which is a universal issue, but it's often solved by manual measuring presently. With the sufficient measurements provided by MoCap, at least all the kinematic parameters would be determined automatically.

Acknowledgments. This work is supported by the National Hi-Tech Project of China under grant 2008AA01Z150 and the Natural Science Foundation of China under grant 60745002 and 61175057, as well as the USTC Key Direction Project and the USTC 985 Project. The authors are grateful to all the other team members, the sponsors and all the people who gives us helpful comments and constructive suggestions.

References

1. Amigoni, F., Bonarini, A., Fontana, G., Matteucci, M., Schiaffonati, V.: Benchmarking through competitions. In: European Robotics Forum-Workshop on Robot Competitions: Benchmarking, Technology Transfer, and Education (2013)
2. Engelhard, N., Endres, F., Hess, J., Sturm, J., Burgard, W.: Real-time 3d visual slam with a hand-held RGB-D camera. In: Proceedings of the RGB-D Workshop on 3D Perception in Robotics at the European Robotics Forum, Vasteras, Sweden, vol. 180 (2011)
3. Fox, D.: Adapting the sample size in particle filters through KLD-sampling. Int. J. Robot. Res. **22**(12), 985–1004 (2003)
4. Grisetti, G., Stachniss, C., Burgard, W.: Improving grid-based SLAM with Rao-Blackwellized particle filters by adaptive proposals and selective resampling. In: Proceedings of the 2005 IEEE International Conference on Robotics and Automation, ICRA 2005, pp. 2432–2437, April 2005
5. Grisetti, G., Stachniss, C., Burgard, W.: Improved techniques for grid mapping with Rao-Blackwellized particle filters. IEEE Trans. Robot. **23**(1), 34–46 (2007)

6. Henry, P., Krainin, M., Herbst, E., Ren, X., Fox, D.: RGB-d mapping: using kinect-style depth cameras for dense 3d modeling of indoor environments. Int. J. Robot. Res. **31**(5), 647–663 (2012)
7. Hornung, A., Phillips, M., Jones, E.G., Bennewitz, M., Likhachev, M., Chitta, S.: Navigation in three-dimensional cluttered environments for mobile manipulation. In: IEEE International Conference on Robotics and Automation (ICRA), pp. 423–429. IEEE (2012)
8. Iocchi, L., Holz, D., del Solar, J.R., Sugiura, K., van der Zant, T.: Robocup@home: analysis and results of evolving competitions for domestic and service robots. Artif. Intell. **229**, 258–281 (2015)
9. Labbé, M., Michaud, F.: Memory management for real-time appearance-based loop closure detection. In: IEEE/RSJ International Conference on Intelligent Robots and Systems (IROS), pp. 1271–1276. IEEE (2011)
10. Schneider, S., Hegger, F., Ahmad, A., Awaad, I., Amigoni, F., Berghofer, J., Bischoff, R., Bonarini, A., Dwiputra, R., Fontana, G., et al.: The rockin@home challenge. In: Proceedings of ISR/Robotik; 41st International Symposium on Robotics, pp. 1–7. VDE (2014)
11. van Beek, L., Chen, K., Holz, D., Matamoros, M., Rascon, C., Rudinac, M., des Solar, J.R., Wachsmuth, S.: RoboCup@Home2015: rule and regulations (2015). http://www.robocupathome.org/rules/_rulebook.pdf
12. van der Zant, T., Wisspeintner, T.: RoboCup X: a proposal for a new league where robocup goes real world. In: Bredenfeld, A., Jacoff, A., Noda, I., Takahashi, Y. (eds.) RoboCup 2005. LNCS (LNAI), vol. 4020, pp. 166–172. Springer, Heidelberg (2006)
13. Wisspeintner, T., Van Der Zant, T., Iocchi, L., Schiffer, S.: Robocup@home: scientific competition and benchmarking for domestic service robots. Interact. Stud. **10**(3), 392–426 (2009)
14. Wurm, K.M., Hornung, A., Bennewitz, M., Stachniss, C., Burgard, W.: Octomap: a probabilistic, flexible, and compact 3d map representation for robotic systems. In: Proceedings of the ICRA Workshop on Best Practice in 3D Perception and Modeling for Mobile Manipulation, vol. 2 (2010)

RoboCup SPL 2015 Champion Team Paper

Brad Hall, Sean Harris$^{(\boxtimes)}$, Bernhard Hengst, Roger Liu, Kenneth Ng,
Maurice Pagnucco, Luke Pearson, Claude Sammut, and Peter Schmidt

School of Computer Science and Engineering, University of New South Wales,
Sydney 2052, Australia
sharris@cse.unsw.edu.au
http://www.cse.unsw.edu.au

Abstract. The Robocup Standard Platform League competition is a
highly competitive league, with very little separating the top teams.
Winning the competition in consecutive years is particularly challenging as other teams look to counter the tactics and game play of the
previous champions. As the reigning champions from 2014, team UNSW
Australia was able to overcome this challenge and win the competition
for a second consecutive year. Although this success is not only related to
developments from this year, this paper focuses on the new innovations
and development by team UNSW Australia for the 2015 Robocup Competition. These innovations include white goal detection, whistle detection, foot detection and avoidance, improved path planning and new
odometry.

1 Introduction

Team *UNSW Australia*, formerly known as *rUNSWift*, has been competing in the
Standard Platform League (SPL) since 1999. We were world champions 3 times
in the years 2000–2003, but were unable to regain that title until 2014, when we
won the competition for the first time in eleven years. In 2015 we successfully
defended our title, to become back-to-back champions for the first time since
2000–2001.

The competition in 2015 introduced new challenges for the teams. The goal
posts changed from being a distinct yellow colour, to being a standard white,
making them significantly less unique and harder to detect. The finals matches
were started by a human referee blowing a whistle, rather than a wifi packet.
Teams were also permitted to wear unique jersey colours, instead of the previous
cyan and magenta standard colours.

In addition to this, the 2015 team also faced UNSW specific challenges. After
a successful 2014 campaign involving a large team of alumni and current students, we had an exceptionally high turnover rate. This lead to almost an entirely
new team, who had little to no experience with the code base before 2015. This
also resulted in a smaller team, and thus limited the scope of new innovations
in 2015.

© Springer International Publishing Switzerland 2015
L. Almeida et al. (Eds.): RoboCup 2015, LNAI 9513, pp. 72–82, 2015.
DOI: 10.1007/978-3-319-29339-4_6

The 2015 UNSW Australia team members are Sean Harris, Roger Liu, Kenneth Ng, Luke Pearson, Peter Schmidt and faculty members Brad Hall, Bernhard Hengst, Maurice Pagnucco and Claude Sammut, many of whom are shown in Fig. 1.

Fig. 1. A subset of the 2015 team. From left to right: Roger Liu, Sean Harris, Kenneth Ng, Luke Pearson, Peter Schmidt. Kneeling: Brad Hall

The success of our campaign cannot be attributed to a single contribution, nor only this year's improvements. Our success in 2015 was a combination of previous success [1], strong team culture and continued innovation in all areas of robot soccer. This paper outlines the innovations of the 2015 team, including new goal detection, whistle detection, foot detection and avoidance, path planning and odometry.

2 Team Organisation and Development Methodologies

Consistent team organisation and routines have played a significant role in the success of the rUNSWift team in recent years. Despite a significant turnover of students moving from 2014 to 2015, the team maintained the regular meeting and testing routines used in 2014.

The team would meet once a week, utilising Google Hangouts when geography made physical meetings challenging. At these meetings each team member would outline their progress over the previous week, ask questions about topics that were challenging and set goals for the next week. This process made all team members accountable for their actions (or lack thereof) over the past week and ensured everyone remained focused. It also allowed team members to ask for assistance on difficulties they had experienced and fostered a culture of support for other team members.

The team also maintained regular testing schedules. The simplest of these is a 'striker test', where a single robot and the ball are placed in set positions around the field, and the time taken to score a goal is measured. This tests the integration of all our modules and improvements over the past week to ensure that changes are actually improving our overall soccer play. We also ran regular practice drills and matches involves multiple robots to test team play, positioning and obstacle avoidance.

3   Whistle Detection

In 2015 the SPL Rules changed to have all finals games start by a human referee's whistle sound. In the past a wifi packet was used to signal the kick off. The wifi packet was not removed in 2015, but instead was delayed 15 s, providing a significant advantage to teams who were able to successfully and reliably detect the whistle. We identified two key problems to solve in successfully implementing a whistle detector. The first was to detect true positive whistles reliably, and the second to avoid false positives from other noises or far away whistles.

To assist with this, the free open source software tool Audacity was used. Figure 2 contains an example Audacity project. Conceptually from Fig. 2 our whistle detection algorithm looks for a white or red rectangle in the 2000–4000 Hz range lasting for at least 250 ms.

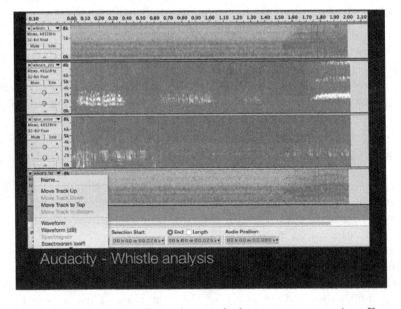

Fig. 2. An Audacity project with tracks switched to spectrogram view. From top to bottom: A quiet whistle in the UNSW Robocup lab, a loud whistle at Hefei competition, a gear noise false positive and a far away whistle in the lab (Color figure online).

To detect a whistle signal, the algorithm integrates 3 main ideas on the NumPy Fast Fourier Transform of raw audio data, henceforth referred to as the spectrum. Firstly, we adaptively grow background noise zones by discarding 200 Hz sections from the low and high end of the spectrum, leaving a smaller signal to focus on. Secondly, we only count the spectrum if the remaining signal is statistically quiet relative to the whole spectrum. Thirdly, we only count the spectrum if it is statistically quiet compared to the spectra recorded over the previous second.

Our whistle detection code runs in a separate Python process, and is fairly resource intensive. It consumes approximately 30–40 % of CPU time on the Nao, which is far too expensive to run at all times. Thus we only trigger the whistle detection once the game state switches to 'Ready', which means we are expecting a whistle shortly. Once the game begins we then turn off the whistle detection until the next time it is required. Figure 3 shows the usage and development cycle for the whistle detection system. All previously recorded whistle sounds are downloaded after a game and the regression test is updated to include the new samples.

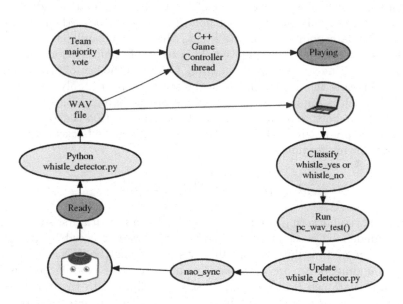

Fig. 3. Whistles: overview of the development and data flows through the Nao. Green circles indicate a game state change. Note that the laptop subsystem is external and occurs after each game (Color figure online).

To improve the reliability of the overall team system, the final decision for whether or not the game has begun is made by a team vote. Once a majority of players agree that the whistle has been blown, the whole team will start playing, regardless of what they voted. This system allowed us to handle infrequent false positives and true negatives by repeating the experiment across all 5 robots and taking the majority decision.

At the competition, our team was perfect at detecting whistles. The team never false started before the whistle had been blown and also never failed to start on a correct whistle. This was despite numerous nearby whistles during games that often fooled human spectators. We were the only team to have a perfect record in detecting the whistle at the 2015 competition.

4   Foot Detection and Avoidance

Foot detection and avoidance was introduced into the 2015 rUNSWift codebase with the goal of reducing the total number of falls that occur as a result of stepping on another robot's foot.

Foot detection works by scanning image for strong edge points. Following this it removes straight lines from the image with RANSAC. It then performs a custom point filtering algorithm which aims to remove points that are unlikely to be part of a robot. The main assumption here is that any vertical rectangular region of non-green coloured pixels is likely to be a robot. Thus we identify any points that have a substantial run of non-green pixels above them and mark them as candidate points. Then any points that aren't candidate points, nor are near candidate points, are removed. This results in points belonging to only field lines being removed, whilst maintaining points on the feet of nearby robots. Following this only the bottom points in the set are kept, i.e. only one point per x location which is minimal in y. This works on the assumption that the foot should be at the bottom of the image. A successful detection is shown in Fig. 4.

Finally the points are bucketed and the midpoint of the bucket is calculated using the following formula:

$$Midpoint = (\frac{\sum_{i=0}^{|B|} B_{xi}}{|B|}, \frac{\sum_{i=0}^{|B|} B_{yi}}{|B|}) \tag{1}$$

Following the bucketing, a region is grown using a breadth first search (BFS) that is bounded by a set of rules. A bounding box is created by taking the minimum x, y, and maximum x, y values visited during the BFS.

The bounding boxes are then passed onto the walk module and used to clip the walk parameters as required. A step is defined with three parameters, *forward*, *left* and *turn* as shown in Fig. 5. *Forward* specifies the intended x location of the foot after the step in coordinates relative to the robot, *left* specifies the intended y location of the foot, and *turn* specifies the rotation of the foot during the step. These parameters are adjusted to be within the physical capabilities of the robot. The parameters are then tested to see whether they fall within a bounding box, which represents the opponent's feet. This is done by intersecting a circle, which represents the toe of our robot, with the set of opponent feet. If the toe lands in any of the boxes the parameters are scaled down. This process is repeated until no intersection occurs.

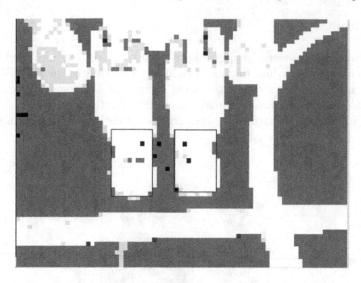

Fig. 4. The colour saliency provided from offnao (our offline debugger) with the feet bounding boxes shown in blue. The image also includes a difficult case of field line intersection which has not been classified as feet (Color figure online).

Foot detection and avoidance proved to be substantially better than having no such system in place. We measured this by examining "danger instances", which represent instances where it was possible to step on, and detect, the opponent's toes. Our 2014 matches had 1 fall every 4.7 danger instances, and our 2015 matches had 1 every 10.1 danger instances. In 2014 there was a 21.2 % chance of stepping on a toe in every danger instance, this percentage is reduced to 9.8 % with foot detection and avoidance.

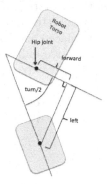

Fig. 5. The walk parameters and the resulting location of the foot.

5 Motion

Competitive bipedal soccer playing robots need to move fast and react quickly to changes in direction while staying upright. An overview of the UNSW motion module is provided in an internal technical report [2]. This report describes our approach to reactive omni-directional locomotion for the Nao robot as used in the RoboCup Standard Platform League competitions in 2014 and 2015 (Fig. 6).

Fig. 6. Nao robot and simulated version. The box-like rendition of the ODE simulated Nao has been programmed with the precise dimensions, masses, and joint-locations from the manufacturer's specification.

We have used reinforcement learning to learn a policy to stabilise a flat-footed humanoid robot using a physics simulator. The learned policy is supported theoretically and interpreted on a real robot as a linearised continuous control function [3].

An interpreter between the behaviour module and the walk engine was introduced in 2015. High level behaviour requests such as lining up to the ball and dribbling are broken down by the interpreter into multiple walk parameters for the walk engine. The interpreter is placed just before the walk engine. Given the state of the walk cycle and the relative ball position, the interpreter ensures precision of steps and a fluid transition between walking and kicking.

Dynamic step size adjustment was improved in 2015. A superellipsoid with axes that represented the maximum forward, side and rotational step was used as a model to constrain directional walk parameters within smooth boundaries.

6 Odometry

A visual odometry module was developed in 2012 that estimated the true change in robot's heading by monitoring changes in pixels between vision frames. This is important during external disturbances such as collision with another robot.

In 2015, z-axis gyroscope measurement was added to Nao H25, which reliably replaced the visual odometry module. The change in heading reported by the walk engine is substituted with z-axis gyroscope measurement. To avoid drift that occurs when integrating biased gyroscope measurement, walk odometry heading is only replaced when the rates are sufficiently different. The resulting odometry is more accurate and robust to external disturbances.

7 Path Planning

Path planning and obstacle avoidance was improved in 2015 by utilising a potential-field based approach, similar to [4]. The idea is that the robot is moved by the vector sums of the attractive force of the target position and repulsive forces from the obstacles. This is achieved by defining potential functions where the obstacles are on "high ground" and the target is at "low ground". The robot follows the negative gradient of the total potential to get to the target. The method uses little computation and is highly effective.

8 Goal Detection

As part of the 2015 rule change, goalposts were converted from yellow to white. This change in colour caused the goalposts to change from being a unique color on the field to becoming the second most common colour, behind only green. The rUNSWift localisation architecture is highly sensitive to false positives, so it was imperative that goal detection for white goals minimised false positive detection. It was also important to still maintain a high detection rate though, so that the robot could remain localised.

The first step in the White Goal detection algorithm is to find regions that may potentially contain goal posts. We call these regions "candidate regions". We identify candidate regions based on an isolated property of the goalposts, which is that regardless of the positioning of the robot relative to a goalpost, the base of the goalpost should always intersect with the field edge. With this in mind, our method for generating candidate regions is to scan along the edge of the field and mark pixels identified as white as starting points for candidate regions. Figure 7 shows an example goal intersecting the field edge.

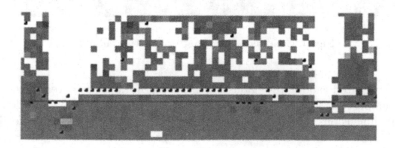

Fig. 7. The detected field edge is shown as a red line (Color figure online)

For each potential starting point, a rectangular region is bound downwards and towards the right whilst the region remains white. This rectangular bound ideally corresponds to the whole base of the goalpost, starting from the field edge to its base.

The candidate region generator uses every white point along the field edge as a starting point to generate rectangular regions. As a result, we are required to cull our candidates such that only the most appropriate regions are left. As such, any candidate region that contains another candidate region, but does not extend further downwards or further outwards (towards the right) are culled from the candidate list. We are then left with the longest candidate regions in each section. An example is provided in Figs. 8 and 9.

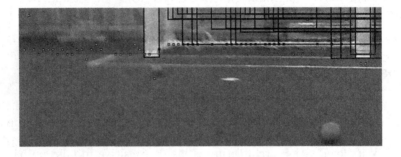

Fig. 8. Candidate regions before selective culling

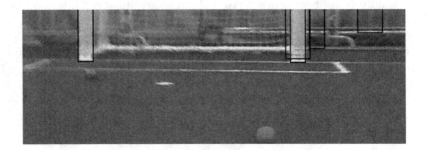

Fig. 9. Candidate regions after selective culling

The generated candidate regions are extended upwards to cover the entirety of the goalpost above the field edge. This is done through prior knowledge of the goal dimensions, in particular the ratio between the width and height of the goal post. As areas on the field are less prone to noise from background, we are able to reliably determine the width of a goalpost based on the amount of obstruction it imposes on the field edge. This width is then multiplied by 9 (the ratio between the height and width of a goalpost) to find the estimated position of the top of the goalpost. The final region is checked against a series of "sanity checks", such as the ratio of colour within the region, to ensure that it meets the criteria for being labelled a goal post.

The effectiveness of the 2015 White Goal Detection was measured through lab testing of placing a Nao at particular points on the field and performing a 360° rotation. The total number of frames for which a goalpost was correctly classified and incorrectly classified were recorded with a 94 % True Positive rate and a 0.4 % false positive rate. These results are shown in Figs. 10 and 11.

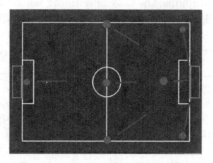

Fig. 10. Goal detection test reference positions

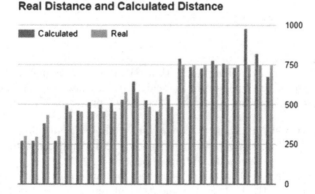

Fig. 11. Calculated distance accuracy

9   Concluding Discussion

After the competition the team members made a list of future developments to overcome weaknesses and add new functionality as a starting point for next year's team. The primary feature identified as needing a big overhaul was the vision system. It is very reliant on accurate colour classification, which can be difficult to achieve with unregulated lighting conditions. The SPL is continuing to move away from distinct colours and regulating the lighting on the field, so we need to make our systems more robust. Our team play is also heavily dependant on reliable wifi connections, which are often not reliable at the competition. This was highlighted by our multiple own goals during the early pool stages when delayed or non-existent wifi packets caused problems for our localisation.

The 2015 team started with a championship quality code base, but also faced a large number of challenges in maintaining and improving the system. With a high student turnover, and many other teams tailoring their strategies to defeat the defending champions, the task of winning back-to-back championships is always challenging. The continued success of team UNSW Australia cannot be attributed to a single factor, or to a single year, but to a combination of integrated software development and team collaboration over several years.

Acknowledgements. The 2015 team wish to acknowledge the legacy left by previous rUNSWift teams and the considerable financial and administrative support from the School of Computer Science and Engineering, University of New South Wales. We wish to pay tribute to other SPL teams that inspired our innovations in the spirit of friendly competition, especially to Thomas Hamboeck and the Austrian Kangaroos for their insights into whistle detection methods.

References identified as UNSW CSE Robocup reports and other Robocup related references in this paper are available in chronological order from: http://cgi.cse.unsw.edu.au/~robocup/2014ChampionTeamPaperReports/.

References

1. Sushkov, O., Ashar, J., Sammut, C., Teh, B., Ashmore, J., Roy, R., Mei (Jacky), Z., Tsekouras, L., Harris, S., Hengst, B., Hall, B., Liu, R., Pagnucco, M.: RoboCup SPL 2014 champion team paper. In: Bianchi, R.A.C., Akin, H.L., Ramamoorthy, S., Sugiura, K. (eds.) RoboCup 2014. LNCS, vol. 8992, pp. 70–81. Springer, Heidelberg (2015)
2. Hengst, B.: rUNSWift Walk 2014 report, University of New South Wales (2014). http://cgi.cse.unsw.edu.au/~robocup/2014championteampaperreports/20140930-bernhard.hengst-walk2014report.pdf
3. Hengst, B.: Reinforcement learning inspired disturbance rejection and Nao bipedal locomotion. In: 15th IEEE RAS Humanoids Conference, November 2015
4. Hwang, Y., Ahuja, N.: A potential field approach to path planning. IEEE Trans. Robot. Autom. **8**(1), 23–32 (1992)

Team Homer@UniKoblenz —
Approaches and Contributions
to the RoboCup@Home Competition

Viktor Seib[✉], Stephan Manthe, Raphael Memmesheimer,
Florian Polster, and Dietrich Paulus

Active Vision Group (AGAS), University of Koblenz-Landau,
Universitätsstr. 1, 56070 Koblenz, Germany
{vseib,smanthe,raphael,fpolster,paulus}@uni-koblenz.de
http://agas.uni-koblenz.de, http://homer.uni-koblenz.de

Abstract. In this paper we present the approaches and contributions
of team homer@UniKoblenz that were developed for and applied during
the RoboCup@Home competitions. In particular, we highlight the different abstraction layers of our software architecture that allows for rapid
application development based on the ROS actionlib. This architectural
design enables us to focus on the development of new algorithms and
approaches and significantly helped us in winning the RoboCup@Home
competition in 2015. We further give an outlook on recently published
open-source software for service robots that can be downloaded from our
ROS package repository on http://wiki.ros.org/agas-ros-pkg.

Keywords: RoboCup@Home · Service robots · homer@UniKoblenz ·
Robot lisa · Robotic architecture

1 Introduction

This paper introduces our team homer@UniKoblenz and its scientific approaches
and contributions to the RoboCup@Home competition. Its predecessor team,
resko@UniKoblenz, was founded in 2006 by the scientific supervisor and the
students of a practical course in the curriculum of the Active Vision Group
(AGAS). The team and its robot *Robbie* participated in the RoboCup Rescue
league. Among others, resko@UniKoblenz won twice the RoboCup Rescue championship in the category "autonomy" (2007 and 2008).

Because of rising demands on hardware engineering in the Rescue league we
put our focus on the @Home league in 2008. This league offered a greater opportunity to follow the original research interests of our group, vision and sensor
data processing, while having significantly less demands on hardware engineering
and mechatronics. While using the rescue robot for the @Home competition in
2008, one year later we built our service robot *Lisa*. In the beginning, Lisa's software was largely based on Robbie's, however, over time many new components

© Springer International Publishing Switzerland 2015
L. Almeida et al. (Eds.): RoboCup 2015, LNAI 9513, pp. 83–94, 2015.
DOI: 10.1007/978-3-319-29339-4_7

were added, specializing on typical at home tasks like human-robot interaction, object perception and manipulation.

Since our first participation in the RoboCup@Home league, Lisa and homer-@UniKoblenz won many awards, among others the Innovation Award (2010) and the Technical Challenge Award (2012). However, the greatest success in the team's history was the 1st place in the RoboCup@Home competition in 2015.

So far, we collaborate with the RoboCup teams Pumas [5] and Golem [9] by organizing research visits and workshops on the Robot Operation System (ROS) [12]. Further, we published some of the integral software components needed to create a service robot on our ROS package repository website[1]. These components are explained in great detail in a contributed chapter in a book on ROS [4].

To this day, the team consists of students pursuing their Bachelor and Master studies and is supervised by a PhD student of the Active Vision Group. Consequently, most of the students leave the team after only one year and new students take their places. Therefore, it is very important to provide an architecture that is easy to learn and to maintain to allow new students to start developing new algorithms very quickly. This paper introduces our approaches to accomplish these goals and is structured as described in the following.

Section 2 shortly introduces our robot Lisa and its hardware components. The software architecture used by our team is presented in Sect. 3. Different abstraction layers are introduced and related to the service robot architecture structure proposed by Pineda et al. [10]. The algorithms and contributions used by our team in the competitions are explained in Sect. 4. This is followed by Sect. 5 that presents recent and ongoing research in the Active Vision Group that is incorporated into our robot's software. Finally, Sect. 6 summarizes this paper and gives an outlook to our future plans and research.

2   Service Robot Lisa

In this year's competition we used two robots (Fig. 1). The blue Lisa is our main robot and is built upon a CU-2WD-Center robotics platform[2]. The old Lisa serves as an auxiliary robot and uses the Pioneer3-AT platform. Every robot is equipped with a single notebook that is responsible for all computations. Currently, we are using a Lenovo Thinkpad W520 equipped with an Intel Core i7-2670QM processor, 12 GB of RAM with Ubuntu Linux 14.04 and ROS Indigo.

Each robot is equipped with a laser range finder (LRF) for navigation and mapping. A second LRF at a lower height serves for small obstacle detection.

The most important sensors of the blue Lisa are set up on top of a pan-tilt unit. Thus, they can be rotated to search the environment or take a better view of a specific position of interest. Apart from a RGB-D camera (Asus Xtion) and a high resolution RGB camera (IDS UI-5580CP-C-HQ), a directional microphone (Rode VideoMic Pro) is mounted on the pan-tilt unit.

---

[1] AGAS ROS packages: http://wiki.ros.org/agas-ros-pkg.

[2] Manufacturer of our robotic platform: http://www.ulrichc.de.

Fig. 1. Lisa (in blue) on the left is our main robot. The Lisa (right) serves as auxiliary robot (Color figure online).

A 6 DOF robotic arm (Neuronics Katana 400HD) is used for mobile manipulation. It is certified for a safe operation around humans and is able to manipulate light-weight objects up to 0.5 kg. The end effector is a custom setup and consists of 4 Festo Finray-fingers.

Finally, a Raspberry Pi inside the casing of the blue Lisa is equipped with a 433 MHz radio emitter. It is used to switch device sockets and thus allows to use the robot as a mobile interface for smart home devices.

3   Software Architecture

Since the foundation of our team we were using a message-based software architecture. In the beginning we used the middleware *Robbie* [22] that was developed in our group. Robbie is a modular architecture that uses messages to exchange data between modules. Each module runs in its own thread and interfaces libraries, hardware drivers or third-party applications. With the advent of ROS we started a step by step conversion of our software modules, which was finished in 2014. Because of the similar design ideas in both architectures, the conversion was easy to manage. Using ROS as middleware facilitates the integration of existing packages, but also sharing our own code with the growing ROS community. Further, we do not need to maintain the middleware and, thus, are able to concentrate on developing new algorithms. This is particularly important, since each year many experienced students are replaced by new ones. However, like in the time of the Robbie architecture, we still adhere to the idea of encapsulating algorithms in stand-alone libraries that are connected by ROS nodes to the application.

In the following we want to highlight some aspects of our architecture and introduce our approaches in implementing RoboCup@Home tests.

3.1   Abstraction Levels

Although our software mainly consists of ROS packages and libraries, several different abstraction layers can be distinguished. We identify the following levels: the functionality level, the behavior level and the application level. Components at each abstraction level are allowed to use only modules from the same or a lower level.

The lowest level is the functionality level. It is formed by nodes that directly access a library or a hardware driver to fulfill one functionality or execute a specific algorithm. Examples are interfacing the robot platform or a sensor, but also executing algorithms like path planning or object recognition. To avoid fragmentation into too small components, these nodes are not limited to exactly interface one algorithm or device. For instance, the node responsible for planning an arm trajectory for grasping also accesses the controller that is responsible to execute the planned path. All these nodes have a rather simple structure, mostly taking a request and providing a response. Because of this type of utilization in most cases it is favorable to use ROS services as interfaces, rather than messages.

The second level is the behavior level. This level is composed by nodes employing the ROS *actionlib*[3]. The actionlib provides standardized interfaces to define and execute (long running) preemptable tasks. While the task is executed, periodic feedback is provided on the state of the task execution. In our system, actions are defined at different levels of granularity. Some actions merely encapsulate one functionality (e.g. voice synthesis), while others contain more complex tasks like grasping an object. The latter example includes detecting and recognizing the object, positioning the robot accordingly and finally moving the arm while avoiding obstacles. System components on this abstraction level are allowed to use other actions (mostly of minor granularity) to facilitate composition of more complex behaviors. In general, complex actions are implemented using state machines that react to different outcomes of sub-task execution. We are currently working on integrating more versatile error handling and feedback for every action.

The third and highest abstraction level is the application level. This level contains predefined or dynamic sequences of behaviors and functionalities, e.g. the RoboCup@Home tests. At this level, a high-level task specification language as proposed in [11] or [24] might be handy to facilitate complex behavior description and application composition. In our case we decided to use the same programming language (C++) as was used in all other abstraction levels, to facilitate the integration of new students. However, we implemented special encapsulating interfaces to be used at this level that significantly facilitate application implementation as compared to pure state machines and ROS interfaces. Subsection 3.3 provides more details on these interfaces.

---

[3] ROS actionlib documentation: http://wiki.ros.org/actionlib.

3.2   Relation to Service Robot Concepts

In [10] Pineda et al. discuss a concept and functional structure of a service robot. They identify the following three system levels in service robot design: system programming, robotic algorithms and the functional specification. These three system levels are related to Newel's levels of computational systems [8], stating that the first two layers can be reduced to more basic components. However, the functional specification level emerges from the lower levels providing additional knowledge about a task structure and, thus, can not be reduced to components of lower levels. For this reason, Pineda et al. propose a specialized task specification language [11].

They further propose two functionality layers: *Application* and *Behaviors*. The Application layer, being the higher layer, defines the task structure and conceptually corresponds to our application level. The Behaviors layer contains structured hierarchical objects corresponding to basic tasks and conceptually resembles our two basic levels: the functionality and the behavior level. In the hierarchical objects on the Behavior layer of Pineda et al., the lowest hierarchy level can be compared to our functional level, whereas the higher hierarchies are similar to our behavior level. However, the exact relation of our system levels to the ones proposed by Pineda et al. is not that clear in this case.

Another crucial component in service robot design identified in [10] is task management. Task management allows the robot to assess its current state, react to unexpected situations and perform proper error handling and recovery. To this date our system architecture lacks a unified task management and task planning component. Although basic task management strategies exist, they are loosely coupled and do not properly relate to the abstraction layers of our system architecture. Since task management is an important component, we are currently focusing on developing a task management concept that best suits our demands.

3.3   Implementing RoboCup@Home Tests

RoboCup@Home tests consist of different behaviors and functionalities that are combined in a specific way to create a desired application. Hence, they reside on the application abstraction level. Pineda et al. [10] identify two different types of RoboCup@Home tests: static and dynamic tasks. While almost all RoboCup@Home tests describe a predefined sequence of actions and are therefore considered static, the *General Purpose Service Robot* (GPSR) test is the only one with a dynamic task structure.

To enable simple static task implementation, we have created a utility module that encapsulates all behaviors our robot can execute. The most simple test is the *Robot Inspection*, where the robot is waiting in front of a closed door. As soon as the door is opened, the robot navigates to a predefined inspection position. After being inspected the robot receives a speech command to exit the arena. In our software architecture, the implementation for this test would look like this:

```
1   ros::NodeHandle n;
2
3   Robot lisa(n);
4   lisa.waitForDoor();
5   lisa.navigateTo("inspection");
6   lisa.speak("I am ready for inspection");
7   lisa.speak("Say 'exit now' if you want me to exit");
8   lisa.waitForCommand("exit now");
9   lisa.navigateTo("exit");
```

Of course this is a simple example where the robot does not have to react to the result of one command to decide which command is next or which parameters to pass to the next command. In practice, even static tasks are rich in control structures like loops and conditional clauses. However, these programs are still easy to read and task oriented, by hiding the implementation details.

Note that all of the behaviors in the above example are blocking. The next command is only executed when the previous one was completed. Nevertheless, sometimes a non-blocking version is required, which is also available in the provided interface. The downside of this approach is that this convenience module gets very complex and requires careful maintenance. We are currently considering several ideas on how to improve these interfaces.

Unlike the other tests, the GPSR test does not have a predefined task sequence. This test is divided into three types of complexity. The first type is an unknown sequence of commands from a known set that need to be executed in the correct order. The second type is similar to the first one, but might contain missing information that the robot must infer on its own or optionally ask for clarification. The third and most complex type deals with incomplete and erroneous information that requires task planning and complex situation analysis.

To solve this test, we use a natural language parser with keyword spotting. Individual tasks and the corresponding parameters are extracted from the given command and executed one after another. Missing information is obtained from a knowledge base containing a hierarchical definition of objects, categories and places. So far all of these information is inserted manually into the knowledge base after being specified by the organizers of the competition. However, this simple approach only allows for solving the first and second type of GPSR commands. The third type requires more complex reasoning and task management which is subject of ongoing work.

4   Approaches and Contribution

In this section we briefly describe the software that we recently contributed to the RoboCup and ROS community. More details are provided in our current team description paper [18] and a detailed tutorial in [4]. Further, we describe other algorithms and approaches that we use to realize the functionalities and behaviors needed for typical @Home tasks.

4.1 Software Contributions

We offer ROS packages for the following tasks: Mapping and Navigation [25] based on [27] (Fig. 2), Object Recognition [15] (Fig. 4), Speech Recognition and Synthesis and a Robot Face [13] (Fig. 3) for human robot interaction. Additionally, a Graphical User Interface provides convenient interfaces to access the provided functionalities. Video tutorials are available online[4].

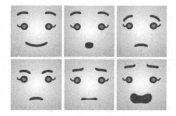

Fig. 2. Mapping and navigation example   **Fig. 3.** Different face expressions

4.2 Approaches and Algorithms

In this Section we introduce our approaches to some important areas of functionalities.

Speech Recognition. For speech recognition we integrated and evaluated different approaches. In the past we used *PocketSphinx*[5] for speech recognition. To improve recognition results, different dictionaries were activated depending on the current state of the task (e.g. if the robot was asking for a drink order only drink names could be understood).

Another approach we tried was the integration of the Android speech recognition. Although an internet connection improves recognition results, it is not mandatory. The main benefit of this approach is that no dictionary needs to be defined and the robot will be enabled to understand everything. This opens different scenarios towards general purpose robotics.

Finally, the approach we are currently using is the grammar based VoCon speech recognition by Nuance[6]. We use the begin and end of speech detection

---

[4] homer@UniKoblenz channel with video tutorials: https://www.youtube.com/user/homerUniKoblenz.

[5] Speech recognition system PocketSphinx:
http://www.speech.cs.cmu.edu/pocketsphinx/.

[6] VoCon speech recognition: http://www.nuance.com/for-business/speech-recognition-solutions/vocon-hybrid/index.htm.

Fig. 4. Object recognition example

to improve recognition results and discard results below a certain confidence threshold. With the Nuance VoCon approach we were able to win the speech recognition benchmark task at the RoboCup 2015 in Hefei.

Multi-robot Coordination. Our first approach to multi-robot coordination aimed at a hierarchical structure, where one robot being the master sent commands to the slave robot [14]. The good speech recognition results we achieved this year allowed for a new experiment. To our best knowledge, we showed the first robot-robot interaction in the RoboCup@Home using natural language. In the final demonstration both robots assigned tasks to one another using synthesized speech and recognizing it. We used a WiFi fallback in case a command was not understood in a certain time. As the human voice is already a general interface for service robots in RoboCup@Home, we showed that it can be used for robots as well. Furthermore, it is imaginable to use it as a general robot-robot interaction interface, even for robots of different teams.

Of course, data that can not be expressed in human language (e.g. object models, slam maps) has to be exchanged in a traditional way. However, if robots communicated using natural language e.g. for task assignment, people would be aware of what the robot is doing and could even provide modifications to previously assigned tasks ("Set the table for one more person, we will have a guest.").

People Detection and Tracking. People are detected by the combination of three sensors. The laser range finder is used to detect legs, while the RGB camera image provides data for face detection. We use the face detection algorithm implemented in the OpenCV library[7]. Finally, the depth camera allows to detect silhouettes of persons [20].

For people tracking we use rich RGB-D data from a depth camera. The sensor is mounted on a pan-tilt unit to actively follow the current person of interest.

---

[7] OpenCV face detection: http://docs.opencv.org/2.4/doc/tutorials/objdetect/cascade_classifier/cascade_classifier.html.

Our people tracker is based on the publicly available 3D people detector of Murano et al. [7] in combination with online appearance learning using adaboost classifiers on color histograms. We estimate the target position and velocity using a linear Kalman filter with a constant velocity motion model. At every time step, we select the detection with highest classification score inside the gating region for target association and update the classifier with positive and negative samples from the set of current detections accordingly. Occlusion detection is based on classification scores as well. We perform Kalman update and appearance learning only if the highest classification score exceeds a given threshold.

Object Detection and Recognition. Objects for mobile manipulation are detected by first segmenting horizontal planes as table hypotheses from the point cloud of the RGB-D camera. Subsequently, all points above the plane are clustered and the resulting clusters considered as objects. We combine the advantages of the point clouds from the RGB-D camera with the high resolution images from the IDS camera by calculating the correspondences between the points in the point cloud and the pixels in the color image. To accomplish this, the relative pose between the depth sensor and the color sensor of the camera was calculated with a stereo calibration algorithm [2]. With the knowledge about the relative poses between the two sensors we were able to project the position of detected objects in the point cloud into the color images and use a high resolution image patch from the IDS camera for object recognition.

Transparent objects (in our case drinking glasses) are detected by making use of one fault of the structured light sensor. The light emitted by the RGB-D camera is scattered in transparent objects providing no valid depth information. We segment areas with no depth data and compare them with drinking glass contour templates. Since the supporting table plane around the transparent objects has valid depth information, a size and location estimation of the transparent objects is obtained and used for grasping. However, we are currently conducting experiments with a deep convolutional neural network to augment recognition and detection results of transparent objects.

Object Manipulation. We use an A* search algorithm within the arm configuration space to plan an arm trajectory for grasping. To avoid collisions, we run an on-the-fly collision detection in which we check a simplified arm model represented by a set of line segments against a kd-tree build on the depth image of the environment. The planned path is smoothed with a recursive line fitting algorithm in the configuration space. The resulting path consists of a minimal set of configurations which are directly connected by linear moves in the configuration space. Although our approach works well, we are not satisfied with the runtime of the planning algorithms, especially if lots of obstacles are present. Therefore, we are integrating the MoveIt library[8] which promises a flexible framework with many different algorithms to choose from.

---

[8] MoveIt library: http://moveit.ros.org/.

5   Current Research

Apart from the various fields mentioned in the previous Sections that we are currently working on the main focus is put into computer vision. The two most important topics are highlighted in this section.

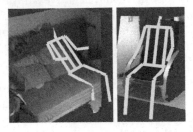  

Fig. 5. Sitting affordance detection with suggested human body pose.

Fig. 6. Recognition result and estimated bounding box.

3D Object Recognition. Leibe et al. presented the Implicit Shape Model (ISM) algorithm as an extension to the popular bag-of-keypoints approach in [6]. Recently, extensions of the ISM approach to 3D data have been proposed [3,26]. Unlike in related work on 3D ISM, we use a continuous Hough-space voting scheme. In our approach [16] (Fig. 6), SHOT features [23] from segmented objects are learned. However, contrary to the ISM formulation, we do not cluster the features. Instead, to generalize from learned shape descriptors, we match each detected feature with the k nearest learned features in the detection step. Each matched feature casts a vote into a continuous Hough-space. Maxima for object hypotheses are detected with the Mean-Shift Mode Estimation algorithm [1]. In a recent work we compared our algorithm with other approaches [17] and achieved best results. Still, the long runtime of our approach is a drawback that we are currently working on. Apart from this purely shape based approach we are also working on combining 2D and 3D features to improve object recognition results based on RGB-D point clouds [19].

Affordance Detection. Affordances have gained much popularity for object classification and scene analysis. Our current research focuses on sitting affordances to analyze scenes regarding sitting possibilities for an anthropomorphic agent. Recently, we introduced the concept of fine-grained affordances [21]. It allows to distinguish affordances on a fine-grained scale (e.g. sitting without backrest, sitting with backrest, sitting with armrests) and thus facilitates the object classification process. Additionally, our approach estimates the sitting pose with regard to the detected object (Fig. 5).

Our current work on that field focuses on adding more affordances, but also experimenting with other agent models. We plan to extend our method to detecting different fine-grained grasping affordances and thus allow for precise grasp planning for mobile manipulation.

6 Summary and Outlook

In this paper we have presented the approaches developed and applied by our team homer@UniKoblenz with robot Lisa that participates in the RoboCup@-Home competition. A special emphasis was put on our software architecture and the therein subsumed abstraction layers that significantly facilitate application development even for new team members without a robotics background.

We further gave a short introduction into our contributed software packages available at our ROS package repository. As we continue developing and improving our software, more packages will be made available to everyone.

Our plans for the future are twofold. On one hand we will continue research in the computer vision field, focusing on 3D object recognition, affordance detection and transparent object detection. On the other hand we will investigate how a meaningful task management can be included into our software architecture. This is an inevitable step towards general purpose service robots which need to be able to handle and recover from errors, be aware of their environment and react to unexpected situations.

References

1. Cheng, Y.: Mean shift, mode seeking, and clustering. IEEE Trans. Pattern Anal. Mach. Intell. **17**(8), 790–799 (1995)
2. Hartley, R., Zisserman, A.: Multiple View Geometry in Computer Vision, 2nd edn. Cambridge University Press, New York (2003)
3. Knopp, J., Prasad, M., Willems, G., Timofte, R., Van Gool, L.: Hough transform and 3D SURF for robust three dimensional classification. In: Daniilidis, K., Maragos, P., Paragios, N. (eds.) ECCV 2010, Part VI. LNCS, vol. 6316, pp. 589–602. Springer, Heidelberg (2010)
4. Koubaa, A. (ed.): Robot Operating System (ROS) - The Complete Reference. Studies in Computational Intelligence, vol. 625. Springer, Switzerland (2016)
5. Bio-Robotics Laboratory. Pumas@home 2015 team description paper (2015). http://biorobotics.fi-p.unam.mx/downloads/finish/3-papers/532-tdp2015
6. Leibe, B., Leonardis, A., Schiele, B.: Combined object categorization and segmentation with an implicit shape model. In: ECCV 2004 Workshop on Statistical Learning in Computer Vision, pp. 17–32 (2004)
7. Basso, F., Munaro, M., Menegatti, E.: Tracking people within groups with RGB-D data. In: Proceedings of the International Conference on Intelligent Robots and Systems (IROS) (2012)
8. Newell, A.: The knowledge level: presidential address. AI Mag. **2**(2), 1 (1981)
9. Pineda, L.A., Rascon, C., Fuentes, G., Estrada, V., Rodriguez, A., Meza, I., Ortega, H., Reyes, M., Pena, M., Duran, J., et al.: The golem team, robocup@home 2014 (2014). http://turing.iimas.unam.mx/golem/pubs/golem-TDP.pdf
10. Pineda, L.A., Rodríguez, A., Fuentes, G., Rascon, C., Meza, I.V.: Concept and functional structure of a service robot. Int. J. Adv. Rob. Syst. **12**, 6 (2015)
11. Pineda, L.A., Salinas, L., Meza, I., Rascon, C., Fuentes, G.: Sitlog: a programming language for service robot tasks. Int. J. Adv. Rob. Syst. **10**, 538 (2013)

12. Quigley, M., Conley, K., Gerkey, B.P., Faust, J., Foote, T., Leibs, J., Wheeler, R., Ng, A.Y.: ROS: an open-source robot operating system. In: ICRA Workshop on Open Source Software (2009)
13. Seib, V., Giesen, J., Grüntjens, D., Paulus, D.: Enhancing human-robot interaction by a robot face with facial expressions and synchronized lip movements. In: Skala, V. (ed.) 21st International Conference in Central Europe on Computer Graphics, Visualization and Computer Vision (2013)
14. Seib, V., Gossow, D., Vetter, S., Paulus, D.: Hierarchical Multi-robot Coordination. In: Ruiz-del-Solar, J. (ed.) RoboCup 2010. LNCS, vol. 6556, pp. 314–323. Springer, Heidelberg (2010)
15. Seib, V., Kusenbach, M., Thierfelder, S., Paulus, D.: Object recognition using hough-transform clustering of surf features. In: Workshops on Electronical and Computer Engineering Subfields, pp. 169–176. Scientific Cooperations Publications (2014)
16. Seib, V., Link, N., Paulus, D.: Implicit shape models for 3d shape classification with a continuous voting space. In: Braz, J., Battiato, S., Imai, F.H. (eds.) VISAPP 2015 - Proceedings of International Conference on Computer Vision Theory and Applications, vol. 2, pp. 33–43. SciTePress (2015)
17. Seib, V., Link, N., Paulus, D.: Pose estimation and shape retrieval with hough voting in a continuousvoting space. In: Gall, J., Gehler, P., Leibe, B. (eds.) GCPR 2015. LNCS, vol. 9358, pp. 458–469. Springer, Heidelberg (2015)
18. Seib, V., Manthe, S., Holzmann, J., Memmesheimer, R., Peters, A., Bonse, M., Polster, F., Rezvan, B., Riewe, K., Roosen, M., Shah, U., Yigi, T., Barthen, A., Knauf, M., Paulus, D.: Robocup 2015 - homer@unikoblenz (Germany). Technical report, University of Koblenz-Landau (2015)
19. Seib, V., Memmesheimer, R., Paulus, D.: Ensemble classifier for joint object instance and category recognition on RGB-D data. In: 2015 22nd IEEE International Conference on Image Processing (ICIP) (2015)
20. Seib, V., Schmidt, G., Kusenbach, M., Paulus, D.: Fourier features for person detection in depth data. In: Azzopardi, G., Petkov, N., Yamagiwa, S. (eds.) CAIP 2015. LNCS, vol. 9256, pp. 824–836. Springer, Heidelberg (2015)
21. Seib, V., Wojke, N., Knauf, M., Paulus, D.: Detecting fine-grained affordances with an anthropomorphic agent model. In: Agapito, L., Bronstein, M.M., Rother, C. (eds.) ECCV 2014 Workshops. LNCS, vol. 8926, pp. 413–419. Springer, Heidelberg (2015)
22. Thierfelder, S., Seib, V., Lang, D., Häselich, M., Pellenz, J., Paulus, D.: Robbie: a message-based robot architecture for autonomous mobile systems. In: INFORMATIK -Informatik Schafft Communities (2011)
23. Tombari, F., Salti, S., Di Stefano, L.: Unique signatures of histograms for local surface description. In: Daniilidis, K., Maragos, P., Paragios, N. (eds.) ECCV 2010, Part III. LNCS, vol. 6313, pp. 356–369. Springer, Heidelberg (2010)
24. Tousignant, S., Van Wyk, E., Gini, M.: Xrobots: a flexible language for programming mobile robots based on hierarchical state machines. In: 2012 IEEE International Conference on Robotics and Automation (ICRA), pp. 1773–1778. IEEE (2012)
25. Wirth, S., Pellenz, J.: Exploration transform: a stable exploring algorithm for robots in rescue environments. In: Workshop on Safety, Security, and Rescue Robotics, vol. 9, pp. 1–5 (2007)
26. Wittrowski, J., Ziegler, L., Swadzba, A.: 3d implicit shape models using ray based hough voting for furniture recognition. In: 2013 International Conference on 3DTV-Conference, pp. 366–373. IEEE (2013)
27. Zelinsky, A.: Environment exploration and path planning algorithms for a mobile robot using sonar. Ph.D. thesis, Wollongong University, Australia (1991)

A Failure-Tolerant Approach for Autonomous Mobile Manipulation in RoboCup@Work

Jan Carstensen[1]([⊠]), Torben Carstensen[1], Simon Aden[1], Andrej Dick[1],
Jens Hübner[1], Sven Krause[1], Alexander Michailik[1], Johann Wigger[1],
Jan Friederichs[2], and Jens Kotlarski[3]

[1] LUHbots, Leibniz Universität Hannover, Hanover, Germany
carstensen@stud.uni-hannover.de, info@luhbots.de
[2] Hannover Centre for Mechatronics, Leibniz Universität Hannover,
Hanover, Germany
friederichs@mzh.uni-hannover.de
[3] Institute of Mechatronic Systems, Leibniz Universität Hannover, Hanover, Germany
jens.kotlarski@imes.uni-hannover.de

Abstract. In this paper we summarize how the LUHbots team was able
to win the 2015 RoboCup@Work league. We introduce various failure
handling concepts, which lead to the robustness necessary to outperform
all the other teams. The proposed concepts are based on failure preven-
tion and failure handling.

1 Introduction

The RoboCup@Work league was established in 2012 [9] to foster the development
and benchmarking of robots in the industrial environment. The main focus of the
league is to improve small but versatile robots capable of doing many different
tasks and, therefore, are interesting not only to huge firms which can afford
many robots, but also to small companies. After introducing the league and
tests performed in 2015, we present our approach, which this year has been
focussed on robustness and failure handling.

2 LUHbots

The LUHbots team was founded in 2012 at the Institute of Mechatronic Systems
Leibniz Universität Hannover, consists of bachelor and master students. Most of
the founding team members have participated in the research inspired practical
lecture RobotChallenge [11]. Nowadays the team is a part of the Hannover Centre
for Mechatronics. The team consists of students from mechanical engineering,
computer science and navigation and environmental robotics (see Fig. 1a). In
2012 the LUHbots team first competed in the RoboCup@Work challenge and
was able to win the competition [10], in 2013 a second place was achieved [1].
In 2015 the LUHbots won both events, the German Open and the RoboCup in
Hefei.

© Springer International Publishing Switzerland 2015
L. Almeida et al. (Eds.): RoboCup 2015, LNAI 9513, pp. 95–105, 2015.
DOI: 10.1007/978-3-319-29339-4_8

(a) The LUHbots Team in Hefei

(b) The RoboCup@Work League

Fig. 1. RoboCup@Work in 2015

3   RoboCup@Work

In this section we introduce the tests of the 2015 RoboCup@Work world championship. The, competition focussed on transportation tasks. In 2015 the rules allowed for picking complexity levels per test [12]. Six teams participating at the world cup in Hefei (see Fig. 1b).

3.1   Tests

In the following we are only going to discuss the complexity levels chosen by the LUHbots.

Basic Navigation Test: The purpose of the Basic Navigation Test (BNT) is testing navigation in a static environment. The arena is initially known and can be mapped during a set-up phase (see Fig. 2). The task consists of reaching and covering a series of markers in a specified orientation and covering a marker completely. In order to increase the complexity, obstacles are positioned in the arena. The position of the unknown obstacles is static.

(a) The arena during testing

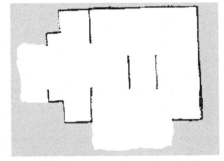

(b) Map used for navigation

Fig. 2. The RoboCup@Work Arena in 2015

Basic Manipulation Test: The Basic Manipulation Test (BMT) focusses on manipulation tasks. The objective is to successfully grasp three objects and place them on a nearby service area. We increased the complexity level by choosing the hardest position, rotation, order and all decoy objects. Thus the complete set-up is defined by chance.

Basic Transportation Test: The Basic Transportation Test (BTT) combines manipulation tasks and navigation tasks. A task description is sent to the robot. The description includes information of starting and end positions for the objects to be transported. The task order and the specific transport tasks have to be determined autonomously by the robot. In order to increase complexity we had to pick objects in a randomly determined order, position and rotation. Furthermore the highest amount of decoy objects was placed on the service areas. We then had to place them according to specification. After placing all objects the robot has to leave the arena.

Precision Placement Test: The Precision Placement Test (PPT) consists of transporting objects and placing them inside small cavities, which are only a few millimetres larger than the object. An initially unknown position of the cavities increased the complexity.

Final: Traditionally the final is a combination of all the above mentioned tests performed on the event. In 2015 the final task consisted of an extended BTT with ten objects. Some objects needed to be placed according to the PPT rules. To further increase the complexity level it was possible to add obstacles. We performed with high manipulation complexity but without additional navigation obstacles.

4   Hardware

Our robot is based on the mobile robot KUKA youBot (see Fig. 3) [2]. The robot consists of a platform with four meccanum wheels [8] and a five degrees of freedom (DoF) manipulator. Additional a gripper is attached at the end of the manipulator (see Fig. 3). The internal computer of the youBot has been replaced by an Intel Core i7 based system. In addition, the robot is equipped with an emergency stop system, allowing for keeping the platform and the manipulator in the actual pose when activated. The manipulator has been remounted to increase the manipulation area. The hardware itself does not offer failure tolerance, this is only achieved in combination with software.

4.1   Sensors

The youBot is equipped with two commercial laser range finders (Hokuyo URG-04LX-UG01) at the platform's front and back. A RGB-D camera (Creative Senz3D) mounted on the wrist of the manipulator (see Fig. 3a).

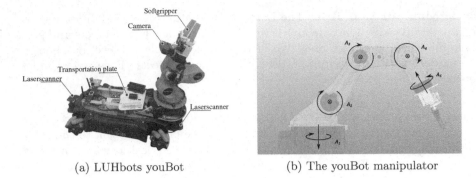

(a) LUHbots youBot

(b) The youBot manipulator

Fig. 3. Hardware overview

4.2 Gripper

One of the major hardware advances performed by the team is the development of a custom gripper. The original gripper has a low speed and stroke. As a result, it is not possible, to grasp all objects defined by the RoboCup@Work rule book, without manually changing the gripper-fingers. Besides the limited stroke, the low speed limit does not allow for an appropriate grasping of moving objects. Even though, in the 2015 competition the Conveyor Belt Test, has not been performed, the hardware design is optimized to meet future requirements. An advancement was to include force feedback into the gripper. Thanks to the integrated feedback within our custom made gripper, we are able to verify performed grasps. If a failure occurs during grasping, we are able to recover. In the current version the gripper uses soft-grippers to allow for a better handling of all objects (see Fig. 4a). A different approach using hall effect sensing has been tested (see Fig. 4b).

(a) Force sensing gripper, without soft-fingers

(b) Advanced sensing gripper

Fig. 4. The LUHbots grippers

5 Approach

We take advantage of an open source software framework called Robot Operating System (ROS) [14]. We used the Indigo release in 2015. Since 2012 we have a series of custom approaches. Each new development is tested extensively in predefined test scenarios before being included in our competition code base. This way we can guarantee a good robustness. Robustness, fail-safes and recovery behaviours are the cornerstone of our development.

5.1 Overview

Since our software architecture is based on ROS, different nodes are used (see Fig. 5). The yellow nodes are drivers they give access to the sensors. The youBot driver in red, can be accessed via the youBot OODL node. The camera data is first processed by the vision node and then filtered and clustered by the observer node, which is triggered by the state machine. The laser scanners are publishing to the navigation stack and the navigation watchdog. The watchdog filters the navigation commands. The task planner and the referee box connection communicate with the state machine. The laser scanner nodes are used unmodified. The ROS navigation stack is used but the global and local planners have been replaced. The youBot OODL driver is heavily modified. All other Nodes are developed entirely by the team.

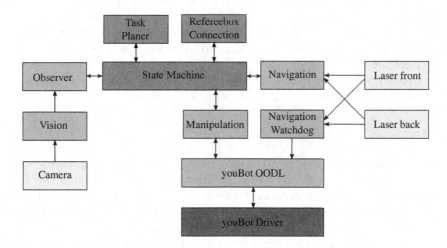

Fig. 5. Overview of the software architecture (Color figure online)

5.2 Manipulation

During the last year we developed a new software system that can be seen as a software development kit (SDK) for manipulation tasks with the youBot. The aim was to facilitate the development of applications for the youBot by

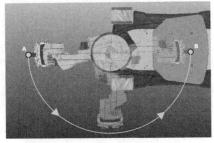
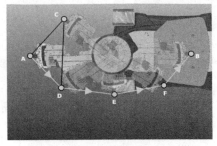

(a) Interpolation based planning   (b) Graph based path planning

Fig. 6. Graph based approach for the path planning, thanks to the proposed approach (b) a shorter motion is executed

providing advanced functionality for the manipulator and the mobile platform combined with user friendly interfaces. Some of the features for the manipulator are: inverse kinematics, path planning, interpolated movement in joint- and task-space, gravity compensation and force fitting. Features for the mobile platform include incremental movement, collision avoidance and movement relative to the environment based on laser scans. The provided interfaces contain a documented API and a graphical interface for the manipulator. In the RoboCup we use this software e.g. to grab objects using inverse kinematics, to optimize trajectories and to create fast and smooth movements with the manipulator. Besides the usability the main improvements are the graph based planning approach (see Fig. 6) and the higher control frequency of the base and the manipulator. Planning on a graph which is based on known and, therefore, valid positions leads to a higher robustness. Using an A*-approach the best path is generated [6]. The higher frequency leads to better executed motion plans and an overall smooth and more accurate motion.

5.3   Navigation

The navigation is based on the ROS navigation stack. The main improvement has been done in the local and global planners. The global planner has been extended to calculate the orientation for each pose of the global plan (see Fig. 7). The plan provided by the global planner is executed by a local planner with high reliance on the global plan. Since the RoboCup@Work arena is mapped before the tests, only a few obstacles are unknown at the beginning of the run. After a short time, the complete arena is mapped including additional obstacles and, therefore, the global plan is very close to the optimal path. Besides improving parts of the navigation stack we implemented a watchdog which operates based on the laser scanner data and is therefore much faster than a costmap-based local planner. The watchdog reduces velocities if an obstacle is too close, or permits the execution of a movement command if a collision would be eminent.

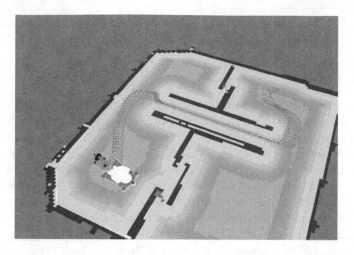

Fig. 7. Global plan with orientations, from start to end pose considering obstacles

5.4   Vision

We use the Creative Senz3D for object recognition, witch has two basic advantages in comparison to similar devices. Firstly it works at close range. Secondly, it is relatively small. Since the camera is not intended to be used in high precision tasks, the obtained 3D points are too noisy to be used directly for object recognition. Instead, we use the 2D images of the infra-red and RGB camera to segment the image, to extract features and to classify the objects. From the infrared image first the objects are separated using the canny algorithm [4]. Then, the objects are then classified using Hu-moments and a random forest classifier [7, 15]. Finally, the 3D points are used to determine the object's position and orientation. In order to get a robust vision system that can handle miss detections and which can memorize detected objects, all detections are clustered using a modified version

Fig. 8. Detected objects, classified and scored

of DBSCAN [5]. Each cluster is weighed, filtered and the positions are averaged. Then, the clusters are classified as objects or as failures (Fig. 8).

5.5   Task Planning

Our task planning is based on a graph based search. In each step all known service areas are used as possible navigation tasks. All objects on the back of the robot (there are up to three allowed) are used as possible placing tasks and the objects on the service area are used as grasping tasks. A greedy-based planning [13] is used up to a max depth and repeated until a complete plan is produced. The greedy algorithm is based on a cost function taking the time to perform the task, the probability to fail and the expected output. For the navigation tasks the distances are precomputed based on the known map. The manipulation time costs are averaged based the last respective manipulation action. When the state machine is not able to successfully recover a failure, the task is rescheduled and replanned increasing the probability to fail.

5.6   State Machine

The state machine is based on SMACH [3]. Which is a python library for building hierarchical state machines in ROS. Due to the capabilities of SMACH our state machine is modular and consists of the main components task planning, task execution, navigation and manipulation (see Fig. 9). The state machine acts as an action client, which sets the goals in navigation and manipulation to accomplish the tasks and receives feedback in case of issues. The state machine is designed for recovery. Each state is analysed. Foreseeable failures are considered. Depending on the failure a direct recovery is applied, the current task will be retried or postponed and tried again later, respectively.

The subroutine for manipulation is basically a linear sequence of actions with several cascaded loops that are repeated if an action fails. The idea is to detect failures as soon as possible and try to recover immediately. If the recovery fails after multiple attempts, the higher level recovery loop is repeated. A manipulation task is defined by a list of objects and an action to be performed with them. For a picking task e.g. the robot first approaches the service area and moves the arm to perform a scan of the objects. The scan is repeated until all objects are found with sufficient certainty or until a maximum number of scan movements is reached. Then, the found objects are grasped and placed on the cargo beginning from left to right. For grasping, the robot moves sideways to the object, then a second close scan is performed to verify the object and to further improve pose estimation. If the certainty of the scan is too low, the scan is repeated from a slightly changed perspective. If the scan fails multiple times, the object is postponed. Otherwise, the object is grasped and the force feedback is evaluated to verify that the grasp was successful. Depending on the result the object is either placed on the cargo or the close scan loop is repeated. If there are any postponed objects left in the end, the outer loop is repeated two times

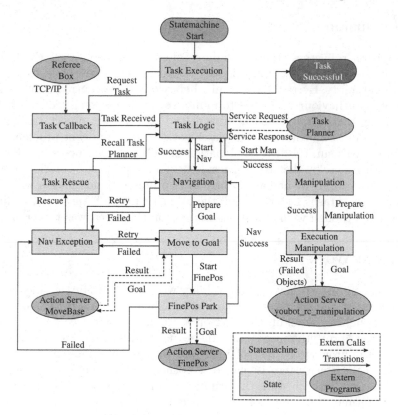

Fig. 9. State machine

with a preceding movement first to the left and then to the right. All postponed objects are reported back to the superior state machine.

6 Results

The results can be seen in Table 1.

Table 1. Results of the RoboCup@Work competition in 2015

Place	Team	BNT	BMT	PPT	BTT	BTT2	Semi Final	Final	Total
1	LUHbots	280	480	630	1265	1920	2050	1840	8465
2	Robo-Erectus	0	480	1405	0	1785	1700	825	6195
3	b-it-bots	575	0	170	420	1020	1105	1615	4905
4	KeJia Worker	260	530	240	270	225	450	600	2575
5	WF Wolves	180	0	0	210	375	600	525	1890
6	robOTTO	195	0	0	0	0	560	375	1130

7   Conclusion

In our opinion robustness through failure tolerant approaches are the key the succeed in RoboCup. Based on different approaches we were able to improve our overall stability. Every failure or miss behaviour ever occurred is either fixed or a recovery behaviour is created to minimise the consequences of the failure. Furthermore, the overall vision approach lead to a robust object recognition. Even though our segmentation and classification had significant problems with the service area's surfaces, our combination of different scan poses, the filtering and clustering done by the observer resulted in an appropriate solution. Besides being able to recover, we had a very fast and optimised manipulation which was able to perform grasps faster then all the other teams and, therefore, giving us an edge. Even though we had a stable navigation the speed was rather slow.

8   Future Work

Even though we already have a good stability we are going to further increase our testing scenarios and the recovery behaviours. The navigation is a topic to work on, even though we are still focussing on a stable collision free navigation, we would like to improve the speed. Since the scenarios are increasing in complexity, we work on improving our task-planner and plan to test different approaches. Changing the vision system is planed. Another major point we will work on will be focussed on improving the gripper we are using. Even though we had a good performance we are looking forward to further increase the speed and robustness. A new gripper is in development, using only one servo motor and including a controller to further increase reliability (see Fig. 10).

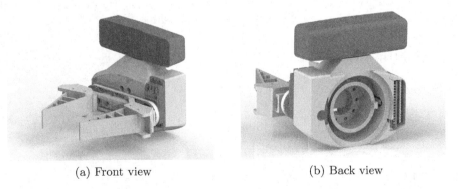

(a) Front view   (b) Back view

Fig. 10. Model of the next generation gripper

Acknowledgements. We would like to thank a couple of institutes and persons supporting our work. The team is supported by the Institute of Mechatronic Systems, the Institute of Systems Engineering Real Time Systems Group, the student affairs office of the faculty of mechanical engineering, the society for the promotion of geodesy and geoinformatics and the Hannover Centre for Mechatronics. The team is being supervised by Jan Friederichs, Johannes Gaa and Daniel Kaczor.

References

1. Alers, S., Claes, D., Hennes, D., Tuyls, K., Fossel, J., Weiss, G.: How to win RoboCup@Work? In: Behnke, S., Veloso, M., Visser, A., Xiong, R. (eds.) RoboCup 2013. LNCS, vol. 8371, pp. 147–158. Springer, Heidelberg (2014)
2. Bischoff, R., Huggenberger, U., Prassler, E.: Kuka youbot - a mobile manipulator for research and education. In: ICRA (2011)
3. Bohren, J., Cousins, S.: The SMACH high-level executive [ROS news]. IEEE Robot. Autom. Mag. 4(17), 18–20 (2010)
4. Canny, J.: A computational approach to edge detection. IEEE Trans. Pattern Anal. Mach. Intell. 8(6), 679–698 (1986)
5. Ester, M., Kriegel, H.P., Sander, J., Xu, X.: A density-based algorithm for discovering clusters in large spatial databases with noise. In: Proceedings of 2nd International Conference on Knowledge Discovery and Data Mining (KDD 1996). pp. 226–231. AAAI Press (1996)
6. Hart, P., Nilsson, N., Raphael, B.: A formal basis for the heuristic determination of minimum cost paths. IEEE Trans. Syst. Sci. Cybern. 4(2), 100–107 (1968)
7. Hu, M.K.: Visual pattern recognition by moment invariants. IRE Trans. Inf. Theory 8(2), 179–187 (1962)
8. Ilon, B.: Directionally stable self propelled vehicle, US Patent 3,746,112, 17 July 1973
9. Nowak, W., Dwiputra, R., Schneider, S., Berghofer, J., Hegger, F., Bischoff, R., Hochgeschwender, N., Kraetzschmar, G.K.: RoboCup@Work: competing for the factory of the future. In: Bianchi, R.A.C., Akin, H.L., Ramamoorthy, S.,Sugiura, K. (eds.) RoboCup 2014. LNCS, vol. 8992, pp. 171–182. Springer, Heidelberg (2015). http://dx.doi.org/10.1007/978-3-319-18615-3_14
10. Kaczor, D., et al.: RoboCup@Work league winners 2012. In: Chen, X., Stone, P., Sucar, L.E., van der Zant, T. (eds.) RoboCup 2012. LNCS, vol. 7500, pp. 65–76. Springer, Heidelberg (2013)
11. Munske, B., Kotlarski, J., Ortmaier, T.: The robotchallenge - a research inspired practical lecture. In: 2012 IEEE/RSJ International Conference on Intelligent Robots and Systems (IROS), pp. 1072–1077, October 2012
12. Nowak, W., Kraetzschmar, G., Hochgeschwender, N., Bischoff, R., Kaczor, D., Hegger, F., Carstensen, J.: Robocup@work rule book (2015). http://www.robocupatwork.org/
13. Papadimitriou, C.H., Steiglitz, K.: Combinatorial Optimization: Algorithms and Complexity. Prentice-Hall Inc, Upper Saddle River (1982)
14. Quigley, M., Conley, K., Gerkey, B.P., Faust, J., Foote, T., Leibs, J., Wheeler, R., Ng, A.Y.: ROS: an open-source robot operating system. In: ICRA Workshop on Open Source Software (2009)
15. Statistics, L.B., Breiman, L.: Random forests. In: Machine Learning, pp. 5–32 (2001)

CMDragons 2015: Coordinated Offense and Defense of the SSL Champions

Juan Pablo Mendoza[✉], Joydeep Biswas, Danny Zhu, Richard Wang,
Philip Cooksey, Steven Klee, and Manuela Veloso

Carnegie Mellon University, Pittsburgh, USA
{jpmendoza,joydeepb,dannyz,rpw,mmv}@cs.cmu.edu,
{pcooksey,sdklee}@andrew.cmu.edu

Abstract. The CMDragons Small Size League (SSL) team won all of its 6 games at RoboCup 2015, scoring a total of 48 goals and conceding 0. This paper presents the core coordination algorithms in offense and defense that enabled such successful performance. We first describe the coordinated plays layer that distributes the team's robots into offensive and defensive subteams. We then describe the offense and defense coordination algorithms to control these subteams. Effective coordination enables our robots to attain a remarkable level of team-oriented gameplay, persistent offense, and reliability during regular gameplay, shifting our strategy away from stopped ball plays. We support these statements and the effectiveness of our algorithms with statistics from our performance at RoboCup 2015.

1 Introduction

The CMDragons 2015 team from Carnegie Mellon University (Fig. 1) builds upon extensive research from previous years (1997–2010 [1–3], 2013–2014 [4]), in areas including computer vision [5], path [6] and dribbling [7] planning, execution monitoring [8] and team architecture [9]. This legacy enables our 2015 team to focus on the problem of coordinated decision-making for a team in a highly dynamic adversarial domain[1]. Furthermore, while previous years have seen a focus on stopped-ball plays [10], this year our team focuses its research efforts on regular gameplay team planning. As a result, our offensive strategy relies heavily on teamwork, with 245 pass attempts and 194 completions over the tournament.

To achieve such coordination, our algorithms are divided into various layers:

1. The *coordinated plays* layer [9] decides how many robots are assigned to a defense subteam and to an offense subteam, based on functions of the state of the game.

---

[1] While this paper focuses on our latest team-decision-making research, our team website http://www.cs.cmu.edu/~robosoccer/small/ provides an overall description of the components of our team.

© Springer International Publishing Switzerland 2015
L. Almeida et al. (Eds.): RoboCup 2015, LNAI 9513, pp. 106–117, 2015.
DOI: 10.1007/978-3-319-29339-4_9

Fig. 1. CMDragons 2015 small size league team. Humans from left to right: Steven Klee, Richard Wang, Juan Pablo Mendoza, Manuela Veloso, Joydeep Biswas, Danny Zhu, and Philip Cooksey.

2. Each of these two subteams creates *coordinated plans* to maximize the team's probability of scoring and not being scored, respectively.
3. Each robot *individually selects actions* that are consistent with the coordinated plan and maximize the probability of success.
4. The robots execute these actions through various new *reusable skills* [10].

This paper focuses on the team coordination layers of items 1 and 2: Sect. 2 discusses the coordinated plays layer, while Sects. 3 and 4 discuss the coordinated plans of offense and defense, respectively. In Sect. 5, we analyze the performance of these algorithms at RoboCup 2015, through various statistics extracted from the tournament. These statistics show how our team (a) maintained persistently offensive gameplay throughout the tournament, (b) successfully executed a team-oriented strategy, and (c) significantly increased its focus on regular gameplay, as opposed to stopped ball gameplay.

2   Coordinated Plays Layer

At the top coordination level, our algorithms divide the team of robots into two subteams: the *defense* robots, whose goal is to minimize the probability of the opponents scoring a goal on our team, and the *offense* robots, whose goal is to maximize the probability of scoring on the opponent team. The number of robots assigned to each of these subteams reflects the desired level of aggressiveness for the current situation.

We specify this subteam division through *plays* [11], each of which is defined by (a) the n roles that the n robots should fill, and (b) the applicability conditions for choosing that play. The CMDragons 2015, playing with n robots, used plays with either 1, $n - 3$, or $n - 2$ offense robots; the team switched among these plays according to the desired level of aggressiveness, as defined by the play applicability conditions. These applicability conditions are based on three functions of the state of the game: ball possession **ballP**, field region **fieldR** in which the ball is located, and the opponent's level of aggressiveness **oppA**.

Ball Possession: Variable **ballP** can take one of four values: ourB, theirB, contendedB, or looseB. We estimate the possession state ballP_t at time t using the length of time $t_{\text{near}}^{\text{us}}$ and $t_{\text{near}}^{\text{them}}$ that the ball has been closer than d_{near} to us and the opponent, respectively, and the time $t_{\text{far}}^{\text{us}}$, and $t_{\text{far}}^{\text{them}}$ that the ball has been farther than d_{far} from us and the opponent, respectively:

$$
\text{ballP}_t = \begin{cases}
\text{ourB} & \text{if } (t_{\text{near}}^{\text{us}} > t_{\text{near}}^{\text{thresh}}) \wedge (t_{\text{near}}^{\text{them}} < t_{\text{near}}^{\text{thresh}}) \\
\text{theirB} & \text{if } (t_{\text{near}}^{\text{us}} < t_{\text{near}}^{\text{thresh}}) \wedge (t_{\text{near}}^{\text{them}} > t_{\text{near}}^{\text{thresh}}) \\
\text{contendedB} & \text{if } (t_{\text{near}}^{\text{us}} > t_{\text{near}}^{\text{thresh}}) \wedge (t_{\text{near}}^{\text{them}} > t_{\text{near}}^{\text{thresh}}) \ , \\
\text{looseB} & \text{if } (t_{\text{far}}^{\text{us}} > t_{\text{far}}^{\text{thresh}}) \wedge (t_{\text{far}}^{\text{them}} < t_{\text{far}}^{\text{thresh}}) \\
\text{ballP}_{t-1} & \text{otherwise}
\end{cases}
\tag{1}
$$

where $t_{\text{near}}^{\text{thresh}}$ and $t_{\text{far}}^{\text{thresh}}$ are time thresholds.

Field Region: Variable **fieldR** attains one of two values: ourH and theirH, according to the half of the field in which the ball is located.

Opponent Aggressiveness: If the opponent has 0 robots in our half of the field, then **oppA** = def; otherwise, **oppA** = off.

Table 1 illustrates how the CMDragons divide the team into offense and defense based on these functions of the state of the world. When the ball is in our half of the field, and we do not have possession of the ball, we enter a very defensive play. When the ball is in the opponent's half of the field or the opponent has no robots in our half, and the opponent does not have possession of the ball, we enter a very aggressive play with $n - 2$ offense robots. In all other circumstances, our team plays with 3 defense robots and $n - 3$ offense robots.

Table 1. Play choice based on the state of the game. Usually, the team is of size $n = 6$, but $n < 6$ is possible due to cautionable offenses or mechanical problems.

Game state	Offense size	Defense size
fieldR = ourH $\wedge$ **ballP** $\neq$ ourB	1	$n - 1$
(**fieldR** = theirH $\vee$ **oppA** = def) $\wedge$ **ballP** $\neq$ theirB	$n - 2$	2
All others	$n - 3$	3

3 Offense Coordination: Zone-Based Selective Reactivity

The CMDragon's multi-robot coordination algorithm, which is in large part responsible for the team's success in 2015, is the product of years of research and change. Here, we describe how our 2015 offense algorithm emerged as a combination of our 2013 and 2014 algorithms, achieving a balance between being *reactive* to the opponent's actions and *assertive* in carrying out its own plans.

3.1 CMDragons 2013: Reactive Offense

Our 2013 offense emphasizes reactivity toward the opponents. Given the set R of offense robots, 1 *Primary Attacker* (PA) robot handles the ball using one of two actions: shoot it on the goal or pass it to a teammate; the remaining $|R|-1$ *Support Attacker* (SA) robots navigate to their estimated optimal location to receive a pass from the PA. All $|R|$ robots are thus highly reactive to the opponents: the PA's decision depends on whether the opponents have left a good shot on the goal open, and on whether the passes to its teammates are likely to succeed; the SAs' receiving location x^* depends on how likely the opponents are to intercept a pass to x^* given their current configuration, and how likely a goal is to be scored from x^*, given the opponents' configuration. Details on these computations can be found in previous work [10].

Such reactivity is a desirable quality, since it enables our robots to choose effective actions in different scenarios. Pure reactivity, however, prevents our robots from generating and carrying out their own plans. Furthermore, since our SAs continuously move in reaction to the opponent to the estimated optimal passing location, and most defending opponents also continuously move in reaction to our robots to prevent them from reaching an effective passing location, the state of the world is constantly and rapidly changing, which complicates the problem of accurately passing and coordinating. Thus, we decided to explore less reactive algorithms that would give our offense more control over the evolution of the state of the world.

3.2 CMDragons 2014: Fixed Zones and Guard Locations

Our 2014 offense emphasizes our team's assertiveness in controlling the evolution of the game in a more predictable way. To this end, we introduced a Zone-based Team Coordination algorithm [12], which we briefly describe here.

The algorithm partitions the field F into $|R|$ zones, such that there exists a one-to-one mapping from robots to zones. Each robot $r_i \in R$ thus gets assigned to a corresponding zone $Z_i \subseteq F$, where $\bigcup_{i=1}^{|R|} Z_i = F$. Figure 2a shows an example of such an assignment, with three offense robots and their respective zones.

Given an assignment of zones to robots, the algorithm determines the behavior of each robot r_i in its zone Z_i. Let X_i^g be a set of guard positions in each zone Z_i. If r_i computes that the ball will enter z_i, then r_i moves to intercept the ball in its zone at the optimal location $x_a(r_i)$; otherwise r_i moves to one of the guard positions $x_g(r_i) \in X_i^g$. The target location $x_t(r_i)$ for r_i is thus given by:

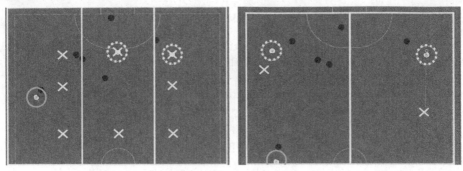

(a) Fixed Zones, Guard Locations (2014)   (b) Zones and Reactive Positioning (2015)

Fig. 2. Fixed zones, guard locations (2014) (Color figure online)

$$\boldsymbol{x}_t(r_i) = \begin{cases} \boldsymbol{x}_a(r_i) \text{ if } \boldsymbol{x}_a(r_i) \in Z_i \\ \boldsymbol{x}_g(r_i) \qquad \text{otherwise} \end{cases} \qquad (2)$$

Figure 2a shows one robot intercepting the ball at its optimal location $\boldsymbol{x}_a$, and two robots placed in their assigned guard positions. These guard positions were determined empirically prior to the games, and were independent of the state of the opponent. Thus, the offense is indifferent to the opponent in its decisions, save for the ball-handling robot, which still decides whether to shoot or pass based on the full state of the world.

This indifference to the state of the opponents gives our 2014 offense more control over the state of the game by enabling pre-defined plans to be carried out, and settling the game into a more slowly-changing pace. However, the lack of reactivity to the opponent has drawbacks, since robots make no effort to improve the probability of scoring by moving to better locations on the field.

3.3   CMDragons 2015: Zones and Reactive Positioning

In 2015, we created an algorithm that combines the strengths of our 2013 offense and our 2014 offense, and awarded the PA more freedom on the field and action options. This coordination algorithm is also zone-based, but the number of zones, their coverage of the field F, and the behavior of each robot given these zones is different. Given R, the algorithm creates $|R| - 1$ zones, which affect the robots' behaviors as described below.

Primary Attacker (PA): One robot is assigned to the role of PA, and is the only robot that handles the ball to shoot on the opponents' goal, pass to a teammate, or keep the ball away from opponents while creating passing or shooting options. The PA is not bound to a zone.

Support Attackers (SAs): Each of the remaining $|R| - 1$ robots r_i is assigned the role of SA within a zone $Z_i \subseteq F$. Similarly to the 2013 algorithm, each r_i estimates the optimal location $\boldsymbol{x}^*(r_i)$ to receive a pass; however, the search is constrained to locations in Z_i, such that $\boldsymbol{x}^*(r_i) \in Z_i$. Similarly to the 2014 algorithm, we define a set of guard positions $\boldsymbol{X}_i^g \subset Z_i$. Then, the target location of robot r_i is given by:

$$\boldsymbol{x}_t(r_i) = \begin{cases} \boldsymbol{x}^*(r_i) & \text{if PA is ready to pass to } r_i \\ \boldsymbol{x}_g(r_i) & \text{otherwise} \end{cases} \tag{3}$$

The evaluation deciding whether the PA is ready to pass to r_i is made using a pass-ahead coordination algorithm [10] that prevents r_i from moving to $\boldsymbol{x}^*$ too early or too late for the pass.

Using this algorithm, our team is able to create and carry out plans independently of the opponent by choosing the zones Z_i and the guard positions $\boldsymbol{X}_i^g$; this enabled our 2015 team to create sequential plans in which the sets of zones assigned to the SAs evolved to move the ball towards the opponents' goal [13]. At the same time, the offense maintains high reactivity to the opponents, since the SAs always search for the best receiving location $\boldsymbol{x}^*$ within their zone Z_i. Importantly, however, the SAs only start moving toward $\boldsymbol{x}^*$ at the last safe moment, according to the pass-ahead coordination; thus, our SAs, and thus usually the opponents marking them, are usually more static and predictable.

Figure 2b shows an example of our 2015 algorithm: The SA on the left half waits in its guard location, since the pass-ahead coordination has computed that it should not start moving yet. On the other hand, the SA on the right half has computed it must start moving to its receive location, and thus has started navigating away from its guard location (yellow circle) toward its receive location (red and black concentric circles).

Figure 3 shows an example of our offense coordination in RoboCup 2015, illustrating the results described in Sect. 5: our coordinated offense exhibits (a) *successful teamwork* through multiple passes, (b) *offensive persistence* by maintaining possession on the opponent's half and repeatedly shooting on open angles on the goal, and (c) *regular gameplay effectiveness* by keeping the ball in play over multiple passes, shots, and a rebound off the opponent.

3.4   Other Offense Contributions

In addition to the offense coordination algorithm presented, CMDragons 2015 includes several other features that greatly supported the success of our team: (a) an extremely dedicated care of our 10-year old robot hardware; (b) scripted plays with dynamic zones tuned by extensive simulation games ahead of the tournament and enabled by our novel autoref algorithm [14]; (c) carefully polished action execution; (d) online adaptation of free kick plays according to observed performance metrics; and (e) novel individual robot skills to enable the dribbling the ball, and intercepting any free ball, independently of its origin.

(a) Play starts with a free kick: chip pass over opponents

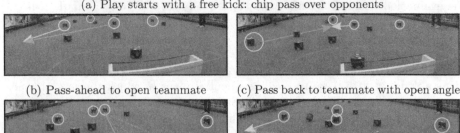

(b) Pass-ahead to open teammate   (c) Pass back to teammate with open angle

(d) Shoot on the open goal angle   (e) Intercept rebound from goalie save

(f) Pass-ahead to open teammate   (g) Shoot and score on open goal angle

Fig. 3. Example of our offense (yellow circles) coordination in RoboCup 2015, illustrating multiple passes (solid yellow and orange arrows), persistent shooting on open angles (green dashed lines), and continuously keeping the ball in play (Color figure online).

4   Defensive Coordination: Threat-Based Evaluation

Our defense relies on a threat-based evaluator [10], which computes the *first-level threat* T_1 and several *second-level threats* T_2^i on our goal: T_1 is the location from which the opponent can most immediately shoot on our goal — i.e., the location of the ball, if it is stationary, or the location of the robot that will receive the ball, if the ball is moving. Threats T_2^i are locations of opponent robots that might receive a pass from the T_1 robot, and then shoot on our goal.

The available defenders are positioned based on the locations of the threats. *Primary defenders* (PDs), of which there are usually one or two, move near the defense area, acting as the line of defense before the goalie; *secondary defenders* (SDs) move further out, intercepting passes and shots by the opponent earlier on. The PDs typically defend against the T_1, blocking open goal angles between the ball and the goal; if there are two PDs, but only one is needed to block a potential shot, the other moves elsewhere on the defense area to guard a T_2. (This situation arises when the ball is close to one of the near corners of the

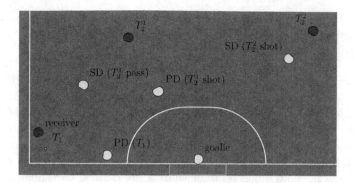

Fig. 4. Example defense (yellow) evaluation and role assignment, as a function of the ball (small orange circle) and opponent (blue) locations. Parenthesized text indicates the type of task assigned to each defender (Color figure online).

field.) The SDs guard against the various T_2s; each one positions itself on a line either from a T_2 to the goal (to block a shot) or from the T_1 to a T_2 (to block a pass). There are never more than two PDs, and there are never any SDs unless there are two PDs. Figure 4 shows the T_1 and T_2^i, and the corresponding PD and SD assignments for a particular state of the world.

The aim of the goalie and the defenders guarding the T_1 is to entirely block the open angle from the threat to the goal; if this is not possible, they position approximately so that as much of it as possible is blocked.

4.1 Defense Behavior Illustration: 5 Defense Robots

We illustrate the defense's coordination behavior when the play system of Sect. 2 has chosen the most defensive play, which has 5 robots assigned to defense. In this case, one defense robot is the goalie, 2 robots are PDs, and 2 are SDs; we illustrate the case in which both PDs are needed to guard T_1. Thus, only the behavior of the SDs varies as a function of the opponent's offense.

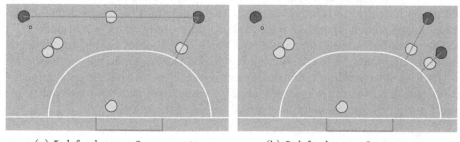

(a) 5 defenders vs. 2 opponents   (b) 5 defenders vs. 3 opponents

Fig. 5. Positioning of the defense in response to attacking opponents. Red lines indicate the geometrical relationships that define the target locations of the SDs (Color figure online).

Two opponents: Figure 5a shows the positioning of a 5-robot defense against 2 offense opponents. Since there is only one T_2, the two SDs block the shot from and the pass to that threat.

Three or more opponents: Figure 5b shows the positioning of a 5-robot defense against 3 offense opponents. Our defense always ranks shot-blocking tasks as more valuable than pass-blocking tasks; thus, both SDs position to block shots from T_2^1 and T_2^2. If there are more than three opponents and hence more than two T_2^i, our algorithm ranks the T_2^i based on the size ϕ of their shooting angle on the goal, ignoring robots, and our SDs mark the highest-ranked threats. If multiple robots have an angle ϕ larger than a pre-defined threshold $\phi^{\max}$, they are ranked based on an estimate of the time t_{goal} it would take to complete a shot on our goal, where lower times are ranked higher. Time t_{goal} is calculated as $t_{\text{goal}} = t_{\text{pass}} + t_{\text{deflect}} + t_{\text{kick}}$, where

$$t_{\text{pass}} = \frac{d(\boldsymbol{x}_b, \boldsymbol{x}_T)}{v_{\text{pass}}}, \quad t_{\text{deflect}} = t_c \cdot \begin{cases} 0, & \theta \leq \theta_{\min} \\ 1, & \theta \geq \theta_{\max} \ , \\ \frac{\theta - \theta_{\min}}{\theta_{\max} - \theta_{\min}}, & \text{else} \end{cases} \quad t_{\text{kick}} = \frac{d(\boldsymbol{x}_g, \boldsymbol{x}_T)}{v_{\text{kick}}}.$$

Time t_{pass} estimates the time to pass to the threat using the distance $d(\boldsymbol{x}_b, \boldsymbol{x}_T)$ between the ball location $\boldsymbol{x}_b$ and the threat location $\boldsymbol{x}_T$, and the estimated opponent passing speed v_{pass}. estimates the time to deflect the ball to our goal based on the angle θ formed by $\boldsymbol{x}_b$, $\boldsymbol{x}_T$, and the goal center point $\boldsymbol{x}_g$: if a one-touch deflection is possible (i.e., $\theta \leq \theta_{\min}$), then $t_{\text{deflect}} = 0$; otherwise, t_{deflect} increases with θ, as defined by constants t_c and $\theta_{\max}$. t_{kick} estimates the shot time similarly to t_{pass}, using the opponent's shooting speed v_{kick}.

4.2   Contributions to Offense

RoboCup 2015 tournament introduced a new rule which prohibited players from entering either team's defense area. As a result, if a team's goalie has control of the ball inside its own team's defense area, it is safe from interference by the other team. We observed that this is effectively the chance to turn a shot on our goal into a free kick; to take advantage of this opportunity, we added the ability for the goalie to behave as a free kick taker under appropriate conditions.

The modularization of our code into *tactics* [9] enables the goalie tactic to simply use the tactic that takes actual free kicks; that tactic handles positioning and pass-ahead coordination with the receiver tactics. Although opportunities for the goalie to take this kind of kick were rare, there were two goals that we scored during the tournament as a direct result of these kicks.

5   RoboCup 2015 Results and Statistics

Our work on defense and offense coordination algorithms resulted in a dominant performance of the CMDragons at RoboCup 2015. The CMDragons played 6

Table 2. Statistics for each CMDragons game in RoboCup 2015. Games RR2 and RR3 ended early due to a Round-Robin 10-goal mercy rule, after 10:20 and 18:25 min respectively (normally, games last 20 min).

Game	Shots					Passes			
	Scored	Missed	Blocked (goalie)	Blocked (other)	Success rate (%)	Success	Missed	Blocked	Success rate (%)
RR1	6	2	4	5	35.3	32	2	7	78.1
RR2	10	2	2	1	66.7	14	4	0	77.8
RR3	10	3	5	7	40.0	30	5	2	81.1
QF	15	8	5	12	37.5	38	2	2	90.5
SF	2	3	6	20	6.5	51	6	6	81.0
Final	5	2	6	7	25.0	29	7	8	65.9
Total	48	20	28	52	—	194	28	23	—
Avg	8	3.3	4.7	8.7	32.4	32.3	4.7	3.8	79.2

official games during the tournament: three Round-Robin games (RR1, RR2, RR3), a Quarter Final (QF), a Semi-Final (SF) and the Final, winning all of the games by scoring 48 goals in total and conceding 0. Table 2 summarizes the offense statistics for shots and passes during each game.

Persistent Offense: First, we highlight the highly offensive gameplay of the CMDragons. As Table 2 shows, our team attempted 148 shots on the opponent's goal throughout the tournament, successfully scoring on 48 opportunities. The CMDragons had an average shooting rate of 3.84 shots per minute of regular gameplay, and a sdcoring rate of 1.25 goals per minute. Figure 6a also demonstrates this highly offensive gameplay: our defensive play, which had only one robot in offense, was used only 21.4 % of the time, while the rest of the time was devoted to more aggressive attack with 3 robots (50.3 %) or 4 robots (28.3 %). Figure 6b shows how this offensive gameplay enabled us to keep the ball mostly on the opponent's half of the field (66 % of the time).

Team Coordination: Table 2 also highlights our team-oriented strategy, with 245 total attempted passes, of which 194 succeeded (79.2 %)[2]. Per minute of regular gameplay, the CMDragons attempted 6.4 passes, and successfully completed 5 passes. Figure 7a shows that this team-oriented strategy played a crucial role in our gameplay: of our 48 goals, 22 were scored immediately after one successful pass, 11 goals after two successful consecutive passes, and 1 goal after three consecutive passes. Thus, team coordination enabled 34 of the 48 goals.

Regular Gameplay Performance: The team coordination algorithms described here shifted the focus of the CMDragons game toward regular gameplay intelligence, rather than stopped ball plays. Figure 7b shows the number of goals

---

[2] As anecdotal reference, the pass completion rates of the human teams in the 2014 World Cup final were 71 % and 80 %.

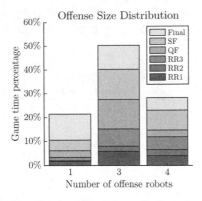
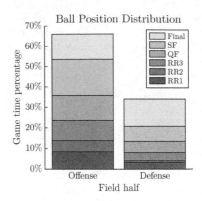

(a) Distribution of number of robots in offense for each game.

(b) Amount of time spent by the ball in each half of the field for each game.

Fig. 6. Measures of gameplay aggressiveness of the CMDragons at RoboCup.

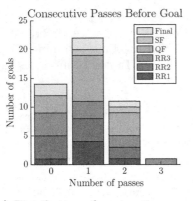
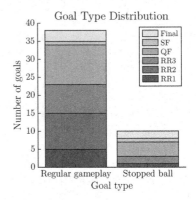

(a) Distribution of consecutive successful passes immediately preceding each goal.

(b) Distribution of regular gameplay goals vs. stopped ball goals

Fig. 7. Passing and goal type statistics for each CMDragons goal in RoboCup.

resulting from regular gameplay vs. those resulting from a stopped ball play. We define stopped ball goals as those that were scored either directly from our stopped ball or after a single pass, since our free kick planners (e.g., [10]) plan one-pass plays. The large majority of our goals were scored during regular gameplay. This distribution is in significant contrast with previous years: while only 56.7 % of our recorded goals in RoboCup 2014 happened during regular gameplay, 79.2 % of our goals in RoboCup 2015 happened during regular gameplay. We believe such shift of focus greatly benefited our team's performance in 2015.

6   Conclusion

Our coordination algorithms in offense and defense are largely responsible for the CMDragons' success at RoboCup 2015. This paper presents the algorithms for (a) assigning our robots to defense and offense subteams, and (b) creating coordinated plans for each subteam. In offense, we present a zone-based coordination algorithm that combines the benefits of planning independently of the opponents with those of appropriately reacting to them. In defense, we describe our threat evaluation algorithm to rank opponents as threats, and to assign defending robots to these threats. We show, via statistics from RoboCup 2015, that our coordination algorithms generated a persistent state of offense, effective teamwork, and a shift from stopped ball plays to regular gameplay. Our future work will focus on greater online autonomous adaptation to different opponents.

References

1. Veloso, M., Stone, P., Han, K.: The CMUnited-97 robotic soccer team: perception and multiagent control. Robot. Auton. Syst. **29**(2–3), 133–143 (2000)
2. Veloso, M., Bowling, M., Stone, P.: The CMUnited-98 champion small-robot team. Adv. Robot. **13**(8), 753–766 (2000)
3. Bruce, J., Zickler, S., Licitra, M., Veloso, M.: CMDragons: dynamic passing and strategy on a champion robot soccer team. In: Proceedings of ICRA (2008)
4. Biswas, J., Mendoza, J.P., Zhu, D., Klee, S., Veloso, M.: CMDragons extended team description paper. In: RoboCup (2014)
5. Bruce, J., Veloso, M.: Fast and accurate vision-based pattern detection and identification. In: ICRA, pp. 1277–1282 (2003)
6. Bruce, J., Veloso, M.: Real-time randomized path planning for robot navigation. In: IROS, pp. 2383–2388 (2002)
7. Zickler, S., Veloso, M.: Efficient physics-based planning: sampling search via non-deterministic tactics and skills. In: AAMAS, pp. 27–33 (2009)
8. Mendoza, J.P., Veloso, M., Simmons, R.: Detecting and correcting model anomalies in subspaces of robot planning domains. In: AAMAS (2015)
9. Browning, B., Bruce, J., Bowling, M., Veloso, M.: STP: skills, tactics, and plays for multi-robot control in adversarial environments. Proc. IME, Part I J. Syst. Control Eng. **219**(1), 33–52 (2005)
10. Biswas, J., Mendoza, J.P., Zhu, D., Choi, B., Klee, S., Veloso, M.: Opponent-driven planning and execution for pass, attack, and defense in a multi-robot soccer team. In: AAMAS (2014)
11. Veloso, M.M., Bowling, M., Chang, A., Browning, B.: Plays as team plans for coordination and adaptation. In: Polani, D., Browning, B., Bonarini, A., Yoshida, K. (eds.) RoboCup 2003. LNCS (LNAI), vol. 3020, pp. 686–693. Springer, Heidelberg (2004)
12. Mendoza, J.P., Biswas, J., Zhu, D., Wang, R., Cooksey, P., Klee, S., Veloso, M.: CMDragons extended team description paper. In: RoboCup (2015)
13. Mendoza, J.P., Biswas, J., Cooksey, P., Wang, R., Zhu, D., Klee, S., Veloso, M.: Selectively reactive coordination for a team of robot soccer champions. In: Proceedings of AAAI-16 (2016, to appear)
14. Biswas, J., Zhu, D., Veloso, M.: AutoRef: towards real-robot soccer complete automated refereeing. In: Bianchi, R.A.C., Akin, H.L., Ramamoorthy, S., Sugiura, K. (eds.) RoboCup 2014. LNCS, vol. 8992, pp. 419–430. Springer, Heidelberg (2015)

UT Austin Villa: RoboCup 2015 3D Simulation League Competition and Technical Challenges Champions

Patrick MacAlpine[✉], Josiah Hanna, Jason Liang, and Peter Stone

Department of Computer Science, The University of Texas at Austin, Austin, USA
{patmac,jphanna,jliang,pstone}@cs.utexas.edu

Abstract. The UT Austin Villa team, from the University of Texas at Austin, won the 2015 RoboCup 3D Simulation League, winning all 19 games that the team played. During the course of the competition the team scored 87 goals and conceded only 1. Additionally the team won the RoboCup 3D Simulation League technical challenge by winning each of a series of three league challenges: drop-in player, kick accuracy, and free challenge. This paper describes the changes and improvements made to the team between 2014 and 2015 that allowed it to win both the main competition and each of the league technical challenges.

1 Introduction

UT Austin Villa won the 2015 RoboCup 3D Simulation League for the fourth time in the past five years, having also won the competition in 2011 [1], 2012 [2], and 2014 [3] while finishing second in 2013. During the course of the competition the team scored 87 goals and only conceded 1 along the way to winning all 19 games the team played. Many of the components of the 2015 UT Austin Villa agent were reused from the team's successful previous years' entries in the competition. This paper is not an attempt at a complete description of the 2015 UT Austin Villa agent, the base foundation of which is the team's 2011 championship agent fully described in a team technical report [4], but instead focuses on changes made in 2015 that helped the team repeat as champions.

In addition to winning the main RoboCup 3D Simulation League competition, UT Austin Villa also won the RoboCup 3D Simulation League technical challenge by winning each of the three league challenges: drop-in player, kick accuracy, and free challenge. This paper also serves to document these challenges and the approaches used by UT Austin Villa when competing in the challenges.

The remainder of the paper is organized as follows. In Sect. 2 a description of the 3D simulation domain is given. Section 3 details changes and improvements to the 2015 UT Austin Villa team including those for variable distance kicks, set plays, and a kick decision classifier for deciding when to kick or dribble the ball, while Sect. 4 analyzes the contributions of these changes in addition to the overall performance of the team at the competition. Section 5 describes and analyzes the league challenges that were used to determine the winner of the technical challenge, and Sect. 6 concludes.

© Springer International Publishing Switzerland 2015
L. Almeida et al. (Eds.): RoboCup 2015, LNAI 9513, pp. 118–131, 2015.
DOI: 10.1007/978-3-319-29339-4_10

2   Domain Description

The RoboCup 3D simulation environment is based on SimSpark,[1] a generic physical multiagent system simulator. SimSpark uses the Open Dynamics Engine[2] (ODE) library for its realistic simulation of rigid body dynamics with collision detection and friction. ODE also provides support for the modeling of advanced motorized hinge joints used in the humanoid agents.

Games consist of 11 versus 11 agents playing on a 30 m in length by 20 m in width field. The robot agents in the simulation are modeled after the Aldebaran Nao robot,[3] which has a height of about 57 cm, and a mass of 4.5 kg. Each robot has 22 degrees of freedom: six in each leg, four in each arm, and two in the neck. In order to monitor and control its hinge joints, an agent is equipped with joint perceptors and effectors. Joint perceptors provide the agent with noise-free angular measurements every simulation cycle (20 ms), while joint effectors allow the agent to specify the speed and direction in which to move a joint.

Visual information about the environment is given to an agent every third simulation cycle (60 ms) through noisy measurements of the distance and angle to objects within a restricted vision cone (120°). Agents are also outfitted with noisy accelerometer and gyroscope perceptors, as well as force resistance perceptors on the sole of each foot. Additionally, agents can communicate with each other every other simulation cycle (40 ms) by sending 20 byte messages.

In addition to the standard Nao robot model, four additional variations of the standard model, known as heterogeneous types, are available for use. These variations from the standard model include changes in leg and arm length, hip width, and also the addition of toes to the robot's foot. Teams must use at least three different robot types, no more than seven agents of any one robot type, and no more than nine agents of any two robot types.

The 2015 RoboCup 3D Simulation League competition included a couple key changes from the previous year's competition. The first of these was a rule change requiring that the ball must either touch an opponent, or touch a teammate outside the center circle, before a team taking a kickoff can score. This rule was put in place to prevent teams from attempting to score directly off a kickoff [3] so as to prevent the competition from potentially devolving into a kickoff taking contest. The second change was to add noise to the beam command used by agents to place themselves at specific positions on the field before a kickoff occurs. Adding noise to the beam command requires agents to use perception when kicking the ball—the added noise prevents agents from beaming to an exact position behind the ball and then blindly executing a kick.

3   Changes for 2015

While many components contributed to the success of the UT Austin Villa team, including dynamic role assignment [5] and an optimization framework used to

---

[1] http://simspark.sourceforge.net/.

[2] http://www.ode.org/.

[3] http://www.aldebaran-robotics.com/eng/.

learn low level behaviors for getting up, walking, and kicking via an overlapping layered learning approach [6], the following subsections focus only on those that are new for 2015. Analysis of the performance of these components is provided in Sect. 4.1.

3.1 Variable Distance Kicks

In 2014 the UT Austin Villa team only had four kicks for its robots to choose from: a fast short kick that goes about 5 m, both a low and a high kick that travel around 15 m each, and a long kick that can propel the ball up to 20 m. This coarse granularity in kick distances limits the set of locations that the ball can be kicked to. Ideally we would like our robots to be able to kick the ball to any precise position on the field such as professional human soccer players are capable of doing. Having the ability to kick the ball to any location opens up possibilities for better passing and teamwork.

For the 2015 competition the UT Austin Villa team added a set of 13 new kicks to its agents with each of the kicks optimized to travel a fixed distance of 3 to 15 m in 1 m increments. This series of new variable distance kicks allow a robot to kick the ball within half a meter of any target 2.5 to 15.5 m away. The kicks, represented as a series of parameterized joint angle poses [7], were optimized using the CMA-ES algorithm [8] and the team's optimization framework incorporating overlapping layered learning [6]. During learning of a d meter kick the robot attempts to kick the ball to a target position d meters directly in front of the robot, and a kick attempt is awarded a negative fitness value equal to the euclidean distance of the ball relative to the target position. Each kick was optimized for 400 generations of CMA-ES with a population size of 150. After optimization of each kick the top 300 highest fitness kick parameter sets were evaluated again over 300 kick attempts each to check for consistency. Finally, a parameter set with both high accuracy and low variance for the target distance was identified from collected data and chosen as the kick to use. This learning process was performed for each kick distance, and run across all five heterogeneous agent types, resulting in a total of $13 \times 5 = 65$ kicks learned.

Variable distance kicks allow for a richer set of passing options as robots can select from many potential targets to kick the ball to as shown in Fig. 1. Each potential kick location is given a score according to Eq. 1, and the location with the highest score is chosen as the location to kick the ball to. Equation 1 rewards kicks for moving the ball toward the opponent's goal, penalizes kicks that have the ball end up near opponents, and also rewards kicks for landing near a teammate. All distances in Eq. 1 are measured in meters.

$$\texttt{score}(target) = \begin{array}{l} -\|opponentGoal - target\| \\ \forall opp \in Opponents, -\max(25 - \|opp - target\|^2, 0) \\ +\max(10 - \|closestTeammateToTarget - target\|, 0) \end{array} \quad (1)$$

Having many available targets to kick the ball to for passing was very important for implementing a keepaway task for the free challenge discussed in Sect. 5.3.

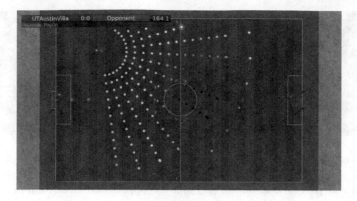

Fig. 1. Potential kick target locations with lighter circles having a higher score. The highest score location is highlighted in red (Color figure online).

Having accurate variable distance kicks was also imperative for doing well during the kick accuracy challenge described in Sect. 5.2.

In addition to allowing for more precise passing, variable distance kicks are also useful for taking shots on goal. Generally speaking, the greater the distance a kick travels the longer and higher the ball may travel in the air, and possibly over the goal, when shooting. To prevent accidentally shooting the ball over the goal, which happened quite frequently during the 2014 RoboCup competition, we limit kicks for shooting on goal to be no more than 7 m in distance beyond the goal line. Having the ability to kick the ball with just the right amount of power such that the ball flies into the goal—but not over it—is a valuable skill during games.

3.2 Set Plays

During the 2014 RoboCup competition the UT Austin Villa team used a multi-robot behavior to score goals immediately off an indirect kickoff. This behavior consisted of having one robot lightly touch the ball before a second robot kicked the ball into the opponent's goal [3]. As rules were changed for the 2015 competition, and now a teammate is required to touch the ball outside of the center circle before a goal can be scored, this kickoff tactic is no longer allowed. Instead the team created legal set plays for kickoffs to try and quickly score.

The first kickoff set play, shown in the left image of Fig. 2, has the player taking the kickoff kick the ball slightly forward and to the left or right side of the field to a waiting teammate ready to run forward and take a shot. The player taking the kickoff chooses which side target to kick the ball to based on which target is furthest from any opponent. If there are opponents near both side targets then the player taking the kickoff instead chooses the kickoff set play shown in the right image of Fig. 2. In this set play the ball is first kicked backwards and to the side to a waiting teammate. The player who receives this backwards pass then kicks the ball forward and across to the other side of the

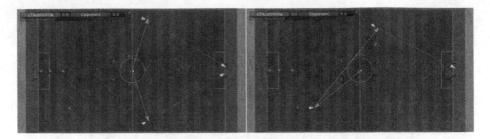

Fig. 2. Kickoff set play to the sides (left image) and pass backwards (right image). Yellow lines represent passes and orange lines represent shots. Dashed lines represent agent movement (red for teammates and blue for opponents) (Color figure online).

field where a teammate is waiting for a pass. It is expected that the player who receives the second pass will be in a good position to take a shot on goal as opponent agents will have been drawn to the other side of the field after the initial backwards pass off the kickoff.

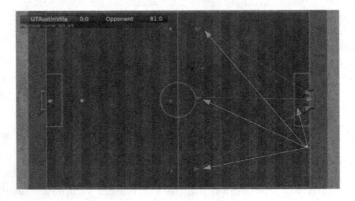

Fig. 3. Corner kick set plays. Yellow lines represent passes and orange lines represent shots. Dashed red lines represent teammate movement. In the example shown the ball would be passed to the teammate waiting for the ball near the bottom of the image as that teammate is most open (Color figure online).

In addition to kickoff set plays the UT Austin Villa team also created set plays for offensive corner kicks. These set plays, shown in Fig. 3, consist of having three teammates move to positions on the midline at the center and both sides of the field. The player taking the corner kick chooses to kick the ball to whichever of these three players is most open. If none of these players are open then the player taking the corner kick just chooses the default option of kicking the ball to a position in front of the goal where several teammates are waiting.

All set plays require passing the ball to specific locations on the field though the use of learned variable distance kicks discussed in Sect. 3.1. Approaching and kicking the ball must be quick as a team has only 15 s to kick the ball once a set play starts.

3.3   Kick Decision Classifier

Before deciding where to kick the ball, first a decision must be made as to whether to kick or dribble the ball. The 2014 UT Austin Villa team chose to always dribble if an opponent is within two meters of the ball—it was assumed that an agent might not have enough time to complete a kick if an opponent is less than two meters from the ball.

Rather than using a hand-picked value to determine if there is enough time to kick the ball, the 2015 UT Austin Villa team decided to train a logistic regression classifier to predict the probability of a kick being successful given the current state of the world. To do so, the team played many games against a common opponent in which agents were instructed to always try and kick the ball. During the course of kick attempts the following state features were recorded and then labeled as positive or negative kick examples based on whether kick attempts were successful.

1. Difference between angle of ball and the orientation of agent
2. Difference between angle of kick target and orientation of agent
3. Angle difference between closest opponent to ball (OPP*) and ball from agent's point of view
4. Difference between angle of ball (from OPP*'s point of view) and the orientation of OPP*
5. Is OPP* fallen or not
6. Magnitude of OPP* velocity
7. Angle between OPP* velocity and ball velocity
8. Distance from agent to ball/OPP* distance to ball
9. Distance from agent to ball/OPP* distance to agent
10. OPP* distance to ball/OPP* distance to agent
11. Distance from agent to ball - OPP* distance to ball
12. Distance from agent to ball - OPP* distance to agent
13. OPP* distance to ball - OPP* distance to agent
14–24. Same features as 3–13 except OPP* is the second closest opponent to ball
25–35. Same features as 3–13 except OPP* is the third closest opponent to ball

The output from a trained classifier is a probability of a kick attempt being successful. A threshold value for this probability, for which kicks are attempted when the probability of a successful kick exceeds this value, was chosen after experimenting with different threshold values while playing 100 s of games against multiple opponents. Metrics monitored during these games were average goal differential, number of kicks performed, goals against, and the probability of a tie or loss.

During the competition the UT Austin Villa team used two different kick classifier models. One was trained against the Apollo3D team which was thought to be one of the fastest teams, and the other was trained against the BahiaRT team which was the opponent UT Austin Villa had the most trouble scoring against at the 2014 competition [3]. Ultimately it was decided to use the more conservative model trained against Apollo3D whenever the team was on defense, so as to be less likely to lose the ball on defense, and then half the time (playing as the left team) switch to the BahiaRT trained model that chose to kick more frequently when on offense and in range of being able to take a shot on goal.

4   Main Competition Results and Analysis

In winning the 2015 RoboCup competition UT Austin Villa finished with a perfect record of 19 wins and no losses.[4] During the competition the team scored 87 goals while only conceding 1. Despite finishing with a perfect record, the relatively few number of games played at the competition, coupled with the complex and stochastic environment of the RoboCup 3D simulator, make it difficult to determine UT Austin Villa being better than other teams by a statistically significant margin. At the end of the competition, however, all teams were required to release their binaries used during the competition. Results of UT Austin Villa playing 1000 games against each of the other 11 teams' released binaries from the competition are shown in Table 1.

Table 1. UT Austin Villa's released binary's performance when playing 1000 games against the released binaries of all other teams at RoboCup 2015. This includes place (the rank a team achieved at the competition), average goal difference (values in parentheses are the standard error), win-loss-tie record, and goals for/against.

Opponent	Place	Avg. Goal Diff.	Record (W-L-T)	Goals (F/A)
FUT-K	2	2.082 (0.036)	927-2-71	2178/96
FCPortugal	3	2.399 (0.040)	945-4-51	2624/225
BahiaRT	4	2.496 (0.044)	944-1-55	2501/5
Apollo3D	5	3.803 (0.046)	995-0-5	3805/2
magmaOffenburg	6	4.167 (0.051)	999-0-1	4171/4
RoboCanes	7	4.187 (0.049)	998-0-2	4235/48
Nexus3D	8	5.571 (0.044)	1000-0-0	5573/2
CIT3D	9	6.321 (0.050)	1000-0-0	6321/0
ITAndroids	11	10.125 (0.041)	1000-0-0	10125/0
Miracle3D	12	10.521 (0.056)	1000-0-0	10521/0
HfutEngine3D	10	11.897 (0.068)	1000-0-0	11897/0

UT Austin Villa finished with at least an average goal difference greater than two goals against every opponent. Additionally UT Austin Villa only lost

---

[4] Full tournament results can be found at http://wiki.robocup.org/wiki/Soccer_Simulation_League/RoboCup2015#3D.

7 games out of the 11,000 that were played in Table 1 with a win percentage greater than 92 % against all teams. This shows that UT Austin Villa winning the 2015 competition was far from a chance occurrence. The following subsection analyzes some of the components described in Sect. 3 that contributed to the team's dominant performance.

4.1 Analysis of Components

Table 2 shows the average goal difference achieved by the following different versions of the UT Austin Villa team when playing 1000 games against the top four teams at RoboCup 2015.

UTAustinVilla Released binary with all features.
NoVarDistKicks No variable distance kicks (except for those used during set plays).
NoSetPlays No set plays.
NoKickClassifier No kick decision classifier.

Table 2. Average goal difference (standard error shown in parentheses) achieved by different versions of the UT Austin Villa team when playing 1000 games against the top four teams at RoboCup 2015.

Opponent	UTAustinVilla	NoVarDistKicks	NoSetPlays	NoKickClassifier
UTAustinVilla	—	−0.042 (0.028)	−1.524 (0.048)	0.060 (0.029)
FUT-K	2.082 (0.036)	1.929 (0.036)	1.467 (0.035)	2.107 (0.036)
FCPortugal	2.399 (0.040)	2.536 (0.044)	2.478 (0.042)	2.181 (0.040)
BahiaRT	2.496 (0.044)	2.407 (0.044)	2.283 (0.040)	2.173 (0.039)

Using variable distance kicks slightly improves performance against most teams except for when playing against FCPortugal. When watching games

Table 3. Scoring percentage of kickoffs and corner kicks achieved by versions of the UT Austin Villa team with and without using set plays while playing 1000 games against the top four teams at RoboCup 2015.

	Kickoff scoring %		Corner kick scoring %	
Opponent	SetPlays	NoSetPlays	SetPlays	NoSetPlays
UTAustinVilla	47.11	0.18	21.32	8.25
FUT-K	54.20	2.65	28.50	12.83
FCPortugal	10.45	6.34	26.03	16.83
BahiaRT	1.25	1.79	33.66	11.79

against FCPortugal it was noticed that the UT Austin Villa team often makes short passes to players who have opponents running toward them and are no longer open by the time the ball is received. This suggests that Eq. 1 in Sect. 3.1 for scoring kick target locations should be improved to take into account the velocity of opponents. Using opponents' velocities, and also performing machine learning to train a function for computing the value of a location to kick the ball to, are future work.

Using set plays really improves performance against UTAustinVilla and FUT-K. Set plays are also beneficial when playing against BahiaRT. Table 3 shows the scoring percentages for kickoffs (measured as having scored within 30 s of a kickoff) and corner kicks (measured as having scored within 15 s of the kick being taken) when both using and not using set plays while playing against the top four teams at RoboCup 2015. Using set plays improves the scoring percentage against all opponents on corner kicks. Kickoffs are not very successful against BahiaRT and FCPortugal with set plays as both teams use formations during kickoffs that are spread out and cover the field well. These spread out formations make it difficult for opponents to find an area with enough free space to receive a pass. Using set plays actually lowers performance against FCPortugal which we attribute to the low kickoff scoring success rate—we find it beneficial to just kick the ball deep into the opponent's side on kickoffs when playing against teams whose formations interfere with our kickoff set plays.

Using the kick decision classifier improves performance against BahiaRT and FCPortugal. This is not surprising as one of the classifiers used was trained against BahiaRT which was built on top of a version of FCPortugal's code base. The kick decision classifier does not help against UTAustinVilla and FUT-K, which are likely the two quickest teams in the league, and are both faster at getting to the ball than either of the teams the kick decision models were trained against. Future work remains to train kick decision classifier models for playing against UTAustinVilla and FUT-K in order to verify if doing so can improve performance against these teams.

5   Technical Challenges

For the second straight year there was an overall technical challenge consisting of three different league challenges: drop-in player, kicking accuracy, and free challenge. For each league challenge a team participated in points were awarded toward the overall technical challenge based on the following equation:

$$\text{points}(rank) = 25 - 20 * (rank - 1)/(numberOfParticipants - 1)$$

Table 4 shows the ranking and cumulative team point totals for the technical challenge as well as for each individual league challenge. UT Austin Villa earned the most points and won the technical challenge by taking first in each of the league challenges. The following subsections detail UT Austin Villa's participation in each league challenge.

Table 4. Overall ranking and points totals for each team participating in the RoboCup 2015 3D Simulation League technical challenge as well as ranks and points awarded for each of the individual league challenges that make up the technical challenge.

Team	Overall		Drop-in player		Kick accuracy		Free	
	Rank	Points	Rank	Points	Rank	Points	Rank	Points
UTAustinVilla	**1**	**75**	**1**	**25**	**1**	**25**	**1**	**25**
FCPortugal	2	61.44	3–6	19.44	2	21	2	21
BahiaRT	3	41.44	3–6	19.44	6	5	3	17
FUT-K	4	37.22	9	7.22	3	17	4	13
magmaOffenberg	5	32.44	3–6	19.44	4	13	—	—
CIT3D	6	31.78	2	22.78	—	—	5	9
HFutEngine3D	7	19.44	3–6	19.44	—	—	—	—
Apollo3D	8	19	10	5	5	9	6	5
Nexus3D	9–10	10.56	7–8	10.56	—	—	—	—
RoboCanes	9–10	10.56	7–8	10.56	—	—	—	—

5.1   Drop-In Player Challenge

The drop-in player challenge,[5] also known as an ad hoc teams challenge, is where agent teams consisting of different players randomly chosen from participants in the competition play against each other. Each participating team contributes two agents to one drop-in player team where drop-in player games are 10 vs 10 with no goalies. An important aspect of the challenge is for an agent to be able to adapt to the behaviors of its teammate. During the challenge agents are scored on their average goal differential across all games played.

Table 5 shows the results of the drop-in player challenge at RoboCup under the heading "At RoboCup 2015". The challenge was played across 8 games such that every agent played at least one game against every other agent participating in the challenge. UT Austin Villa used the same strategy employed in the 2013 and 2014 drop-in player challenge [9], and in doing so was able to win this year's drop-in player challenge.

Drop-in player games are inherently very noisy and it is hard to get statistically significant results when only playing 8 games. In order to get a better idea of each agents' true drop-in player performance we replayed the challenge with released binaries across all $(\binom{10}{5} * \binom{5}{5})/2 = 126$ possible team combinations of drop-in player games ten times each. Results in Table 5 of replaying the competition over many games show that UT Austin Villa has an average goal difference more than five times higher than any other team, thus validating UT Austin Villa winning the drop-in player challenge.

---

[5] Details of the drop-in player challenge at http://www.cs.utexas.edu/~AustinVilla/sim/3dsimulation/2015_dropin_challenge/3D_DropInPlayerChallenge.pdf.

Table 5. Average goal differences for each team in the drop-in player challenge when playing all possible parings of drop-in player games ten times (1260 games in total) and at RoboCup 2015.

| Team | Avg. Goal Diff. | At RoboCup 2015 | |
		Rank	Avg. Goal Diff.
UTAustinVilla	**1.823 (0.045)**	1	**1.625**
FCPortugal	0.340 (0.067)	3–6	−0.125
BahiaRT	0.182 (0.068)	3–6	−0.125
magmaOffenburg	−0.039 (0.068)	3–6	−0.125
FUT-K	−0.052 (0.068)	9	−0.625
RoboCanes	−0.180 (0.068)	7–8	−0.375
CIT3D	−0.361 (0.067)	2	1.125
HfutEngine3D	−0.501 (0.067)	3–6	−0.125
Apollo3D	−0.593 (0.066)	10	−0.875
Nexus3D	−0.620 (0.066)	7–8	−0.375

5.2 Kick Accuracy Challenge

For the kick accuracy challenge[6] robots are asked to kick a ball to the center point of the field from ten different starting ball positions ranging in distances to the center of the field of 3–12 m in 1 m increments. Having already optimized variable distance kicks for each of the integer distances in the 3–12 m range as described in Sect. 3.1, the UT Austin Villa agent participating in the challenge simply chose to execute the appropriate kick for the distance the ball was from the field center during each kick attempt.

Table 6. Average kick distance error in meters for each of the participating teams in the kick accuracy challenge.

Team	Error
UTAustinVilla	**0.200**
FCPortugal	0.282
FUT-K	0.963
magmaOffenberg	1.302
Apollo3D	2.714
BahiaRT	4.652

---

[6] Details and framework for the kick accuracy challenge at https://github.com/ magmaOffenburg/magmaChallenge#kick-challenge.

Results of the kick accuracy challenge are shown in Table 6. UTAustinVilla won the challenge by having the lowest average kick distance error due to its very accurate learned kicks. It is worth noting that the FCPortugal team, who also had a very low average error and took second place in the challenge, used a different strategy of learning a single general kicking skill for different distances using contextual policy search [10].

5.3   Free Challenge

During the free challenge, teams give a five minute presentation on a research topic related to their team. Each team in the league then ranks the top five presentations with the best receiving 5 votes and the 5th best receiving 1 vote. Additionally several respected research members of the RoboCup community outside the league vote, with their votes being counted double. The winner of the free challenge is the team that receives the most votes. Table 7 shows the results of the free challenge in which UT Austin Villa got first place.

UT Austin Villa's free challenge submission[7] focused on describing the team's approach, incorporating machine learning, for kicking and passing the ball to team-mates. This included the following topics: how to approach and kick the ball to different targets (learning skills for walking up to and kicking the ball discussed in [6] as well as learning variable distance kicks presented in Sect. 3.1), where to kick the ball (using a kick location scoring function detailed in Sect. 3.1), when to kick the ball (by querying the kick decision classifier described in Sect. 3.3), and how to have teammates move to receive a pass (using kick anticipation explained in [3]).

UT Austin Villa's free challenge presentation culminated in the demonstration of a keepaway task in which one team attempts to maintain possession of the ball and keep it away from another team for as long as possible. During the demonstration a team was shown to be able to maintain possession and keep the ball away from the 2014 RoboCup champion UT Austin Villa team for over two minutes.[8]

Table 7. Results of the free challenge.

Team	Votes
UTAustinVilla	**79**
FCPortugal	67
BahiaRT	49
FUT-K	32
CIT3D	29
Apollo3D	13

---

[7] Free challenge entry description at http://www.cs.utexas.edu/~AustinVilla/sim/3dsimulation/AustinVilla3DSimulationFiles/2015/files/UTAustinVillaFreeChallenge.pdf.

[8] A video of the keepaway task can be found at http://www.cs.utexas.edu/~AustinVilla/sim/3dsimulation/#2015challenges.

6 Conclusion

UT Austin Villa won the 2015 RoboCup 3D Simulation League main competition as well as all technical league challenges.[9] Data taken using released binaries from the competition show that UT Austin Villa winning the competition was statistically significant. The 2015 UT Austin Villa team improved dramatically from 2014 as it was able to beat a version of the team's 2014 champion binary (the NoScoreKO agent in [3] that does not attempt the now illegal behavior of scoring on a kickoff) by an average of 1.838 ($+/-0.047$) goals across 1000 games.

A large factor in UT Austin Villa's success in 2015 was due to improvements in kicking and the coordination of set plays. In order to remain competitive, and challenge for the 2016 RoboCup championship, the team will likely need to improve multiagent team behaviors such as passing. Additionally, as other teams in the league advance their own passing capabilities, UT Austin Villa will look to implement marking strategies to account for opponents' offensive strategies.

Acknowledgments. Thanks to Klaus Dorer for putting together the kicking accuracy challenge. This work has taken place in the Learning Agents Research Group (LARG) at UT Austin. LARG research is supported in part by NSF (CNS-1330072, CNS-1305287), ONR (21C184-01), AFRL (FA8750-14-1-0070), and AFOSR (FA9550-14-1-0087).

References

1. MacAlpine, P., Urieli, D., Barrett, S., Kalyanakrishnan, S., Barrera, F., Lopez-Mobilia, A., Ştiurcă, N., Vu, V., Stone, P.: UT Austin Villa 2011: a champion agent in the RoboCup 3D soccer simulation competition. In: Proceedings of 11th International Conference on Autonomous Agents and Multiagent Systems (AAMAS 2012) (2012)
2. Stone, P., Collins, N., MacAlpine, P., Lopez-Mobilia, A.: UT Austin Villa: RoboCup 2012 3D simulation league champion. In: Chen, X., Stone, P., Sucar, L.E., van der Zant, T. (eds.) RoboCup 2012. LNCS, vol. 7500, pp. 77–88. Springer, Heidelberg (2013)
3. Depinet, M., Liang, J., MacAlpine, P., Stone, P.: UT Austin Villa: RoboCup 2014 3D simulation league competition and technical challenge champions. In: Bianchi, R.A.C., Akin, H.L., Ramamoorthy, S., Sugiura, K. (eds.) RoboCup 2014. LNCS, vol. 8992, pp. 33–46. Springer, Heidelberg (2015)
4. MacAlpine, P., Urieli, D., Barrett, S., Kalyanakrishnan, S., Barrera, F., Lopez-Mobilia, A., Ştiurcă, N., Vu, V., Stone, P.: UT Austin Villa 2011 3D Simulation Team report. Technical Report AI11-10, The University of Texas at Austin, Department of Computer Science, AI Laboratory (2011)
5. MacAlpine, P., Price, E., Stone, P.: SCRAM: scalable collision-avoiding role assignment with minimal-makespan for formational positioning. In: Proceedings of the Twenty-Ninth AAAI Conference on Artificial Intelligence (AAAI 2015) (2015)

---

[9] More information about the UT Austin Villa team, as well as video highlights from the competition, can be found at the team's website: http://www.cs.utexas.edu/~AustinVilla/sim/3dsimulation/#2015.

6. MacAlpine, P., Depinet, M., Stone, P.: UT Austin Villa 2014: RoboCup 3D simulation league champion via overlapping layered learning. In: Proceedings of the Twenty-Ninth AAAI Conference on Artificial Intelligence (AAAI 2015) (2015)
7. MacAlpine, P., Depinet, M., Stone, P.: Keyframe sampling, optimization, and behavior integration: towards long-distance kicking in the RoboCup 3D simulation league. In: Bianchi, R.A.C., Akin, H.L., Ramamoorthy, S., Sugiura, K. (eds.) RoboCup 2014. LNCS, vol. 8992, pp. 571–582. Springer, Heidelberg (2015)
8. Hansen, N.: The CMA Evolution Strategy: a Tutorial (2009). http://www.lri.fr/hansen/cmatutorial.pdf
9. MacAlpine, P., Genter, K., Barrett, S., Stone, P.: The RoboCup 2013 drop-in player challenges: experiments in ad hoc teamwork. In: Proceedings of the IEEE/RSJ International Conference on Intelligent Robots and Systems (IROS) (2014)
10. Kupcsik, A.G., Deisenroth, M.P., Peters, J., Neumann, G.: Data-efficient generalization of robot skills with contextual policy search. In: AAAI (2013)

RoboCup 2015 Humanoid AdultSize League Winner

Seung-Joon Yi[1]([✉]), Stephen McGill[1], Heejin Jeong[1], Jinwook Huh[1],
Marcell Missura[1], Hak Yi[2], Min Sung Ahn[2], Sanghyun Cho[2], Kevin Liu[2],
Dennis Hong[2], and Daniel D. Lee[1]

[1] GRASP Lab, Engineering and Applied Science, University of Pennsylvania,
Philadelphia, USA
{yiseung,smcgill3,heejinj,jinwookh,missura,ddlee}@seas.upenn.edu
[2] University of California, Los Angeles, CA 90095, USA
{yihak,aminsung,albe1022,kevinliu676,dennishong}@ucla.edu
http://www.seas.upenn.edu/~robocup

Abstract. Major rule changes for the RoboCup Humanoid League in
2015 pose significant vision and locomotion challenges for disambiguat-
ing similarly colored objects and navigating soft terrain. These significant changes highlight the need for applying general purpose humanoid
robotics approaches that can handle abrupt environment modifications,
and we utilize the general purpose THOR (Tactical Hazardous Operations Robot) series of robot from the recent DARPA Robotics Challenge
(DRC). Specific techniques for vision, kicking and autonomy complement
software developed for robust deployments in the DRC. In this paper,
we present these soccer playing techniques, which were validated in the
Humanoid AdultSize league in Hefei.

1 Introduction

Team THORwIn represents the second iteration of University of Pennsylvania-
UCLA joint RoboCup Humanoid AdultSize league team. First competing in
the KidSize class, we have competed in the Humanoid League since 2010 as
Team DARwIn, Team DARwIn-XOS and Team THORwIn. We have been for-
tunate to achieve first place results in many years as well as work on general
purpose humanoids outside of RoboCup. In the AdultSize league, we introduced
the general purpose THOR humanoid robot platform to RoboCup in 2014. Most
AdultSize league robots are designed to be very lightweight and compliant. They
are well optimized for soccer playing but their features are too limited for general
applications. On the other hand, THOR is a heavier, more powerful and more
versatile robot designed for disaster response scenarios in unstructured environments [1]. Using a heavy, general purpose humanoid robot for autonomous soccer
playing poses a number of unprecedented challenges, which have been our main
research focus in AdultSize league.

Additionally, the year 2015 has brought major changes in the rules for the
Humanoid leagues. The field is now a soft surface covered by artificial grass that

L. Almeida et al. (Eds.): RoboCup 2015, LNAI 9513, pp. 132–143, 2015.
DOI: 10.1007/978-3-319-29339-4_11

Fig. 1. The THOR robot for DRC Finals and the modified THOR robot for the RoboCup 2015.

makes both locomotion and kicking much more challenging. New white colored goalposts and balls pose a large burden on perception systems to disambiguate vision cues. In this paper, we present in detail how we utilized the general purpose THOR humanoid robot, developed by Robotis Co, Ltd.[1], to overcome all those challenges with RoboCup 2015 (Fig. 1).

2   THOR Robot Platform

First used in the DARPA Robotics Challenge (DRC) [2], the THOR robot successfully completed many manipulation tasks with its hardware and software features for disaster response situations. The robot was upgraded in computation power, kinematics, and joint torque capabilities for the DRC Finals. We modified this robot to compete in RoboCup 2015 with similar qualities as the RoboCup 2014 entry.

The newer version developed for DRC Finals event held in June 2015 has a number of improvements over the older THOR robot [3]. Unlike the former DRC Trials event where each tasks are tried separately with own time limits and the robot is protected by overhead gantry system, the DRC Finals event consists of a single run where the robot has to complete tasks sequentially without protection. The new robot has completely redesigned legs with more powerful actuators and more rigid structural parts. To lower the center of mass for increased stability,

---

[1] http://www.robotis-shop-en.com/.

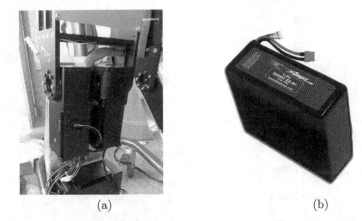

(a)   (b)

Fig. 2. (a) Parallel knee actuators of THOR-RD robot (b) 488 Wh LiPo battery pack housed in the thigh section.

now the leg houses the main battery packs which have been increased in capacity as well. These are shown in Fig. 2.

Both THOR humanoid robot versions use off-the-shelf Robotis Dynamixel Pro series actuators for most joints that are connected into kinematic chains via standardized structural components. With standardized dimensions, the robot is reconfigured quickly in case of damage or feature change needs.

To support its heavier weight while traversing uneven terrain, all leg actuators are increased in power; the most demanding knee joint is now powered by two custom 200 W actuators with right angle gearboxes. They are connected with special cable to form a master-slave pair, and are controlled as a single actuator. Battery compartments have also moved into leg cavities to lower the center of mass as much as possible. This helped the robot in a number of demanding manipulation tasks on uneven surfaces at the DRC Finals event. Lastly, the hip is redesigned to provide a wider range of motion for difficult whole body motions such as car egress. This leg redesign greatly increased the mass and inertia of the legs, so we adopt an additional balancing algorithm that considers the masses of the legs as well.

2.1   Modification for RoboCup

The THOR robot originally had two asymmetric seven degree of freedom (DOF) arms with grippers, providing a large workspace and high dexterity required to accomplish all the DRC Finals tasks. However, we could not use the stock arms as they are too long to conform with the RoboCup rules. Long arms also increase the chance of touching an obstacle, and can obstruct the main camera. We have prototyped a lightweight two DOF arm design with a thin aluminum truss structure that can pick up and throw a ball. However, we eventually discarded it as the throw-in challenge has been removed this year, which had been the sole reason to have actuating arms. Instead, we use a very thin and short one DOF arm design, which further decreases the footprint of the robot and has minimal weight.

Another modification we introduced last year was the enlarged foot size for providing more stability, especially during strong kick motions. This year, we have found that the new robot has a much lower center of mass height due to the large mass of the redesigned legs, and this limits the maximum size of the feet. Instead of designing custom feet for RoboCup, we use an adjustable foot design developed for the DRC Finals competition that is smaller than the feet of last year's RoboCup. To prevent possible backspin of the kicked ball, we added toe bumpers.

3   Walk Controller

For the first time, RoboCup has introduced a realistic field surface covered with artificial grass. This change has made the surface soft and uneven, which means that locomotion becomes a much harder task. We have found that due to the soft and sometimes non-uniform terrain, the support foot can dig into the grass and tilt the robot during walking. Although we use the ankle push recovery strategy and landing timing adaptation to keep the robot balanced, we have found that the frontal tilt of the robot while walking forward make the robot kick the ground hard and quickly lose balance. Increasing the step height can provide more ground clearance, but it will make the landing harder, which will worsen the effect of early landing. To solve this issue, we implement a human like heel-strike, toe-off walk gait to provide toe clearance and smooth the landing.

3.1   Toe and Heel Lift Controller

The DRC Finals competition requires the capability to traverse over uneven terrain and staircases. The kinematic dimensions are too large for the THOR robot to handle with quasi-stationary locomotion. To handle this traversal without sacrificing the stability, we have developed a walk controller that utilizes foot tilt around the toe and heel edges to extend the effective length of the legs [4]. As the RoboCup environment assumes a flat surface, we can use a simplified version that has a closed form solution (Fig. 3).

Given the hip position x_{hip} and the projected ankle position over the landing surface x_F, the position of the ankle is

$$x_{ankle} = T(x_F)T([0, 0, h_{ankle}])[0, 0, 0, 1]^T \qquad (1)$$

where T is the translation matrix, R_y the rotation matrix around the y axis, h_{ankle} the ankle height. Then we have the following kinematic constraint:

$$||x_{ankle} - x_{hip}|| \leq \text{Leg}_{\text{MAX}} \qquad (2)$$

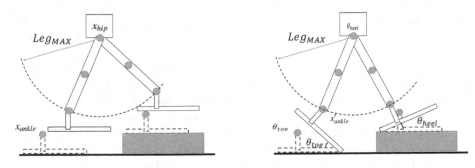

Fig. 3. Toe and heel lift calculation

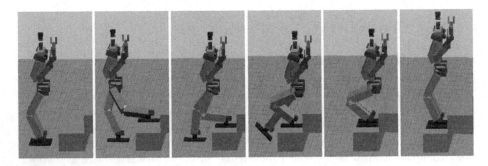

Fig. 4. The THOR-RD robot climbing on the staircase using the toe and heel lift controller in simulation.

In case (2) does not hold, we can lift the heel by θ_{heel} around the toe contact position to increase the effective length of the leg. This leads to the new ankle transform

$$x^H_{ankle} = T(x_F)T([F_T, 0, 0])R_y(\theta_{heel})T([-F_T, 0, h_{ankle}])[0, 0, 0, 1]^T \qquad (3)$$

where F_T is the distance between the projected ankle axis to the toe contact position. Likewise, we can lift the toe by θ_{toe} around the heel contact position, resulting in the ankle transform

$$x^T_{ankle} = T(x_F)T([-F_H, 0, 0])R_y(-\theta_{toe})T([F_H, 0, h_{ankle}])[0, 0, 0, 1]^T \qquad (4)$$

We can solve the following equations to calculate the minimum toe and heel lift angles that satisfy the kinematic constraints

$$||x^H_{ankle} - x_{hip}|| = \text{Leg}_{MAX} \qquad (5)$$

$$||x^T_{ankle} - x_{hip}|| = \text{Leg}_{MAX} \qquad (6)$$

which have closed form solutions. Once the lift angles are found, the type and amount of lift angle is determined based on the foot configuration and the range of motion of the ankle joint. Figure 4 shows the THOR-RD robot climbing over a staircase utilizing toe and heel lifting.

3.2 Heel Strike Toe Off Walk

The Toe and Heel Lift Controller automatically generates foot tilt motion when the target feet positions are beyond the kinematic limits, which can happen when the robot takes a large stride or walks over an uneven surface. As we want to make the robot always perform heel strike toe off gait, we provide a target foot tilt angle to the Toe and Heel Lift Controller. To support the heel-strike, toe-off walking gait on flat surfaces, we extend our piecewise linear ZMP trajectory so that the ZMP moves linearly from heel to toe in single support. Once the reference ZMP and foot trajectories are generated, we use our hybrid walk controller [5] to generate the torso trajectory which both uses a ZMP preview controller and an analytic ZMP controller (Fig. 5).

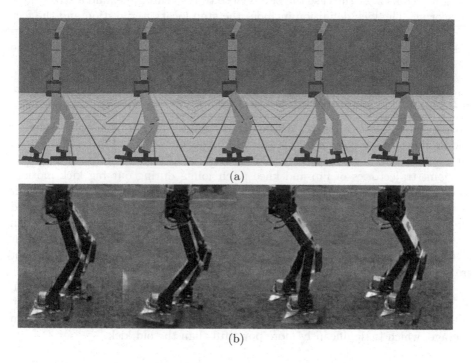

Fig. 5. Heel strike Toe off walk. (a) simulation (b) on RoboCup field

4 Kick Controller

The artificial grass covered field has not only made locomotion harder, but also made our previous kick much weaker due to the increased surface friction. We could consistently kick the ball more than eight meters on smooth surfaces with our old kick, but we have found that the same kick can reach only three meters on the grass covered field in the worst case. In this section we present how we have improved our kick controller to be more powerful, while keeping the tight integration with the locomotion engine that allows for rapid kicking without fully stopping.

4.1 Cartesian Space Kick Generation

We use a kick engine based on our hybrid walk controller, that uses online switching of walk controllers and custom step sequence and feet trajectory to generate kick motion [6]. Our kick engine is fast and very flexible. It is very easy to design a number of different kicks, and the same kick definition can be used to initiate kick during either walking or standing state. Since last year, we use three different type of kicks; the quick, non-stopping kick for the first kick, the low powered kick for the case the ball is close to the center line, and a strong kick for scoring. The kick definition includes the step location, duration and the feet trajectory which we define using spline curves in Cartesian space. For RoboCup 2015, we added the heel pitch trajectory to prevent the toe from touching the ground.

4.2 Hybrid Kick Generation

Although the Cartesian space definition of the kick motion is simple to design and easily can be integrated with walking, linear movement of feet can result in undesirable joint angle trajectory that reduces the kick power. Figure 6(a) shows the joint trajectories of hip and knee pitch joints during our big kick motion, where the knee joint does not move in synchronization with the hip pitch joint at the time of impact. This problem can be solved by utilizing a hand designed joint trajectory, which is shown in Fig. 6(b), but such a joint level trajectory can be harder to integrate with walk controller, which handles the feet movement in Cartesian space. We solved this problem by linearly interpolating two joint trajectories at the beginning and the end of the kick.

Figure 7 shows the comparison of the old kick using Cartesian space definition and the new hybrid kick that uses joint space trajectory as well. On the RoboCup field, we have found that the new kick can consistently move the ball to 5 m on average, which is significantly more powerful than the old kick.

4.3 Whole Body Balancing Control

Unlike the Linear Inverted Pendulum Model (LIPM) we use for dynamics calculation, the physical robot has distributed mass which can seriously affect the stability of the robot if not correctly accounted for. When humanoid robots are used for manipulation tasks, the different arm configuration affects the overall center of mass location, and the robot must compensate for it to keep balanced. For the DRC Trials, we calculated the upper body COM location based on current arm configuration, and shifted the torso position so that the upper body COM stays fixed. Leg masses are not considered, as we assume that all manipulation takes place when robot is standing still with preset leg stance. However, as

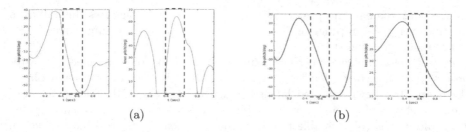

(a)   (b)

Fig. 6. Joint trajectories of leg pitch joints, where the red box shows the possible ball impact. (a) Cartesian space kick (b) Joint space kick

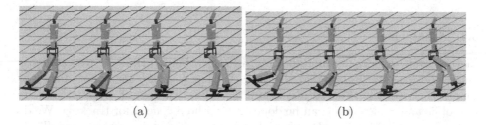

(a)   (b)

Fig. 7. Comparison of kick sequences. (a) Cartesian space kick (b) Hybrid kick

the new version of THOR robot has much heavier legs and lighter upper body, we use a fine grained mass model to calculate the COM location of the whole robot, and iteratively update the torso position to compensate the COM error. The balancing controller is now always active, using the reference COM position from the walk controller instead of the middle point of the two foot positions as the target COM position. As a result, the robot can now correctly balance in a slow single support phase while performing arm motions, which helps balancing during kicks.

5   Vision and Localization

In the AdultSize league, the robot must be aware of its surroundings. Two stationary obstacles are avoided while planning, while a moving goalie should be avoided while kicking. The robot also detects a white ball, lines, and goal posts on the green turf field. We tackle these challenges using color classification, bitwise preprocessing operations, and strict shape filters. Our locomotion strategy depends on accurate goal post detection, calibrated locomotion odometry, and some line information.

A single Logitech C920 HD camera provides color pixels for processing. We train a Gaussian mixture model for applying maximum likelihood estimation of the color class of a pixel – white, green, off-white. The maximum likelihood estimate is encoded in a look-up table to categorize pixels into one of three color labels. We use two white color classes, where one class models the lines and one class models the ball and posts. We found that this disambiguates the ball from the lines very well in practice, but required a highly trained human supervisor in the model training stage.

The labeled image from the lookup table is then down-sampled for faster processing before being fed into high level blob detection routines and object classifiers. The down-sampled labeled image is a bitwise OR concatenation of the color classes of the pixels from the higher resolution image in the subsampled 2×2 block. The robot is able to detect the ball, goal posts, and lines based on their color, size, and shape information. Given the exact dimensions of objects, the distance and angle of detected objects are obtained by projection from the image frame to the robot's egocentric coordinates.

5.1   Obstacles

In the AdultSize League in RoboCup 2015, two black cylinders are randomly placed between the robot and the goal. During a game, the robot must not touch these obstacles. Placed just before the "Play" command, the obstacles are immobile and so detection can be done once without a need for tracking. We do not model explicitly the color black, but only as "uncategorized colors." Thus, blue or red colors are just as uncategorized as the black color of the obstacles. We apply blob detection used for the similarly shaped posts, but we add additional checks for surrounding green, much larger size. Also, we only consider pixels below the horizon by a certain angle. In this way, uncategorized pixel blobs off the field pose no issues.

The labeled image is scanned horizontally and the position of the lowest black pixel (*depth* of black) in each column is marked. The columns with black *depth* difference within a threshold are stitched together to be connected regions, and those are the candidates of obstacles which then go through a filtering process including height check, width check, on-the-ground check, and within-the-field check. A final filtering process is performed after transforming the positions of the detected obstacles from image frame to the robot's egocentric, polar coordinates. The potential obstacles are clustered into groups based on their angles to the robot, and statistical information is calculated for each group providing the certainty and weighed average position of each obstacle. Finally, the closest two obstacles are registered for later trajectory planning.

5.2   White Objects

When the ball lies on a line, detection becomes difficult as the color classification leads to continuous, indiscriminate, white pixels. We built a pipeline of bitwise region erosion and dilation to eliminates weak pixel connections between a ball and lines/posts. The number of erosion steps and dilation steps are tuned to eliminate background noise – false positive ball and line pixels. We model erosion as bitwise AND on 2×2 blocks of pixels, while dilation is modeled as a bitwise OR operation on 2×2 blocks.

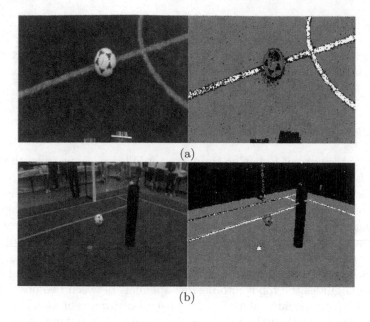

(a)

(b)

Fig. 8. Ball-on-the-line cases. (a) Nearby ball. (b) Faraway ball.

Distant balls and close balls require different erosion and dilation steps, but in practice, it is difficult to make these step choices scale invariant. While we used an additional color class for the particular shade of white on the ball, we found that bitwise erosion and dilation is useful for removing noisy color classifications (especially off the field) for field and line colors. With less noisy pixels, object checks for the ball can be tighter. To detect white field lines, we implement a decision tree for color class edges, where 2×2 blocks of pixels are mapped to a coordinate in Hough transform space. We modulate weights and coordinates for Hough transform counts. This more stringent guideline using edges for line population finds Green/white/neither edges on 2×2 blocks. We are able to find good line statistics even with a ball on a line far away, as shown in Fig. 8.

5.3   Localization

In previous years, we used a particle filter to track the three-dimensional robot pose (x, y, θ) on the field, in which orientation is reset when the robot falls down. The particles are probabilistically updated using a motion model based on the odometry of the robot and yaw angle based on inertial measurement, as well as vision information that takes into account pre-specified positions of landmarks such as goal posts, lines and corners.

This year we could further simplify this framework due to the uniqueness of the AdultSize league: the robot always starts at exactly the same initial position. It is never supposed to fall down, and each trial has quite a short time limit of 90 s. This makes the inertial tracking of the robot heading angle θ very precise during the match, so we assume that the current estimate of heading direction is correct. Then we discard the distance information from goalpost observations, which can be quite noisy for distant goalposts, to handle the redundant information. This simplification greatly helped the localization during the whole competition, where the localization error was hardly visible.

6   High Level Behavior

According to the current RoboCup Humanoid AdultSize league rules, the attacking robot is only allowed to score after crossing the centerline. The attack ends if the attacking robot touches the obstacle or the kicked ball stops inside the penalty box. So the best tactic for the attacking robot is to move the ball to just cross the centerline and powerfully kick the ball to score. We have used a discrete search tree that evaluates possible kick paths that considers obstacles, the border distance and goal angle. Due to the newly introduced grass covered surface, kick power has decreased significantly and we have found that our old planner often guides the robot into narrow openings between two obstacles. We have increased the penalization term for obstacles to prevent this issue, which made the robot move the ball to a less populated area before trying the big kick. Figure 9 shows an example of a global plan that avoids both obstacles.

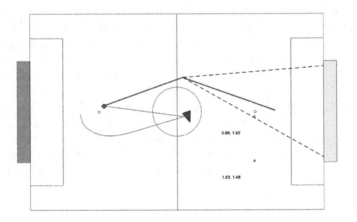

Fig. 9. An example of high level kick planning.

7   Results

We have found that the new version of the THOR robot with new heel strike toe off walk controller can reliably walk over the new grass covered soccer field without any issues. The higher friction of the surface rendered our old kicks much weaker. We have utilized the hybrid walk controller which also uses joint level target trajectories to make the kick harder while being tightly integrated with a walk engine. To make up for a weaker kick, we have increased the maximum stride length to 40 cm, which allowed the robot to try up to four kicks to score within the time limit of 90 s. As the AdultSize league has only one attacking robot on the field, our vision algorithm had no problem differentiating various localization cues of the same color. Throughout the matches, the new hierarchical planner reliably found good kick sequences that lead to scores.

8   Conclusions

In this work, we described our methods for executing soccer tasks on a general purpose full sized humanoid robot for the RoboCup 2015 competition. To overcome the new challenges of the 2015 rule changes, we utilized a heel strike toe off walk controller to improve ground clearance for the swing foot and a new hybrid kick controller to compensate for the kick distance loss due to the high friction of the surface. We also present the vision and planning systems we developed for the RoboCup competition, which worked robustly throughout the competition.

Acknowledgments. We acknowledge the Defense Advanced Research Projects Agency (DARPA) through grant N65236-12-1-1002. We also acknowledge the support of the ONR SAFFIR program under contract N00014-11-1-0074.

References

1. Yi, S.-J., McGill, S., Vadakedathu, L., He, Q., Ha, I., Rouleau, M., Hong, D., Lee, D.D.: Modular low-cost humanoid platform for disaster response. In: Proceedings of the 2014 IEEE/RAS International Conference on Intelligent Robots and Systems (2014)
2. Yi, S.-J., McGill, S.G., Vadakedathu, L., He, Q., Ha, I., Han, J., Song, H., Rouleau, M., Zhang, B.-T., Hong, D., Yim, M., Lee, D.D.: Team THOR's entry in the DARPA robotics challenge trials 2013. J. Field Rob. **32**(3), 315–335 (2014)
3. McGill, S., Yi, S.-J., Lee, D.: Team THORs adaptive autonomy for disaster response humanoids. In: 15th IEEE-RAS International Conference on Humanoid Robots (Humanoids). IEEE (2015)
4. Yi, S.-J., Hong, D., Lee, D.: Heel and toe lifting walk controller for traversing uneven terrain. In: 15th IEEE-RAS International Conference on Humanoid Robots (Humanoids). IEEE (2015)
5. Yi, S.-J., Hong, D., Lee, D.D.: A hybrid walk controller for resource-constrained humanoid robots. In: 13th IEEE-RAS International Conference on Humanoid Robots (Humanoids), October 2013
6. Yi, S.-J., McGill, S., He, Q., Hong, D., Lee, D.: Walk and kick motion generation for a general purpose full sized humanoid robot. In: Workshop on Humanoid Soccer Robots, IEEE-RAS International Conference on Humanoid Robots (Humanoids). IEEE (2014)

Fuzzy Inference Based Forecasting in Soccer Simulation 2D, the RoboCup 2015 Soccer Simulation 2D League Champion Team

Xiao Li$^{(\boxtimes)}$ and Xiaoping Chen

Department of Computer Science,
University of Science and Technology of China, Hefei, China
runtu@mail.ustc.edu.cn, xpchen@ustc.edu.cn

Abstract. WrightEagle is the champion of RoboCup 2D simulation league, the latest technic used in WrightEagle will be presented in this paper. This paper introduces the application of fuzzy inference system in WrightEagle. The fuzzy inference system is built without knowledge of the mathematical description of the opponents. With the fuzzy inference system, we are able to forecast the actions of the opponents in some aspects. Once we can forecast the actions of the opponents, we can make corresponding decisions.

1 Introduction

In RoboCup 2D simulation league, every team consists of 12 autonomous agents (11 players and an online coach). For each agent, its goal is to make the best decision with the information observed. Up to now, Markov Decision Processes framework [1], multi-step decision-making framework [2], MAXQ hierarchical structure [3] and heuristic online planning [4,5] techniques have been applied on WrightEagle to make decisions. To make better decisions, we want to forecast the actions or decisions made by the opponents. Apparently, as in a competition, if the agent is able to forecast the actions or decisions made by the opponents, it will be very helpful for the agents to make corresponding decisions. To this end, an intuitive method is to build a mathematical model for the opponent. But the scale of 2D simulation is so large that it is nearly impossible to build a mathematical model for the opponent completely. Then we come up with the fuzzy inference system. The fuzzy inference system makes it possible to map inputs (observations) to outputs (actions) without knowledge of the mathematical description of the opponent. With the rules or knowledge we concluded in the past, we can build some fuzzy inference systems. These systems can help us to forecast opponents' actions in some aspects.

2 Fuzzy Inference System (FIS)

Fuzzy inference is the process of formulating the mapping form a given input to an output using fuzzy logic [6]. In fuzzy logic, a statement is able to assume

© Springer International Publishing Switzerland 2015
L. Almeida et al. (Eds.): RoboCup 2015, LNAI 9513, pp. 144–152, 2015.
DOI: 10.1007/978-3-319-29339-4_12

any real value between 0 and 1, representing the degree to which an element belongs to a set. So it is able to work with human inputs. On the other hand, fuzzy logic does not need complicated mathematical models. So the FIS is easy to implement.

Mamdani-Type Fuzzy Inference and Sugeno-Type Fuzzy Inference are two of the most common types of fuzzy methodology. In the following sections, we will use these two methods to deal with a simple and specific scene in the 2D simulation.

2.1   Mamdani-Type Fuzzy Inference

The Mamdani's fuzzy inference method is the most common fuzzy methodology. It was first proposed by Ebrahim Mamdani [7] in 1975. A Mamdani's fuzzy inference system is generally composed of five steps:

1. Determine a set of fuzzy rules,
2. Fuzzify inputs with the input membership functions,
3. Obtain each rule's conclusion,
4. Aggregate conclusions obtained in step 3,
5. Defuzzification.

The detailed description of these steps will be introduced in the following part.

Determine a Set of Fuzzy Rules. The form of fuzzy rules in Mamdani's fuzzy inference system is always like this:

$$If\ x\ is\ A\ and/or\ y\ is\ B,\ then\ z\ is\ C.$$

where A, B and C are linguistic values defined in the fuzzy set.

For such a scene in the 2D simulation (Fig. 1), we assume that opponent1 (opp1) is attending to pass the ball to opponent2 (opp2) to point A. Our team-mate1 (tm1) is marking on opponent2. The goal of tm1 is to judge if opp1 would pass the ball to point A. Here, we assume the possibility of opp1 to pass the ball successfully to be the evaluation of opp1's action.

It is intuitive that the possibility of opp1 to pass the ball to opp2 successfully (P) is positively related to $\Delta c1 = c1 - c2$ and negatively related to $\Delta c2 = c1 - c3$. So two simple rules can be concluded here:

$$rule1:\ If\ \Delta c1\ is\ large\ and\ \Delta c2\ is\ small,\ then\ P\ is\ high.$$

$$rule2:\ If\ \Delta c1\ is\ small\ or\ \Delta c2\ is\ large,\ then\ P\ is\ low.$$

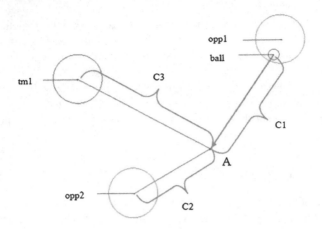

Fig. 1. A specific scene in 2D simulation. c1 is the cycle needed by the ball to get to point A; c2 is the cycle needed by opp2 to get to point A; c3 is the cycle needed by tm1 to get to point A.

Fuzzify Inputs with the Input Membership Functions. For a fuzzy set A, its membership function is defined as $\mu_A : X \to [0,1]$. Set X is the universe of discourse. For each $x \in X$, the value $\mu_A(x)$ is called the degree of membership of x to fuzzy set A. For example, if $\mu_A(x) = 0.7$, we can say that x belongs to A to a degree of 0.7.

With the fuzzy set, we can describe fuzzy concept like *"large"* and *"small"*. Therefore, the FIS is able to deal with human inputs. For the *"large"* and *"small"* in the rules, their membership functions are defined in Fig. 2.

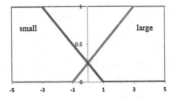

Fig. 2. The membership functions of *"small"* and *"large"*.

Using the fuzzy sets defined in Fig. 2, we can fuzzify the inputs. For example, with the inputs $\Delta c1 = 1$, $\Delta c2 = -2$, we can obtain the membership values for each rule (Fig. 3).

The fuzzy operators *"and"* (min) and *"or"* (max) are used in Fig. 3, then we obtain a single membership value for each rule.

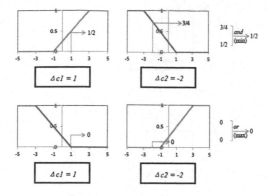

Fig. 3. Obtain a single membership value for each rule.

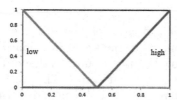

Fig. 4. The membership functions of *"low"* and *"high"*.

Obtain Each Rule's Conclusion. To obtain each rule's conclusion, we should define the membership functions of the outputs first. For the *"high"* and *"low"* in the rules, their membership functions are defined in Fig. 4.

Then, with these membership values obtained in step 2, we can use a fuzzy implication operator (min) to obtain a fuzzy set for each rule (Fig. 5).

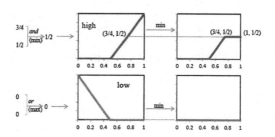

Fig. 5. Obtain a fuzzy set for each rule.

Aggregate Conclusions. We choose the max to be the aggregation operator, then we can combine the results obtained in last step (Fig. 6).

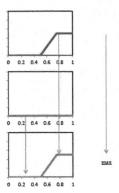

Fig. 6. Aggregate conclusions.

Defuzzification. The result of last step is a fuzzy set, it is not we want when we need to make a decision. As in the scene mentioned above, we do not want to know the P is high or low but the accurate value of P. Therefore, we need to transform the fuzzy set into an accurate value. The process of this transformation is called defuzzification. Here, we use the centroid method, which is one of the most common used methods for defuzzification. Centroid method is to compute the center of the mass. Here, we take some sample points to calculate the center (Fig. 7).

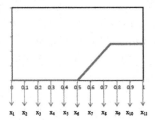

Fig. 7. Aggregate conclusions.

$$P = \frac{\sum\limits_{i=1}^{11} x_i \times \mu(x_i)}{\sum\limits_{i=1}^{11} \mu(x_i)} = 0.83 \tag{1}$$

Now, the accurate value of P is obtained. So we can say that the possibility of opp1 to pass the ball to opp2 successfully is 0.83. Then, we can forecast that the possibility of opp1 to pass the ball to point A is 0.83. In other words, now we are able to forecast the possibility of an opponent to pass the ball to a certain point. Therefore, we can forecast to which point an opponent will pass the ball through enumeration.

2.2 Sugeno-Type Fuzzy Inference

The Sugeno fuzzy inference method was first proposed in 1985 [8], it is very similar to the Mamdani's method. Generally, the Sugeno fuzzy inference system is composed of four steps:

1. Determine a set of fuzzy rules,
2. Fuzzify inputs with the input membership functions,
3. Obtain each rule's conclusion,
4. Aggregate conclusions obtained in step 3.

Determine a Set of Fuzzy Rules. The form of the rules in Sugeno method is like this:

$$\text{If } x \text{ is } A \text{ and/or } y \text{ is } B, \text{ then } z \text{ is } f(x, y).$$

Unlike the Mamdani's method, the output of Sugeno method is no longer a fuzzy set. It is a function of the inputs. So there is no need of defuzzification. For the same scene mentioned above, we can conclude two rules:

$$rule1: \text{If } \Delta c1 \text{ is large and } \Delta c2 \text{ is small, then } P = f(\Delta c1 - \Delta c2).$$
$$rule2: \text{If } \Delta c1 \text{ is small or } \Delta c2 \text{ is large, then } P = g(\Delta c1 - \Delta c2).$$

where $f(x)$ and $g(x)$ are defined in Fig. 8.

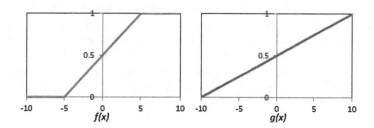

Fig. 8. $f(x)$ and g(x).

Fuzzify Inputs with the Input Membership Functions. In this step, the Sugeno method is exactly the same with Mamdani's method (Fig. 9).

Obtain Each Rule's Conclusion. Science the output is a function of the inputs, it is much simpler to obtain each rule's conclusion.

$$rule1 : P = f(\Delta c1 - \Delta c2) = 0.8$$
$$rule2 : P = g(\Delta c1 - \Delta c2) = 0.65$$

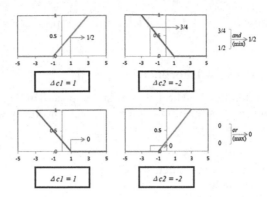

Fig. 9. Obtain a single membership value for each rule.

Aggregate Conclusions. In Sugeno method, aggregation is to compute the weighted average of the outputs obtained is last step. The membership value obtained in step 2 is used to be the weight of each rule ($w_1 = 1/2, w_2 = 0$).

$$P = \frac{\sum\limits_{i=1}^{2} w_i \times p_i}{\sum\limits_{i=1}^{2} w_i} = 0.8 \tag{2}$$

2.3   Comparison Between Mamdani and Sugeno

Based on the form of the rules, we can find out that the Mamdani's method is more intuitive. But in 2D simulation, every decision should be made within 100 ms. If the number of the rules is large, it will take much time to compute. So we should apply the Mamdani's method on scenes with small number of rules. Science the output is a function of the inputs in Sugeno method, it is computationally efficient. So it is able to handle with scenes with large number of rules.

3   Experiment

We have implemented a FIS and applied it on the marking module of WrightEagle. To test the FIS, we played 2000 full games (1000 before we implemented the FIS and 1000 after) with each team (Helios2014, Yushan2014, Gliders and Oxsy). The binary code of Helios2014, Yushan2014, Gliders and Oxsy we used were released by RoboCup 2014. These games were played on the same hardware and software environment. For each team, we take account average goal and win rate. The results are shown in Tables 1 and 2.

Table 1 shows that, after we implemented the FIS, the win rate against opponent teams increased. The increase of win rate shows that the FIS works well.

What's more, Table 2 shows that the average goal of opponent teams decreased, from which we can conclude that the FIS did increase the defensive ability of WrightEagle.

Table 1. Win rate against each team

	Before	After
Helios2013	58.3 %	66.6 %
YuShan2013	84.8 %	88.2 %
Gliders	71.4 %	82.2 %
Oxsy	90.3 %	94.9 %

Table 2. Average goal against each team

	Before	After
Helios2013	2.38:1.47	2.41:1.15
YuShan2013	3.30:0.77	3.43:0.64
Gliders	2.34:0.77	2.57:0.74
Oxsy	4.64:0.81	4.63:0.79

4   Related Work

As for a FIS, how to get rules is the most important challenge. Soccer is the most popular sport in the world. The sport has developed for hundreds of years. There are so many matches from which we can conclude rules. The human players and coaches have come up with numbers of strategies. So we can conclude some general rules from these strategies. On the other hand, the 2D simulation has developed for decades, we can conclude rules from different teams and different matches. The machine learning techniques and big data techniques also can be used here.

The FIS we talked above is a multi-input, single-output model. There are some scenes in 2D simulation in which we need use multi-input, multi-output models. Actually, a multi-input, multi-out model can be separated into some multi-input, single-output models.

5   Conclusion

For the purpose of forecasting actions of opponents in soccer simulation 2D, we introduced Mamdani-Type Fuzzy Inference and Sugeno-Type Fuzzy Inference to deal with a specific scene in soccer simulation 2D. Then we implemented and applied a FIS to the marking module of WrightEagle. Based on the result of the experiment, we conclude that the FIS improves the defensive ability of our team. It did work like what we expected. The most important advantage of FIS is that it can work well without the knowledge of the mathematical description of the opponent. Another advantage of the FIS is that, when we want to take account more things, what we need to do is to add new rules or edit old rules. Then we do not have to take much time to write a new algorithm.

References

1. Puterman, M.L.: Markov Decision Processes: Discrete Stochastic Dynamic Programming, vol. 414. Wiley, Hoboken (2009)
2. Zhang, H., Chen, X.: The decision-making framework of WrightEagle, the RoboCup 2013 soccer simulation 2D league champion team. In: Behnke, S., Veloso, M., Visser, A., Xiong, R. (eds.) RoboCup 2013. LNCS, vol. 8371, pp. 114–124. Springer,Heidelberg (2014)
3. Dietterich, T.G.: The maxq method for hierarchical reinforcement learning. In: ICML, pp. 118–126. Citeseer (1998)
4. Bai, A., Wu, F., Chen, X.: Online planning for large mdps with maxq decomposition. In: Proceedings of the Autonomous Robots and Multirobot Systems Workshop, at AAMAS 2012, June 2012
5. Bai, A., Wu, F., Chen, X.: Towards a principled solution to simulated robot soccer. In: Chen, X., Stone, P., Sucar, L.E., van der Zant, T. (eds.) RoboCup 2012. LNCS(LNAI), vol. 7500, pp. 141–153. Springer, Heidelberg (2013)
6. Zadeh, L.A.: Fuzzy sets. Inf. Control **8**(3), 338–353 (1965)
7. Mamdani, E.H., Assilian, S.: An experiment in linguistic synthesis with a fuzzy logic controller. Int. J. Man-Mach. Stud. **7**(1), 1–13 (1975)
8. Sugeno, M.: Industrial Applications of Fuzzy Control. Elsevier Science Pub. Co., Amsterdam (1985)

CIT Brains KidSize Robot: RoboCup 2015 KidSize League Winner

Yasuo Hayashibara[1](✉), Hideaki Minakata[1], Kiyoshi Irie[1],
Daiki Maekawa[1], George Tsukioka[1], Yasufumi Suzuki[1],
Taiitiro Mashiko[1], Yusuke Ito[1], Ryu Yamamoto[1], Masayuki Ando[1],
Shouta Hirama[1], Yukari Suzuki[1], Chisato Kasebayashi[1],
Akira Tanabe[1], Youta Seki[1], Moeno Masuda[1], Yuya Hirata[1],
Yuuki Kanno[1], Tomotaka Suzuki[1], Joshua Supratman[1],
Kosuke Machi[2], Shigechika Miki[3], Yoshitaka Nishizaki[4],
Kenji Kanemasu[5], and Hajime Sakamoto[6]

[1] Chiba Institute of Technology, 2-17-1 Tsudanuma, Narashino, Chiba, Japan
`yasuo.hayashibara@it-chiba.ac.jp`
[2] Sagawa Electronics, Inc., Matsudo, Chiba, Japan
[3] Miki Seisakusyo Co., Ltd., 1-7-28 Ohno, Nishiyodogawa, Osaka, Japan
[4] Nishizaki Co., Ltd., 1-7-27 Ohno, Nishiyodogawa, Osaka, Japan
[5] Yosinori Industry, Ltd., 1-1-7 Fukumachi, Nishiyodogawa, Osaka, Japan
[6] Hajime Research Institute, Ltd., 1-7-28 Ohno, Nishiyodogawa, Osaka, Japan
`sakamoto@hajimerobot.co.jp`

Abstract. In this paper, we describe technologies of the robot Accelite that is the winner of the RoboCup2015 Humanoid KidSize League. In 2015, the rules of the humanoid league were changed dramatically. The ball was changed from a monochromatic orange ball to a real soccer ball. Furthermore, the goal color was also changed from yellow to white. Therefore, the soccer robots in the humanoid league were not able to detect the ball based on just colors. We developed some detection methods and verified their effectiveness. Consequently, our robots detected the ball robustly and we succeeded in the technical challenge "Goal-Kick from Moving Ball". In RoboCup 2015, the field of play was also changed from a punch carpet to an artificial grass. Because the pile length of the artificial grass in RoboCup2015 was long, our robots were not able to walk robustly when we started to use the field. We also explain how we solved the problem. We also discuss how we succeeded in the technical challenges.

Keywords: Humanoid robots · Image processing · RoboCup

1 Introduction

In this paper, we describe the technologies of the robot Accelite that is the winner of the RoboCup2015 Humanoid KidSize League. In RoboCup 2015 Hefei, we won the first prize of the 4on4 soccer and the technical challenge. The results are indicated in the Table 1. Since 2013, we have used the robot Accelite, which is the fourth generation robot in our team CIT Brains [1–3].

© Springer International Publishing Switzerland 2015
L. Almeida et al. (Eds.): RoboCup 2015, LNAI 9513, pp. 153–164, 2015.
DOI: 10.1007/978-3-319-29339-4_13

Table 1. Results of CIT Brains in RoboCup2015

Soccer 4 on 4	8 wins – 0 loss
	Total goals: 18 goals - 0 losses
Technical challenge	Push recovery: 5 points
	Goal-Kick from moving ball: 10 points
	High jump: 5 points
	High-Kick challenge: 10 points
	Total: 30 points

CIT Brains is a joint team consisting of Hajime Research Institute and Chiba Institute of Technology (CIT). The aim of this development is not only research, but also education. Almost all members of our team are undergraduate students. They do not have enough knowledge and experience to understand our robot system in its entirety. Through improving our robot system, they have acquired a lot of hands-on experience in robot technologies.

In 2015, the rules of the humanoid league were changed dramatically. The ball was changed from a monochromatic orange ball to a real soccer ball. Furthermore, the color of the goal was changed from yellow to white. Therefore, the robots were not able to detect the ball by a simple image processing using only color features. For detecting the ball robustly, we built new image processing methods and verified their effectiveness. Consequently, we developed a robust detection system and we succeeded in the new technical challenge "Goal-Kick from Moving Ball". To accomplish the challenge, the robot should detect a "moving" ball. Our developed method is able to detect a moving ball robustly.

In RoboCup2015, the field of play was also changed from a punch carpet to an artificial grass. We started to develop a stable walking on an artificial grass in 2014, because one of the technical challenges in RoboCup2014 is "Artificial Grass Challenge". Our robot walked on the artificial grass robustly since 2014. However, when we start to use the soccer field in RoboCup2015, the robot could not walk robustly, because the length of the artificial grass was long. To solve this problem, we tried to find a solution heuristically in the venue. Consequently, we replaced the soles of our robots extemporaneously, and the robots were able to walk more robustly and speedily than before.

2   Conventional Ball Detection Method

Figure 1 shows our conventional method to detect the ball. The sequence is the following:

(1) Categorizing each pixel to a color cluster according to a preset color lookup table
(2) Calculating areas of the ball color
(3) Distinguishing whether some conditions are satisfied or not
(4) Registering as the ball if all conditions are satisfied.

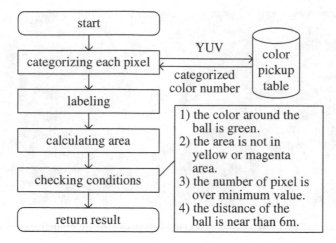

Fig. 1. Sequence of the conventional image processing

Because the CPU of our robot is Intel Atom D525 (dual core, 1.8 GHz), we could not apply an image processing algorithm that demands a high performance CPU. In RoboCup2014, the ball was painted monochromatic orange and it was prohibited to put objects with same orange color on the green field. Therefore, we applied a simple algorithm to detect the ball using only color feature. The method was effective for a low performance CPU, and it detected the ball up to every 20 frames per second robustly.

This method can be applied for another colored ball if the ball has an individual color against other objects on the field. However, the ball color in RoboCup2015 is almost white and there are many white objects on the field. Therefore, we could not apply the conventional method to detect the ball in RoboCup2015.

3 Ball Detection Method Using Color Histogram

Figure 2 shows a soccer ball, which was provided as a sample of a game ball preliminarily. In the rule of humanoid league, it is described that the color of the ball should be over 50 % white. The sample ball contains many colors, so we try to detect the ball by using a color histogram. Figure 3 shows the developed image processing algorithm to detect the ball. The sequence is the following:

(1) Calculating search points around the previous ball position using a particle filter
(2) Calculating a color histogram for every search point
(3) Evaluating a histogram intersection for every search point
(4) Registering as the ball if the maximum value is over a threshold

First, the robot calculates search points using a particle filter as the following:

(1) If the ball is found,
        distributing 50 particles around the ball position according to Gaussian distribution and distributing 50 particles uniformly in the captured image

Fig. 2. A sample of a game ball (Color figure online)

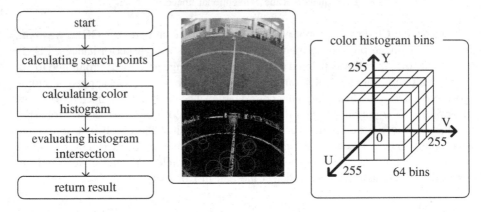

Fig. 3. Sequence of the image processing using color histogram, and the histogram bins

If the ball is not found,
    distributing 500 particles uniformly in the captured image

(2) Calculating diameters of circles in the captured image correspond to the outer shapes of the balls
    The diameter is changed according to the distance to the ball. The distance is calculated based on the position in the captured image and the tilt angle of the camera.

Second, random-sampled 100 points in the circle are categorized to color histogram bins. The camera of the robot outputs YUV image data, whose values are from 0 to 255. Each value of Y, U or V is categorized for 4 clusters as shown in the Fig. 3. Therefore, the total number of the histogram bin is 64. The histogram is calculated as the following:

(1) Picking 100 points up randomly in the circle correspond to the outer shape of the ball

Here, the number of the points is decided with consideration for the performance of the computer heuristically.

(2) Categorizing YUV data of each point P_{ki} in the k-th circle to a color histogram bin according to the following equation:

$$B(P_{ki}) = \text{floor}\left(\frac{Y(P_{ki})}{64}\right) \times 16 + \text{floor}\left(\frac{U(P_{ki})}{64}\right) \times 4 + \text{floor}\left(\frac{V(P_{ki})}{64}\right) \quad (1)$$

Here, B is the color histogram bin whose range from 0 to 63. Y, U and V are the values of YUV data whose ranges are from 0 to 255. Using the floor function, each Y, U or V value is categorized into one of four bins.

(3) Calculating the histogram for each search point

The value of the histogram bin H_{kj} is calculated by the following equation:

$$A_{kj} = \{P_{ki}|B(P_{ki}) = j\}$$
$$H_{kj} = \text{n}(A_{kj}) \quad (2)$$

Here, the function $\text{n}(A_{kj})$ implies the number of elements of the set A_{kj}.

Third, the histogram intersection is calculated at each search point [4]. In order to evaluate the degree of the coincidence of the histogram, we chose the histogram intersection as a simple methods. In advance, a reference color histogram of the ball is memorized manually. The histogram intersection M_k between the reference data R and the color histogram H_k at the k-th search point is calculated according to the following equation:

$$M_k = \frac{\sum_{j=0}^{63} \min(H_{kj}, R_j)}{\sum_{j=0}^{63} R_j} \quad (3)$$

Finally, if the maximum value of the histogram intersections is over a threshold, the corresponding search point is registered as the ball position.

The sequence of this method is more complex than the conventional method and it needs more computational time. So, the frame rate of the image processing was down to 5–10 frames per second. Using this method, the robot was able to detect the ball within a 3 m radius robustly. Consequently, we got the first prize in RoboCup2015 Japan Open Fukui.

4 Ball Detection Method Using a Template

We planned to use the above mentioned method using the color histogram in Robo-Cup2015 Hefei. However, the ball in RoboCup2015 was changed from the previous one. Figure 4 shows the game ball in RoboCup2015. The colors of the ball are just

Fig. 4. A sample of a game ball in the RoboCup2015

white and gray. The gray of the ball color is too similar to a shadow color of the goal or white line. Therefore, it was hard to distinguish the ball from other objects like goals and white lines using color histogram. Then, we tried to apply another detection method using the template of the ball as the Fig. 5. The sequence is the following:

(1) Categorizing each pixel to a color cluster according to a preset color lookup table
(2) Calculating search points around the previous ball position using a particle filter
(3) Evaluating the degree of coincidence by the following method
(4) Registering as the ball if the maximum value is over a threshold

As the Fig. 5 shows, the template implies the shape of the ball. The size of the template is changed according to the distance to the ball. The degree of coincidence at each search point is calculated as the following:

(1) Picking 100 points up randomly in a rectangle around a search point
The rectangle corresponds to the template and the size of the rectangle is changed according to the distance of the ball.

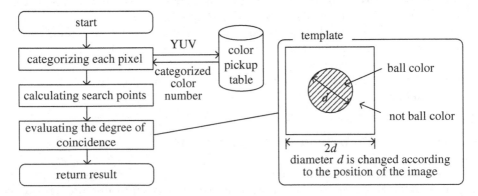

Fig. 5. Sequence of the image processing using the template

(2) Calculating the degree of coincidence Doc_k by comparing a template data of each point P_{ki} in the k–th rectangle as the following equation:

$$X_k = \{P_{ki}|C(P_{ki}) \text{ is ball color}, P_{ki} \text{ is inside circle}\}$$
$$Y_k = \{P_{ki}|C(P_{ki}) \text{ is ball color}, P_{ki} \text{ is outside circle}\}$$
$$Doc_k = \frac{n(X_k) - n(Y_k)}{n(P_{ki}|P_{ki} \text{ is inside circle})}$$

(4)

Here, C is the categorized color that is derived by using the lookup table. If the sampled image is completely the same as the template, the degree of coincidence will be 1.0. Before using this method, we should set the lookup table for categorizing the ball. All colors of the ball are set as a just ball color. By using this method, the robot could detect the ball within a 3 m radius robustly in our experimental soccer field.

Here, for detecting the ball robustly, it is desirable that the lookup table is set as distinguishing the ball's white and other objects' white. The rate of this procedure was 5–10 frames per second, it was almost same as the method using the color histogram.

5 Moving Ball Detection

The robot could detect the stationary ball using the method described in the previous section. However, if the ball moved quickly, it would get lost. Specially, it is a serious problem for a goal keeper. The main causes of losing the ball are a blurring of the ball image and a distribution of the search points. The problem of the blurring could be solved easily. It is solved by adjusting the shutter speed. Another problem about the distribution is more complex. If the number of search points and the distribution will be increased, detecting the moving ball will be possible. However, the cost of the calculation will also be increased. Here, a coming ball detection is more important for the goal keeper. Therefore, we modified the distribution of the particles like the Fig. 6. The particles are distributed in accordance with a long vertical ellipse whose center position is slightly shifted to the bottom. By this modification, the robot could detect a moving ball robustly. Consequently, we succeeded in the technical challenge "Goal-Kick from Moving Ball".

6 Modification in the RoboCup2015

As mentioned above, we developed the robust ball detection method. However, in the venue of RoboCup2015, we faced some new problems. The color of the white line was too similar to the ball color. Then, the robot could not detect the ball robustly in the play field in RoboCup2015. To solve this problem, we modified the template as Fig. 7.

The new template contains the colors of the play field and the robot. The new degree of coincidence Doc' is calculated as the following equation:

Fig. 6. Modified distribution of particles for detecting a coming ball

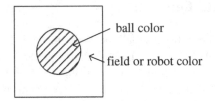

Fig. 7. The template for detecting the ball in RoboCup2015

$$Z_k = \{P_{ki} | C(P_{ki}) \text{ is field or robot color}, P_{ki} \text{ is outside circle}\}$$

$$Doc'_k = \frac{\mathrm{n}(X_k) - \mathrm{n}(Y_k)}{\mathrm{n}(P_{ki} | P_{ki} \text{ is inside circle})} \frac{\mathrm{n}(Z_k)}{\mathrm{n}(P_{ki} | P_{ki} \text{ is outside circle})} \tag{5}$$

Using this method with utilizing the play field color and the robot color outside the circle, the robot was able to distinguish the ball robustly. The green color of the field is able to detect more easily than the white color of the ball. That might contribute the robustness to detection. However, when the ball is located on the white line as the Fig. 8, it is slightly hard to detect the ball because the green area is decreased by the white line. Then, the robot sometimes lost the ball on the goal line. Furthermore, the place near the goal line was also the edge of the field, so the green area would be decreased more.

7   Discussion of a Walking on the Artificial Grass

In 2015, the soccer field of humanoid league was changed from a punch carpet to an artificial grass. Our robot attended the technical challenge "Artificial Grass" in 2014 and acquired the second prize of the challenge. Therefore, the robot has already walked

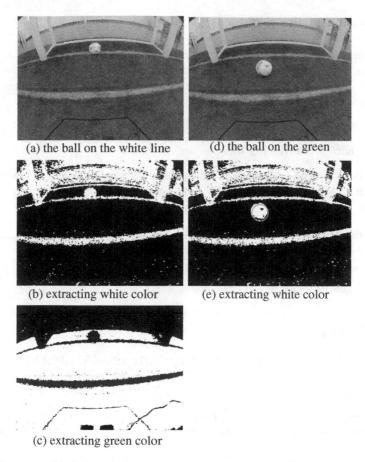

(a) the ball on the white line   (d) the ball on the green

(b) extracting white color   (e) extracting white color

(c) extracting green color

Fig. 8. The cases of losing the ball on the white line (left) and finding the ball (right) (Color figure online)

on an artificial grass robustly since 2014. However, because the length of the artificial grass in RoboCup2015 was long (35 mm), the robot could not walk robustly. Some teams used bolts attached to the sole for a stable walking. We also tried to use bolts like other teams. Consequently, the stability of the walking was somewhat improved. However, it became more difficult to kick the ball. When the bolts touch the field during kicking the ball, a strong force is applied to the foot. For solving this problem, we assembled soles of the robot upside down and made the sole longer back and forth. We had already made long soles for the technical challenge "High-Kick Challenge" as shown in the Fig. 9. That helps to find the solution. The bent edges are attached to not step on its feet. Using the edges under the sole, the robot could walk robustly. Consequently, the walking speed on the artificial grass could be increased.

8   Discussion of the Technical Challenges

In 2015, three new technical challenges were introduced. Our robot succeeded in all challenges. Especially, the technical challenges "High-Kick Challenge" and "Goal-Kick from Moving Ball" were difficult to achieve. We explain about how to achieve the challenges.

8.1   Goal-Kick from Moving Ball

In our system, we write a sequence of a behavior in Python language. By changing the program, the robot can execute various tasks. A sequence of the technical challenge "Goal-Kick from Moving Ball" is following:

(1)   Rotating the neck in the direction of the ball
(2)   Play a kick motion if the ball comes in a certain area

   By only the above sequence, we achieved the challenge at the first trial. Figure 10 shows the result of the technical challenge.

8.2   High-Kick Challenge

We also succeeded in the "High-Kick Challenge". The robot kicked the ball over a wall of 220 mm height as shown in the Fig. 11. The robot is controlled according to the following sequence.

(1)   Behaving as a normal forward player
(2)   Switching the kick motion if the robot enters inside a certain area in front of the goal

Fig. 9. Modified soles (left) and conventional soles (right)

Fig. 10. Result of Goal-Kick from Moving Ball

Fig. 11. Result of High-Kick Challenge

For succeeding in this challenge, the position of the ball from the wall is important. The challenge will not be succeeded when the ball is too near or far from the wall of the goal. For kicking higher, we made the long sole. As already mentioned, it helped to find the robust walking.

9   Conclusion

In this paper, we described the technologies of the humanoid robot Accelite in RoboCup2015. For detecting the ball, we developed image processing methods using the color histogram and the template. Consequently, our robots detected the moving ball robustly. In RoboCup 2015, the field was also changed from a punch carpet to an artificial grass. To walk stably, we assembled soles of the robot upside down and made the sole longer back and forth. We also discussed how we succeeded in the technical challenges.

References

1. Hayashibara, Y., et al.: CIT brains KidSize robot: RoboCup 2014 best humanoid award winner. In: Bianchi, R.A., Akin, H., Ramamoorthy, S., Sugiura, K. (eds.) RoboCup 2014. LNCS, vol. 8992, pp. 82–93. Springer, Heidelberg (2015)
2. Hayashibara, Y., Minakata, H.: The research development of autonomous soccer humanoids through the RoboCup. J. Soc. Instrum. Control Eng. **52**(6), 487–494 (2013)
3. Hayashibara, Y., Minakata, H., Irie, K., Sakamoto, H.: Development of a humanoid "Dynamo2012" for RoboCup. In: Proceedings of JSME Robotics and Mechatronics Conference 2013, 1P1-A12 (2013)
4. Swain, M.J., Ballard, D.H.: Color indexing. Int. J. Comput. Vis. **7**(1), 11–32 (1991)

The Goalkeeper Strategy of RoboCup MSL Based on Dual Image Source

Xueyan Wang[1(✉)], Peng Zhao[2(✉)], Shiyu Mou[1], Di Zhu[7],
Jieming Zhou[8], Dongbiao Sun[1], Song Chen[3(✉)], Zhe Zhu[4(✉)],
Ye Tian[5], Zongyi Zhang[5], Lv Ye[6], Xinxin Xu[6], Wanjie Zhang[6],
Kang Wu[6], Binbin Li[5], Yuan Li[1], Abudouyimujiang Kader[1],
Yiheng Su[1], and Chenyu Wang[6]

[1] Beijing Information Science and Technology University,
No. 12, East Qinghexiaoying Rd, Beijing 100192, China
342429193@qq.com
[2] No. 21 Research Institute of China Electronics Technology
Group Corporation, Shanghai 200233, China
peng.zhao@ia.ac.cn
[3] Beijing Aerospace Guanghua Electronics Technologies Limited Corporation,
Beijing 100854, China
25127601@qq.com
[4] China Telecom, Beijing 100032, China
luke.zhezhu@yahoo.com
[5] Shanghai Yingchuang Limited, Shanghai 201100, China
[6] Beijing Waterplus Limited, Beijing 100000, China
[7] University of Science and Technology Beijing, Beijing 100083, China
[8] Australian National University, Canberra 00120C, Australia

Abstract. Gatekeeping is one of the most important tasks in MSL, especially defending aerial shots. This paper proposes a goalkeeper strategy based on combining two image sources. The goalkeeper uses the Omni-vision system to adjust heading orientation of the Kinect sensor for capturing both depth and RGB images of the ball. The depth information is used to detect the spatial position of the ball and the RGB image ensures the validity of the recognition. Once opponents shoot, the proposed system uses the first few detected positions of the ball to predict the interception point by the goalkeeper based on the least square method. The accuracy and effectiveness of the proposed strategy are verified by experimentation.

Keywords: Goalkeeper strategy · Dual image source · Kinect sensor · Depth camera · RGB camera · Defensive position

1 Introduction

RoboCup Soccer has been one of the competitions of RoboCup ever since. The rules used in the middle size league (MSL) are gradually approaching the rules of human soccer competition of FIFA. The standard ball has been introduced into the middle size league and the field has been extended to 12 m*18 m. The current technical level in

© Springer International Publishing Switzerland 2015
L. Almeida et al. (Eds.): RoboCup 2015, LNAI 9513, pp. 165–174, 2015.
DOI: 10.1007/978-3-319-29339-4_14

MSL makes it a very exciting competition in RoboCup, attracting large audiences and even investors given its high commercial potential [1–3]. Moreover, MSL covers several scientific and technological domains from computer vision [4] to localization and navigation [5], multi-agents cooperation [6], communications, etc., making it an important research platform. Team Water from Beijing Information Science and Technology University is a regular team in RoboCup MSL and has performed well in the competition in recent years [7]. Based on the Omni-vision system and local map matching technology, the team Water keeps improving the precision and robustness of robot self-localization, navigation and detection of targets and obstacles. Last year, the Microsoft Kinect depth camera [8] was introduced to improve the performance of objects detection, which played an important role for goalkeeping in RoboCup2015.

The goalkeeper ability to prevent the opponent team to score is called interception. The performance of the interception behavior has a significant impact on the competition. Based on the ball coordinates got from the robot's Omni-vision system, two main strategies are used to intercept the ball. One is keeping the goalkeeper standing on the line connecting the ball to the center of the goal. Another is fitting the ball trajectory according to the records of recent past ball coordinates and then predicting the point of interception. Both methods perform well when the ball rolls on the floor. Facing high balls, however, these strategies fail due to the distortion introduced by the convex mirror of the Omni-vision system and the impossibility to fit the flying trajectory causing the goalkeeper to miss the actual interception point.

To improve the ball interception with aerial shots we propose a dual image source strategy for the goalkeeper. We use the omnidirectional mirror image and the RGB primary color and depth images from Kinect. Then we reduce the dimension of the ball trajectory to that of a top view straight line using the least squares method and then we predict the closest position to intercept the ball.

The structure of this paper is as follows. The next section introduces the structure of the goalkeeper. Section 3 establishes the coordinates conversion. Section 4 details the strategy of the dual image sources. Section 5 shows the experiments designed to test the strategy proposed in Sect. 4. Section 6 presents the conclusion.

2   Structure of the Goalkeeper Robot

Figure 1 shows the detection system of goalkeeper, which includes an Omni-vision system, an IMU sensor and a Kinect. The Omni-vision system collects the image from all around the robot using a convex Omni-mirror facing down, projecting the 360° image on a vertically mounted camera in the same axis. The Kinect sensor is installed on the front of the robot. This sensor has an RGB camera and a depth camera. The RGB camera can collect the color image and the depth camera can acquire the depth information associated to each pixel.

As shown in Fig. 2, the robot system mainly includes a laptop, a control board and servo drivers. The laptop receives the commands such as "Start", "Pause" and "Stop" from the Referee Box based on the TCP/IP protocol through the laptop wireless LAN connection. Then, combining the information of the Omni-vision system, the Kinect and an electronic compass, the robot can locate itself using a local matching method,

Fig. 1. The goalkeeper physical structure.

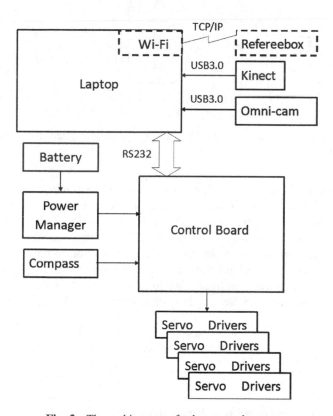

Fig. 2. The architecture of robot control system.

detect the ball and distinguish barriers. Based on this information, we compute the speed and target coordinates for the robot so that it intercepts the ball and we send such set-points to the robot control board. This board converts the received commands to speed set-points for the wheels motor servo drives. In addition, the power unit consists of Li-batteries and a voltage monitor module for power management.

3   Coordinates Transformation

This section presents the notation used and the conversion between different coordinate systems, namely the World $\{W\}$, the Robot $\{R\}$ and the Kinect $\{K\}$ coordinate systems (Fig. 3). The world coordinate system $\{W\}$ takes the center of the field as the origin, the X-axis is parallel to the field midline and pointing left when facing the opponent's half field. The Y-axis is perpendicular to the midline and points to the goal. The Z-axis is perpendicular to the ground and pointing upward.

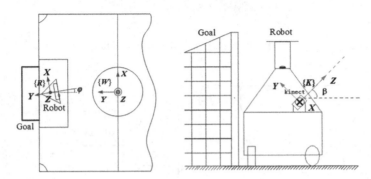

Fig. 3. The world coordinate system $\{W\}$, robot coordinate system $\{R\}$ and the kinect coordinate systems $\{K\}$.

The robot coordinate system $\{R\}$ has the origin on the center of the projection of the robot on the field. The axis toward the right side of the robot, when facing it, is the X-axis. The axis toward the back of the robot is the Y-axis. The Z-axis points upwards, vertically. Based on the position of the field lines using the Omni-vision system and a map matching method, the robot gets its location information in the world coordinate system $\{W\}$. The location information includes the robot coordinates $(x_R, y_R, 0)$ and heading angle φ. The conversion from robot coordinates in $\{R\}$ to the world coordinate system $\{W\}$ is achieved with the transformation matrix $^{W}T_R$ (Eq. 1).

$$^{W}T_R = \begin{bmatrix} \cos\varphi & -\sin\varphi & 0 & x_R \\ \sin\varphi & \cos\varphi & 0 & y_R \\ 0 & 0 & 1 & 0 \\ 0 & 0 & 0 & 1 \end{bmatrix} \tag{1}$$

The RGB and depth cameras of Kinect share the same lens. Hence, they share the coordinate system $\{K\}$ whose origin corresponds to the camera focal point, the X-axis is parallel to the ground and points to the right side of the robot, when facing it, the Z-axis points outwards along the lens axis and the Y-axis follows the right-hand rule. The conversion from the Kinect coordinate system $\{K\}$ to robot coordinates $\{R\}$ is achieved with the transformation matrix $^{R}T_K$ (Eq. 2) where (x_K, y_K, h) is the position of the Kinect origin in the robot coordinate system $\{R\}$ and β represents its tilt angle with respect to the horizontal plane.

$$
{}^{R}T_{K} =
\begin{bmatrix}
1 & 0 & 0 & x_K \\
0 & \sin\beta & -\cos\beta & y_K \\
0 & \cos\beta & \sin\beta & h \\
0 & 0 & 0 & 1
\end{bmatrix}
\tag{2}
$$

Therefore, a generic point ${}^{K}P_i$ $({}^{K}x_i,\ {}^{K}y_i,\ {}^{K}z_i)$ in the Kinect system is represented in the world coordinate system with ${}^{W}P_i$ $(x_i,\ y_i,\ z_i)$ as given by Eq. 3.

$$
{}^{W}P_i =
\begin{bmatrix}
x_i \\
y_i \\
z_i \\
1
\end{bmatrix}
= {}^{W}T_R * {}^{R}T_K *
\begin{bmatrix}
{}^{K}x_i \\
{}^{K}y_i \\
{}^{K}z_i \\
1
\end{bmatrix}
\tag{3}
$$

4 Goalkeeper Strategy Based on Dual Image Source

Following the dual image source strategy the goalkeeper keeps the ball within the Kinect field of view using the images captured by the Omni-vision system. The ball 3D coordinates in {W} are then obtained from the depth information captured by the Kinect while the RGB information is used to remove false identifications. Once the goalkeeper finds that the ball is flying towards to the goal, it fits its trajectory projection on the ground into a straight line, determines the intersection of that line with the goal line and moves to that defensive position.

4.1 Adjust Goalkeeper Heading with Omni-Vision System

The image acquired by the Omni-vision system is an ordinary RGB image, thus the pixels of the ball in the image can be conventionally recognized by the threshold segmentation and the region-growing method [9]. Firstly, the threshold range parameters of the ball in HSV space are calibrated by the calibration toolbox. Based on the range parameters, the image is processed with the banalization method. Then, the scattered pixels in the binary image are clustered together into different connected regions by the region-growing method. The biggest connected region in the image is taken as the ball.

Figure 4 shows an image captured by the Omni-vision system. The U-axis points to the right of the goalkeeper while the V-axis points towards the front. The central pixel of the ball image (u_b, v_b) is used to get the angle of the ball relative to the robot (θ_b), as described by Eq. 4, where (u_0, v_0) is the fiducial point of the Omni-vision system camera. Knowing θ_b the goalkeeper can adjust its orientation to track the ball angle, thus keeping it within the field of view of the Kinect.

$$
\theta_b = \tan^{-1} \frac{v_b - v_0}{u_b - u_0}
\tag{4}
$$

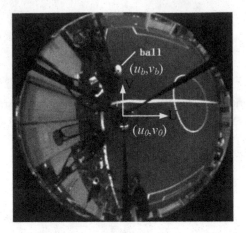

Fig. 4. The ball in the image captured by the Omni-vision system.

4.2 Recognize the Ball's Imaging Blob in the Depth Image

The pixels value ^{H}z in the depth image represent the distance from the camera to the target surface along the Z-axis of the Kinect coordinate system. However, the ball is a regular small sphere resulting in different depth values along its surface that are close to each other. Thus, these ball pixels in the depth image can be taken as a blob with similar depth value. The whole ball detection process is described in Fig. 5.

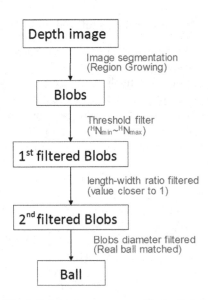

Fig. 5. The ball detection process with depth information.

Firstly, the blobs from the depth image with different depth values are segmented. Then, the blobs of the wall and surrounding robots that are oversize are filtered out. At the same time, the blobs of image noise that are undersize are also removed. The rest of the blobs are referred as available blobs.

Secondly, the blobs whose ratio between length and width is close to 1 are extracted. Thirdly, using Eq. 5 we estimate the corresponding diameters in the world coordinates system (D_b) and we take the blob that has a diameter closer to that of the real ball. Note that $\overline{^HZ_b}$ is the blob mean depth value, ^HD_b is the diameter of the blob and ^Hk is the amplification factor of the depth camera.

$$D_b = \overline{^HZ_b} * \frac{^HD_b}{^Hk} \tag{5}$$

4.3 Check the Imaging Blob and Measure the Ball's Spatial Location

The result of the detection process described in the previous section is the blob barycenter pixel coordinates $(^Hu_b, {}^Hv_b)$. Based on these coordinates, the corresponding RGB image is used to check whether the surrounding pixels belong to the calibrated threshold range in HSV space. A match means that the detection is valid and the ball coordinates in the Kinect coordinate system ^KP_b can be computed with Eq. 6. Note that $(^Hu_0, {}^Hv_0)$ is the fiducial point of depth camera. Finally, the ball position in the world coordinate system (^WP_b) can be determined using Eq. 3.

$$^KP_b = \begin{bmatrix} ^Kx_b \\ ^Ky_b \\ ^Kz_b \end{bmatrix} = \overline{^HZ_b} \begin{bmatrix} ^Hk & 0 & ^Hu_0 \\ 0 & ^Hk & ^Hv_0 \\ 0 & 0 & 1 \end{bmatrix} \begin{bmatrix} ^Hu_b \\ ^Hv_b \\ 1 \end{bmatrix} \tag{6}$$

4.4 Determining the Goalkeeper Defensive Position

Once the goalkeeper detects that the distance from the ball to the goal is decreasing fast, which means the ball is flying towards to the goal, it tracks the ball trajectory with just the first two dimensions of the 3D coordinates ^WP_b (its ground projection or top view) [8]. Then, it applies the Least Squares Method to the ground projected ball path with n (x_i, y_i) points to derive a straight line. The line parameters $\{a, b\}$ are computed as in Eq. 7 where $\bar{x}$ and $\bar{y}$ are the average values of the respective coordinates.

$$\begin{cases} b = \frac{\sum_{i=1}^{n} x_i y_i - n\bar{x}\bar{y}}{\sum_{i=1}^{n} x_i^2 - n\bar{x}^2} \\ a = \bar{y} - b\bar{x} \end{cases} \tag{7}$$

Finally, we determine the intersection (x_d, y_d) between the linearized ground projected ball path and the goal line using Eq. 8, where L is the length of the field. This intersection is our defensive point to where we drive the goalkeeper.

$$\begin{cases} y_d = \frac{L}{2} \\ x_{d.} = \frac{L/2 - a}{b} \end{cases} \tag{8}$$

4.5 Two Measurement Methods Based on Single Image Source

To show the advantages of the proposed dual image based approach, we define here the two single image source based methods which we use for comparison. The first method uses the RGB camera of the Kinect, only. It recognizes the ball region in the image based on calibrated threshold range in HSV color space and estimates the distance between the ball and camera according to the size of the region. The spatial position is calculated with the camera pin-point model. The second method uses the depth camera of the Kinect, only, which essentially does the recognition process described in Sect. 4.3 without cross checking with RGB information.

5 Experiments

In order to verify the performance of our proposed strategy we carried out several experiments that are described next. Firstly, the goalkeeper is placed in a fixed location. Then, the ball is thrown at the goal from three different distances with 20 times at each spot. For each spot we also determine the straight line between the throwing point and the landing point, and we record the intersection between this line and the goal line as the desired defensive point. At the same time, we also determine the defensive point using the Kinect system according to the strategy proposed in Sect. 4. Finally, we record the errors between the two defensive points, the observed and the computed one (Fig. 6).

Fig. 6. The experimental scenario

The above experimental process was repeated for the dual image source method (RGB-D) as well as for the methods that separately use RGB and depth images. The statistics of the measured errors (average/standard deviation) for these three methods are shown in Table 1, for the three different distances.

Table 1. Error statistics for 3 methods (cm)

	Distance		
Method	2.5 m	4.5 m	6.5 m
RGB-D	10.42/5.88	17.65/12.81	38.74/10.37
RGB	27.83/14.37	42.50/27.94	112.07/82.75
Depth	13.42/6.75	18.65/13.09	28.47/7.279

In order to ensure the image quality captured by the Kinect, the goalkeeper will remain still for a while once the ball leaps up from the field. Additionally, the goalkeeper needs time to move to the interception point, thus, it will not observe the ball for a long time. Therefore, in the experiments the goalkeeper captures only a few frames of information from the Kinect to compute the defensive point for every shot.

The experimental results show that the proposed dual image source method is better than the other two methods for short distances. For longer distances, just using depth seems to yield good results. Conversely, the defensive points measured by the RGB camera alone show significantly larger errors resulting in a bad performance in handling aerial shots. In general, as expected, the results also show that the error increases as the shooting distance increases.

6 Conclusion

This paper proposed a goalkeeper strategy based on dual image sources combining the Omni-vision system and the Kinect sensor. It uses the Omni-vision system to get orientation of the ball and direct the Kinect to it. The Kinect utilizes the depth information to measure the spatial positions of the ball and the RGB image to ensure the validity of the recognition. The Least Square Method is applied to the ground projection of the ball path to derive a straight line whose intersection with the goal line determines the defensive position of the goalkeeper. The experimental results show the accuracy and effectiveness of this strategy.

Acknowledgment. This work was supported by the project of Beijing Science and Technology Educational Committee (KM 201511232003).

References

1. Min, T., Shuo, W., Zhiqiang, C.: Multi-robot Systems. Tsinghua University Press, Beijing (2005)
2. Shi, L., Xuming, X., Qin, Y., Zengqi, S.: International robot soccer tournament and correlative technique. Robot **05**(5), 420–426 (2000)
3. Stone, P.: Layered Learning in Multi-agent Systems: a Winning Approach to Robotic Soccer. MIT Press, Cambridge (2000)
4. Fei, L.: Research on key issues of color omni-directional vision for soccer robots. Doctor of Engineering in Control Science and Engineering, National University of Defense Technology (2007)
5. Yonghui, H.: A method for the middle-sized mobile robot's self-location based on mark lines in robocup enviroment. Microcomput. Appl. **10**, 71–75 (2011)
6. Fangyi, S.: Research on behavior control of soccer robot with reinforcement learning. Master's Degree, National University of Defense Technology (2008)
7. Chen, S., et al.: Team water: the champion of the RoboCup middle size league competition 2013. In: Behnke, S., Veloso, M., Visser, A., Xiong, R. (eds.) RoboCup 2013. LNCS, vol. 8371, pp. 37–48. Springer, Heidelberg (2014)
8. Ribeiro Neves, A.J., Trifan, A., Dias, P., Azevedo, J.L.: Detection of aerial balls in robotic soccer using a mixture of color and depth information. In: 2015 IEEE International Conference on Autonomous Robot Systems and Competitions (ICARSC), 8–10 April 2015, pp. 227–232 (2015). doi:10.1109/ICARSC.2015.13
9. Huimin, L.: The research on the omnidirectional vision system for autonomous mobile robots. Doctor of Engineering in Control Science and Engineering, National University of Defense Technology (2010)

Oral Presentations

Regression and Mental Models for Decision Making on Robotic Biped Goalkeepers

Joseph G. Masterjohn[1]([✉]), Mihai Polceanu[2], Julian Jarrett[1],
Andreas Seekircher[1], Cédric Buche[2], and Ubbo Visser[1]

[1] Department of Computer Science, University of Miami,
1365 Memorial Drive, Coral Gables, FL 33146, USA
{joe,julian,aseek,visser}@cs.miami.edu
[2] National Engineering School of Brest (ENIB),
LAB-STICC, 25 rue Claude Chappe, 29280 Plouzané, France
{polceanu,buche}@enib.fr

Abstract. We investigate the decision-making and behavior of robotic biped goalkeepers, applied to the RoboCup 3D Soccer Simulation League. We introduce two approaches to the goalkeeper's behavior: first a heuristics-based approach that uses linear regression and Kalman filters for improved perception, and another based on mental models which uses nonlinear regression for ball trajectory filtering. Our experiments consist of 30,000 kick-and-save tests, using 100 random angle and distance kicks from six distance categories and four angle categories repeated 30 times. Our benchmark results show that both proposed approaches bring significant improvements for the goalkeeper's save success rates ($>200\%$) and validate the applicability of the novel mental model based decision-making process.

1 Introduction

The RoboCup 3D simulation league soccer is an international contest promoting artificial intelligence and autonomous robotics. Virtual replicas of the Aldebaran NAO robot make teams of 11 players and compete against each other in a virtual game of soccer.

With the aim to improve our team's performance in upcoming contests, the work presented in this paper focuses on an important challenge: the goalkeeper. The agent playing this role relies greatly on anticipation techniques. The first, most important step is to accurately track the ball and be able to predict its trajectory. This prediction should account for the time needed by the agent to perform a correct set of actions that would deflect the ball from the goal gate. Second, performance of attackers has improved such that the ball can be shot towards the goal with high velocities that force available reaction time below one second. This leaves little time to react and thus the goalkeeper must take advantage of every available millisecond in order to save the goal. Finally, noisy perception contributes to difficulties in accurate action planning. Therefore, the goalkeeper should begin moving to intercept the predicted ball trajectory while

© Springer International Publishing Switzerland 2015
L. Almeida et al. (Eds.): RoboCup 2015, LNAI 9513, pp. 177–189, 2015.
DOI: 10.1007/978-3-319-29339-4_15

also constantly improving its plan using recent perceptions. In the ideal case, it should also be able to perform last-moment adjustments using fast hand or foot movements to save goals.

The use of scripted behavior in the NAO goalkeeper, as well as in agents in general, can be highly efficient in many scenarios where a certain solution can be applied directly to the context at hand. For example, in this case, one can instruct the robot to position itself on the "goalkeeper's arc" (an imaginary arc 2–6 yards from the goal) to maximize its range, or that it should move to a future location of the ball as soon as it is launched. However, in the cases where a decision has to be made along the way such as whether or not to dive, extend arms or feet in order to block the ball, scripted behavior becomes difficult, especially without the knowledge of whether or not it is necessary to do so and which action would give the best results.

The contributions of this paper consist in the description and evaluation of two approaches to improve the anticipation and decision performance of the goalkeeper agent, an important aspect of the RoboCup competition which has been relatively under-represented in research. We first discuss relevant work in the next section and describe the two approaches to goalie anticipation in Sect. 3. Our experimental setup and the conducted robot tests are explained in Sect. 4. Finally we discuss the pros and cons of our results and outline future work in the remaining Sect. 5.

2   Related Work

Multi-behavior goalkeepers have been studied in the realm of the RoboCup mid sized league. Although the Standard Platform League (SPL) and simulation league face unique challenges that are not present in the mid sized league, the study of behavior control within the mid sized league is still applicable. Menegatti et al. [13] were the first to create an *ad hoc* model for behavior and motion control. Their goalkeeper sweeps an arc in front of the goal and tries to intercept shots coming towards the goal. This work was furthered and refined by Lausen et al. [9]. Their goalkeeper is implemented by a 2-level hierarchical state machine that coordinates primitive tasks and behaviors. Complex motion within a task is carried out by a non-linear control algorithm.

Garcia et al. [7] implemented an ethological inspired architecture to generate autonomous behaviors, and focused on modeling a humanoid goalkeeper according to the rules of the SPL. The architecture implements the goalkeeper behavior using a deliberative structure for planning (self-positioning) and a reactive control mechanism for goal saving. They report being able to track the ball 100 % of the time, positioning itself properly 84 % of the time, and saving goals from 62 % of the trajectories tested.

Bozinovski et al. [3] have presented an approach for an evolving behavior model for the role of a goalkeeper. Using an emotion based self-reinforcement learning algorithm, they present a learning curve as result from training experiments.

Birbach *et al.* [2] have implemented a realtime perception system for catching flying balls with DLR's humanoid *Rollin' Justin*. The system tracks and predicts the trajectory of balls thrown towards the agent using a Multiple Hypothesis Tracker that uses Unscented Kalman Filters to track the different hypotheses for different balls. We expand upon the use of Kalman filters for state estimation and introduce regression methods for trajectory fitting.

Adorni et al. [1] used the current ball positions obtained from images frame by frame, then used a look-up table mechanism for an internal local representation of where the ball is relative to the agent and which eases computation and then compared two subsequent frames to estimate ball position, motion direction and speed. This straight-forward method worked on a Middle-Size League (MSL) agent in the beginning of RoboCup. Since then, research in soccer agents and robots has focused on several critical aspects of gameplay such as efficient omnidirectional movement [10,11] and accurate perception [16].

Human goalkeepers have been shown to use visual cues in the shooter's body posture to predict the direction of the ball [5], before the shot has been performed. However, in the context of the RoboCup 3D soccer competition, goalkeeper agents rely solely on the ball trajectory after shooting due to the constraints (ex. noisy perception, limited computational resources) posed by perceiving the opponent robot's posture.

Despite abundant research into the various challenges posed by the robot soccer context, work on the goalkeeper agent is limited to simple rules that are acceptable for increasing chances of saving a goal. Moreover, goalkeepers use the same perception methods as field player agents, although they require a much more precise, three-dimensional ball localization for brief periods of time.

Goalkeepers also require special skills as they are the only player agents which are allowed to intentionally dive and use their hands to block the ball. Diving is the most popular goalkeeper skill present in the competition, but deciding to activate this skill relies on handcrafted rules. Such rules include instructing the goalie to move towards the position where the ball will be found before entering the goal and if the distance between the goalie and the ball is too large for a direct block, the goalie would perform a dive [12]. However, there are exceptions where after the dive the ball bounces over the fallen goalie resulting in a goal. Another example is when diving happens too slowly to block the ball and a slight leg movement would be more efficient instead.

3   Approach

We investigate two approaches to improve goalkeeper efficiency in the RoboCup 3D simulation competition. First, a rule-based approach which uses linear regression and Kalman filters to filter the ball trajectory. Second, a novel approach to decision-making that utilizes nonlinear regression for ball perception. In the following, we discuss the details of each approach and provide insight into their strengths and drawbacks.

3.1   Thiel-Sen and Kalman-Filter

In our first approach, we filter the ball position and velocity using linear regression, to estimate the location where the ball will pass the goal line. Based on this prediction, the goalie employs a deterministic behavior in order to improve its success in obstructing the ball. In the following we describe the logic behind this behavior and how the filtering has been implemented.

Behavior Logic. We describe the rational stages of behavior of the goalkeeper agent, first actions for the initial positioning of the agent, moving to obstruct the ball, final actions including decisions and diving and other auxiliary checks and behavior to maintain stable execution of the agent.

The agent acting as the goalkeeper first analyzes how to position itself in the goal based on the initial ball position of the free kick to be taken. The goalkeeper has one of five positions on the goal line to choose from. The rational for making these options available for the goalkeeper are motivated from human soccer, watching various free kicks and observing the behavior of professional goalkeepers. The options in initial position are as follows: (a) the center of the goal if the kicker is directly in front of the goal. (b) between the center and the right goal post if the kick is coming from the right and is a short kick. (c) between the center and the left post if the kick is coming from the left and is a short kick. (d) close to the right post if the kick is coming from the right and is a long kick. (e) close to the left post if the kick is coming from the left and is a long kick. A "short" vs. "long" kick is determined by a distance threshold.

The goalkeeper, based on perceptions of his own position and the position of the ball, calculates the distance between himself and the ball as well as the angle between the vector from goalkeeper to ball and the center line. This first perception indicates the predicted direction of the free kick, the vector from the opponent to the center of the goal.

The goalkeeper uses the calculated angle to access a set of behaviors followed by the calculated distance to decide on a specific behavior within this set. (e.g. - an angle $\theta \in [\frac{\pi}{3}, \frac{\pi}{2}]$ is indicative of a kick coming from the left side, and a distance $d < threshold$ indicates a short kick.)

The goalkeeper receives filtered perceptions of the balls position and velocity from a provider module, the details of which are discussed later. The goalkeeper projects the ball along the velocity vector onto the goal line to determine the position at which it crosses the goal line. The goalkeeper characterizes a kick by observing the velocity and detecting when it accelerates past a certain threshold (i.e. - a kick has been performed). This sets off the rational actions to obstruct the path to goal. Once the kick has been taken, the goalkeeper uses the predicted goal line crossing to move towards that position.

Once the predicted ETA of the ball on the goaline becomes less than the estimated time needed to perform a dive action, the goalkeeper makes its final decisions. (1) If he is close enough to the predicted goal line crossing, he will remain in place to obstruct the ball (2) If he is too far from the crossing to obstruct with his feet, he will dive in the direction of the goal line crossing.

Diving is actuated by a predetermined and hand tuned motion defined by an interpolated sequence of joint angle values, for each actuated joint in the agent.

Methods and Implementation. Upon initial analysis of the accuracy of the existing ball position in the software, we decided that the first and most elementary method to predicting the ball velocity would be to compute a simple linear regression on the existing ball positions. This method operates under a few assumptions that are not accurate in the realm of the simulator:

1. $|v| \neq 0$, for v the velocity of the ball
2. $\forall \, \delta t \in \mathbb{R}, \; x'([t, t + \delta t]) = c, \; c \in \mathbb{R}$ (i.e. - constant velocity)

This model does not take into account the complex motion of the ball, but it does serve as a starting point for an efficient way to model the ball velocity. In order to do the initial regression, the Thiel-Sen algorithm was used. This method for calculating the linear trend of a set of points is not as sensitive to outliers as other methods [17]. The full algorithm for determining the velocity follows:

> **function** LINEAR REGRESSION$(S \leftarrow [(x_1, y_1), ..., (x_n, y_n)])$
> $\quad M \leftarrow$ median of all unique slopes $m = \frac{y_j - y_i}{x_j - x_i} \mid i \neq j$
> $\quad v \leftarrow (1, M)$ or $(-1, -M)$ depending on the rough direction of the samples.
> $\quad v \leftarrow$ unitized v scaled by distance between p_n and p_1 divided by $t_n - t_1$
> $\quad$ **return** v
> **end function**

While the first method succeeded in giving an estimate of the ball velocity accurate enough to predict the goal line crossing and ETA, a second, more complex and contemporary method (Kalman Filter) was used for comparison. This decision was made with the additional foresight that the success of the linear regression model might be dependent on the magnitude of error realized in the simulation. Thus, a method robust enough to predict the state of the ball for a real world environment would then be integral. The Kalman Filter is a well known and studied method of estimating variables that cannot be directly measured through a series of noisy measurements, observed over time [6]. We formulate the Kalman Filter matrices as follows:

$$X = \begin{bmatrix} x \\ y \\ x' \\ y' \end{bmatrix} \qquad F = \begin{bmatrix} 1 & 0 & \delta t & 0 \\ 0 & 1 & 0 & \delta t \\ 0 & 0 & 1 & 0 \\ 0 & 0 & 0 & 1 \end{bmatrix} \qquad P = \begin{bmatrix} 0 & 0 & 0 & 0 \\ 0 & 0 & 0 & 0 \\ 0 & 0 & 1000 & 0 \\ 0 & 0 & 0 & 1000 \end{bmatrix}$$

$$Q = \begin{bmatrix} 0.01 & 0 & 0 & 0 \\ 0 & 0.01 & 0 & 0 \\ 0 & 0 & 0.01 & 0 \\ 0 & 0 & 0 & 0.01 \end{bmatrix} \qquad H = \begin{bmatrix} 1 & 0 & 0 & 0 \\ 0 & 1 & 0 & 0 \end{bmatrix} \qquad R = \begin{bmatrix} 0.95 & 0 \\ 0 & 0.95 \end{bmatrix}$$

P is the a posteriori error covariance matrix (a measure of the estimated accuracy of the state estimate); F is the state transition matrix; Q is the covariance of the process noise; H is the observation model; R is the covariance of the observation noise; X is the state vector.

The state transition matrix based on the kinematic equations:

$$x_{i+1} = x_i + v_{ix}\delta t \qquad y_{i+1} = y_i + v_{iy}\delta t$$
$$v_{(i+1),x} = v_{ix} \qquad v_{(i+1),x} = v_{iy}$$

Here we see some simplifications that will inevitably affect the accuracy of this estimation. The velocity is assumed to be constant between small time steps, as in the regression model. If we assume deceleration due to friction is negligible, especially during a hard kick or a kick that spends a considerable amount of its trajectory in the air, the accuracy will be acceptable for the duration of a kick. Another simplification is that this model neglects the z movement of the ball. The dynamics of the ball are drastically different in the air vs. on the ground and if the goalie decides to dive when the ball enters the goal at a high position he will miss the ball.

3.2   Nonlinear Regression and Experimental Anticipation

Our second approach consists of enhancing the goalkeeper with a mental model of itself that it can simulate ahead of time to anticipate the efficiency of its current actions and eventual changes to current behavior. We first give a brief introduction to the concepts of this approach, the requirement of more precise velocity estimation and finally how the goalkeeper can use this cognitive layer to improve its results.

Orpheus, a Generic Mental Simulation Framework. To enable the goalkeeper agent to predict its environment, itself included, we use a generic open source framework for mental simulation entitled ORPHEUS[1] which has been successfully applied in different contexts requiring anticipative agents [14,15].

The ORPHEUS architecture provides an agent with an "imaginary world" which can be used to evaluate the outcomes of its actions in accordance with other objects (e.g. ball) and agents that populate the environment it is placed in. To clarify, the goalkeeper has its own functional representation of itself and the ball, and is able to manipulate it towards improving its behavior in the game. Providing the agent with models of itself and its environment enables it to evolve and evaluate various courses of actions in its imagination, based on which it can make decisions. This cognitive layer functions in parallel to the main agent behaviors, which makes it well suited in the context of robot soccer agents where interruptions in sending motor instructions cause unwanted results.

Two important aspects lead to a successful use of the mental simulation paradigm: perception and mental model accuracy. Specifically, the position and velocity of the ball and robot as well as the models of environment physics and the robot's own body motion must be accurate enough so that future events can be successfully predicted. Hence our focus is on improving velocity estimation and constructing the mental models for the robot.

---

[1] ORPHEUS source code: https://bitbucket.org/polceanum/orpheus.

Nonlinear Regression for Ball Position and Velocity. In the robot soccer simulation contest, as well as for real settings, the trajectory of the ball is shaped by damping factors. However, we can safely assume that, as long as the ball does not collide with any object as it's traveling, it will describe a smooth non-linear trajectory. Unlike linear models, using a curve to approximate the trajectory, even on the X and Y axes where gravity does not intervene, can account for the damping factors.

In this approach, we use the Levenberg-Marquardt algorithm as implemented in Dlib-ml [8] to fit a curve on the raw online perception data. The choice of algorithm was based on its good performance in practice, efficiency on small input datasets and that it allows describing the curve model, as is the case for the goalie's perception of the ball trajectory. We develop the damped motion equations $x_n = x_{n-1} + v_{n-1}\delta t$ and $v_n = \zeta v_{n-1}$ over multiple time steps, resulting in the model to be used in the algorithm:

$$\frac{p_1 \delta t (1 - \zeta^n)}{1 - \zeta} + p_0$$

where δt is the time difference between two perception frames, ζ is the damping factor, n is the step number, p_0 and p_1 are the position and velocity parameters respectively. With this approach, we store the ball perception frames after the kick and refit the model parameters at each step to gradually improve accuracy. For the Z axis, we extend the model to account for gravitational acceleration:

$$p_2 \delta t^2 (n - \frac{1 - \zeta^n}{1 - \zeta}) + \frac{p_1 \delta t (1 - \zeta^n)}{1 - \zeta} + p_0$$

The obtained velocity estimation enables more accurate prediction of the ball for default behavior such as walking towards the ball but also for prediction using mental simulations.

Learning Body Movement Models and Imaginary Reenactment. Once the ball can be predicted, the second step in this approach is to have the robot learn the effects of its actions and be able to reenact them in its imaginary world. This produces a set of mental models which are then used to perform mental simulations.

To this end, we rely on querying the location of each of the robot's body parts which are represented with primitives in the imaginary world. Various actions can be learned by having the robot perform it, while a model training module provided by the ORPHEUS framework distributes the data to the models which are being trained. For the goalkeeper's actions, we use the K-Nearest Neighbors (KNN) algorithm as implemented in MLPack [4]. The dataset is constructed with the position, velocity and rotation in function of time, of each body part at each time step, relative to the goalie's location at time zero.

Reenactment of each learned action with KNN uses the time and position of the robot's body parts and finds the closest corresponding velocities and rotations. From this point forward, ORPHEUS can perform mental simulations on demand, using the learned models through the process briefly described in the following. Perception data (robot, body part and ball position, velocity and rotation) at a given moment in time is submitted to the cognitive layer. From this data, the mental simulation process constructs subsequent mental images by employing the available mental models. This process leads to obtaining a future state of the environment, to a certain accuracy, ahead of time. By evaluating this future state, the robot can make a range of decisions, as detailed in the following.

Self Monitoring and Deciding to Change Behavior. There are limiting factors on how the goalkeeper agent can behave to deflect the ball. The main such factor is the high ball velocity achieved by other teams in the RoboCup 3D Simulation League, which gives little time to react. Moreover, the robot does not have the ability to move fast enough to reach the ball in many situations. There are however situations when extending its limbs or diving may help deflecting the ball.

The goalie can assesses whether or not to perform an action and if so, which one is more favorable, by running mental simulations where it approximates the ball trajectory and its own movements and evaluate the results. Testing showed that simply moving towards the predicted position of the ball when it would reach the defense zone performs well, given satisfactory estimations of the ball position and velocity. Therefore, starting from a reactive decision to move towards this location of interest, we use mental simulations to predict whether or not simply walking would enable the robot deflect the ball. As soon as the kick is detected, a coarse future location of the ball is computed and the robot starts moving towards that location. In parallel, it uses perceptual information to imagine the success or failure of this strategy.

We studied outcome prediction accuracy (Table 1) with a set of 100 penalty shots from random locations, 5 to 10 meters away from the goal gate. This resulted in an average of $\sim$0.5 s in which the goalie could apply a different decision, after subtracting the time required to obtain reasonable ball velocity accuracy ($\sim$0.1 s) and the time required for mental simulations to finish; in our implementation, mental simulation speed ratio is $\sim$3x, i.e. the robot can simulate 1 s of real events in $\sim$0.33 s.

As shown in Table 1, the accuracy of predicting the outcome of a set of actions rises above 50 % (random choice) as soon as valid data is acquired from the environment, and improves with time. Prediction failure is caused by the difference between mental models and reality, which could be later improved by learning more accurate models.

Concurrently with self monitoring, the goalkeeper also imagines and evaluates a set of actions that can be performed. This results in a set of solutions in the case in which it decides the current set of actions will not be effective. If a failure is predicted, the goalkeeper is able to choose one of the previously imagined solutions that may result in a last-moment save.

Table 1. Outcome prediction accuracy (% correct predictions) based on mental simulation, in function of the time before the ball passes the goalie's location.

Time left	0.5 s	0.4 s	0.3 s	0.2 s	0.1 s
Correctly predicted outcome	69.1 %	80.1 %	84.8 %	89.5 %	93.2 %

This experimental approach is limited by the set of possible actions that the robot can perform, and the time available to evaluate the outcomes of these actions. However, more efficient actions can be later added to increase the success rate. We investigate whether this approach is feasible for full integration in the goalkeeper. Therefore, during the tests we enable the complete reasoning process to verify that no negative effects are introduced.

4 Results

We empirically evaluated the two approaches – with Linear Regression/Kalman-Filter (LR/KF) and Nonlinear Regression (NLR) – on a range of kick distance intervals. For the second approach, we developed two versions so that the experimental mental model based prediction (ORPHEUS) can be evaluated separately (Fig. 1).

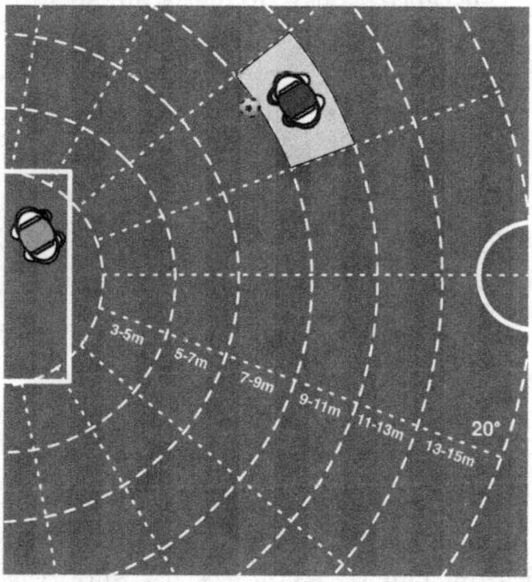

4.1 Experiment Setup

The evaluation consisted in running a goalkeeper with an attacker that will kick the ball from a random distance within a specified interval, at a given angle. As a benchmark, we tested the naive goalkeeper behavior, used so far by the Robo-Canes team in previous competitions.

Fig. 1. Setup for our experiments: six distance categories, 4 × 2 angle categories, 100 random kicks repeated 30 times on five agents.

There would be a large variance in the kicks of the striker, even if the striker kicks from the same distance and angle. Therefore, we reproduce the exact same kicks by assigning a velocity to the ball in the beginning of the kick. Thus, we are able to create a random sequence of kicks and repeat the exact same kicks multiple times using different goalies. Variances in the results are only caused by variance in the goalkeepers behavior, but not by variances in the kicks.

To better estimate the relationship between distance, shot angle and performance, we devised 2 m intervals starting from 3 m away (which is roughly the margin of the goalkeeper's area) to 15 m away (midfield). The obtained results for each interval (Table 2) are the success rates for 100 different kicks repeated 30 times and averaged out (3,000 kicks per interval).

The experiment has been performed on an Intel Core i7-5930K machine with 6 cores. Results were also verified on two less powerful machines – a Quad Core Intel Xeon and a Dual Core Intel Pentium – and replicate the main experiment.

4.2 Goalkeeper Performance

Results of distance-based evaluation (Table 2) show that both approaches bring significant improvements over Naive goalkeeper which was used by the Robo-Canes team in previous competitions.

Table 2. Goalkeeper success rates for 100 random kicks from different distances. The kick angles are chosen randomly from −45 to 45°. Each rate is averaged over 30 repetitions of the 100 kicks (standard deviation shown in parentheses).

Distance	Naive goalie	Kalman Filter	LR	NLR	ORPHEUS
3–5 m	29.8 % (2.2)	35.8 % (3.6)	35.6 % (3.4)	46.1 % (2.8)	45.8 % (2.3)
5–7 m	22.0 % (1.9)	31.8 % (3.4)	31.7 % (4.0)	50.4 % (2.9)	50.2 % (2.3)
7–9 m	19.5 % (1.9)	31.8 % (4.6)	30.1 % (3.0)	48.6 % (3.0)	48.0 % (3.0)
9–11 m	17.8 % (1.9)	23.9 % (3.5)	23.4 % (2.9)	47.5 % (3.4)	46.9 % (3.6)
11–13 m	17.0 % (2.4)	21.5 % (3.3)	22.1 % (2.9)	45.0 % (3.5)	45.1 % (3.8)
13–15 m	14.8 % (2.0)	18.6 % (2.7)	19.5 % (2.8)	43.9 % (3.5)	44.8 % (3.4)

The first important difference between the naive goalkeeper and the two approaches presented in this paper is that previously, the naive keeper was instructed to place itself at the middle between the ball position and the goal, while the proposed goalkeeper's behavior uses the goalkeeper's arc. Midway positioning can be exploited by shooting the ball over the goalkeeper, resulting in easy scoring, illustrated by the constant drop in success rates with respect to distance. This effect is significantly less prominent in the LR/KF and NLR approaches, due to superior initial positioning of the goalkeeper.

The increased success rates obtained by the LR/KF and NLR approaches are also attributed to the improvements in ball tracking (Fig. 2). In the context of the RoboCup 3D soccer simulation league, the ball does not have an ideal trajectory due to the realistic simulation performed by the physics engine which includes friction and damping. This leads to better results with NLR caused by a better fit of the damped trajectory.

Results of the experimental ORPHEUS version of the goalie do not vary in a statistically significant manner from the NLR version, which shows that it is feasible to integrate prediction in the goalkeeper without suffering a trade-off.

In order to more extensively evaluate our goalies' performance, we also varied the angle from which the ball is shot and measured success rates (Table 3).

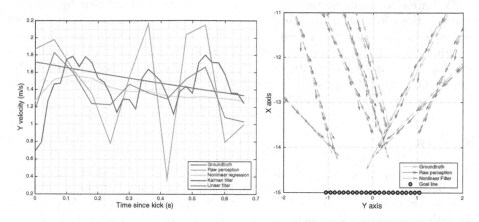

Fig. 2. Comparison of ball velocity values for Y axis over time since the kick (left) and 2D velocity perception during kicks (right).

Table 3. Average goalkeeper success rates for 100 random kicks as in Table 2, here grouped by angle (using the distance 5–7 m)

Angle °	Naive goalie	Kalman Filter	LR	NLR	ORPHEUS
0–20	18.3 % (2.1)	26.8 % (3.7)	26.1 % (3.8)	43.6 % (3.3)	43.4 % (2.8)
20–40	21.5 % (2.0)	31.9 % (3.5)	30.0 % (3.1)	50.0 % (2.9)	49.8 % (2.7)
40–60	33.1 % (2.7)	38.4 % (3.7)	38.7 % (4.5)	65.0 % (3.3)	64.7 % (3.1)
60–80	63.3 % (3.6)	32.1 % (2.9)	37.0 % (3.3)	76.0 % (3.5)	78.1 % (3.5)

Results show that the increase in success rates is directly proportional with the shooting angle for the Naive, LR, NLR and ORPHEUS versions.

5   Conclusion and Future Work

We have presented two approaches to improving the goalkeeper agent of the RoboCanes team: one based on linear regression, Kalman Filter and scripted decision making, and another based on nonlinear regression and an experimental prediction capability based on internal mental models. The proposed approaches have been extensively evaluated along with the naive version used in previous competitions as benchmark, showing significant improvements over all kicking conditions (distance intervals and angles from the goal).

Design features of each approach were correlated with the results they obtained, and concluded the importance of accurate perception and fast reaction in the goalkeeper's efficiency. We have also provided a novel mechanism for the goalkeeper to perform auto-evaluation which has the potential to further improvement of our goalkeeper's performance.

We intend to continue the improvement of the goalkeeper agent by further studying the behavior of the proposed approaches in exceptional cases such as collisions of the ball with the ground or other objects. Future work will also include efforts on improving the prediction ability of the ORPHEUS version by learning better models and integrating it into the decision process of the goalkeeper. The aim of this approach is to enable the goalkeeper to adjust its scripted motion to save more seldom and more difficult shots.

References

1. Adorni, G., Cagnoni, S., Mordonini, M.: Landmark-based robot self-localization: a case study for the RoboCup goal-keeper. In: Proceedings of the 1999 International Conference on Information Intelligence and Systems, pp. 164–171. IEEE (1999)
2. Birbach, O., Frese, U., Bäuml, B.: Realtime perception for catching a flying ball with a mobile humanoid. In: 2011 IEEE International Conference on Robotics and Automation, pp. 5955–5962 (2011)
3. Bozinovski, S., Jäger, H., Schöl, P.: Engineering goalkeeper behavior using an emotion learning method. In: Proceedings of the RoboCup Workshop, KI 1999: Deutsche Jahrestagung für Künstliche Intelligenz, pp. 48–56 (1999)
4. Curtin, R.R., Cline, J.R., Slagle, N.P., March, W.B., Ram, P., Mehta, N.A., Gray, A.G.: MLPACK: a scalable C++ machine learning library. J. Mach. Learn. Res. **14**, 801–805 (2013)
5. Diaz, G.J., Fajen, B.R., Phillips, F.: Anticipation from biological motion: the goalkeeper problem. J. Exp. Psychol. Hum. Percept. Perform. **38**(4), 848 (2012)
6. Faragher, R., et al.: Understanding the basis of the Kalman filter via a simple and intuitive derivation. IEEE Sig. Process. Mag. **29**(5), 128–132 (2012)
7. García, J.F., Rodríguez, F.J., Fernández, C., Matellán, V.: Design an evaluation of RoboCup humanoid goalie. J. Phys. Agents **4**(2), 19–26 (2010)
8. King, D.E.: Dlib-ml: a machine learning toolkit. J. Mach. Learn. Res. **10**, 1755–1758 (2009)
9. Lausen, H., Nielsen, J., Nielsen, M., Lima, P.: Model and behavior-based robotic goalkeeper. In: Polani, D., Browning, B., Bonarini, A., Yoshida, K. (eds.) RoboCup 2003. LNCS (LNAI), vol. 3020, pp. 169–180. Springer, Heidelberg (2004)
10. Liu, J., Liang, Z., Shen, P., Hao, Y., Zhao, H.: The walking skill of Apollo3D – the champion team in the RoboCup2013 3D soccer simulation competition. In: Behnke, S., Veloso, M., Visser, A., Xiong, R. (eds.) RoboCup 2013. LNCS, vol. 8371, pp. 104–113. Springer, Heidelberg (2014)
11. MacAlpine, P., Barrett, S., Urieli, D., Vu, V., Stone, P.: Design and optimization of an omnidirectional humanoid walk: a winning approach at the RoboCup 2011 3D simulation competition. In: AAAI (2012)
12. MacAlpine, P., Collins, N., Lopez-Mobilia, A., Stone, P.: UT Austin Villa: RoboCup 2012 3D simulation league champion. In: Chen, X., Stone, P., Sucar, L.E., van der Zant, T. (eds.) RoboCup 2012. LNCS, vol. 7500, pp. 77–88. Springer, Heidelberg (2013)
13. Menegatti, E., Nori, F., Pagello, E., Pellizzari, C., Spagnoli, D.: Designing an omnidirectional vision system for a goalkeeper robot. In: Birk, A., Coradeschi, S., Tadokoro, S. (eds.) RoboCup 2001. LNCS (LNAI), vol. 2377, pp. 81–91. Springer, Heidelberg (2002)

14. Polceanu, M., Buche, C.: Towards a theory-of-mind-inspired generic decision-making framework. In: IJCAI 2013 Symposium on AI in Angry Birds (2013)
15. Polceanu, M., Parenthoen, M., Buche, C.: ORPHEUS: mental simulation as support for decision-making in a virtual agent. In: 28th International Florida Artificial Intelligence Research Society Conference (FLAIRS-28), pp. 73–78. AAAI Press (2015)
16. Seekircher, A., Abeyruwan, S., Visser, U.: Accurate ball tracking with extended kalman filters as a prerequisite for a high-level behavior with reinforcement learning. In: The 6th Workshop on Humanoid Soccer Robots at Humanoid Conference, Bled (Slovenia) (2011)
17. Sen, P.K.: Estimates of the regression coefficient based on Kendall's tau. J. Am. Stat. Assoc. **63**(324), 1379–1389 (1968)

A Fast Method for Adapting Lookup Tables Applied to Changes in Lighting Colour

Trent Houliston$^{(\boxtimes)}$, Mitchell Metcalfe, and Stephan K. Chalup

School of Electrical Engineering and Computer Science,
The University of Newcastle, Callaghan, NSW 2308, Australia
{trent.houliston,mitchell.metcalfe,stephan.chalup}@newcastle.edu.au

Abstract. This paper proposes a simple and fast method for adapting colour lookup tables to lighting changes in real-time. The method adjusts the classified colour space regions keeping both their surface area and volume constant. Two variations of the method were compared and tested in a RoboCup soccer setting. Detection success rate was measured as a function of the speed and magnitude of hue change to the lighting environment. Compared to a static lookup table, these experimental results show improved robustness against lighting changes for detection of coloured objects.

Keywords: Color lookup table · Illumination invariance · Color space · Computer vision · Robotics

1 Introduction

Mobile robots competing in the RoboCup humanoid and standard platform leagues use vision as their primary sensor. Colour perception is useful as the typical RoboCup soccer field environment has several specified colours. These included an orange ball and blue and yellow coloured goal posts. Consequently, most teams at RoboCup investigate methods to implement stable and efficient colour vision, including colour segmentation in the presence of illumination changes on their robots [4,6–8]. This is in contrast to general mobile robots that avoid colour perception issues by using laser range finders and sonar not unlike those used for the DARPA challenges. Sridharan and Stone [14] have provided a review of some of the most significant work on illumination invariance on mobile robots up until 2008. The challenge for robots that rely on video sensors is that they have to be able to deal with shadows, reflections, natural light changes and other artefacts caused by robot motion in RoboCup and real-world domains. For example when a robot approaches a ball it often ends up in a shadow cast by another robot. Another challenge is that many methods require significant human supervision [11] or substantial computational resources and time [2,12,13]. Only recently more appropriate solutions accessible to small robots such as the DARwIn-OP or Nao have been developed [9].

© Springer International Publishing Switzerland 2015
L. Almeida et al. (Eds.): RoboCup 2015, LNAI 9513, pp. 190–201, 2015.
DOI: 10.1007/978-3-319-29339-4_16

This paper proposes a fast hands-on implementation of a basic procedure to adapt lookup tables to varying lighting conditions. This implementation improves an initial lookup table and ensures its robustness to illumination changes in a RoboCup environment with a limited number of uniformly coloured objects.

2   Adaptive Lookup Tables

A Lookup Table (LUT) for colour classification is a map from the colour space of an input image to a set of colour classes. The LUT can be represented as a cube of discrete voxels, indexed by the three components of the pixel colours in the input image. For example, y, u, and v, for the YUV colour space.

Robots at RoboCup often have difficulty classifying colour when the field lighting is not uniform or changes during a game. This paper outlines a method of dynamically adapting a LUT to the current lighting conditions in order to resolve this problem. This is achieved through continually updating a LUT using feedback from object detectors, like a ball detector. Each object detector recognises a single type of object that is coloured with a single class of colour, like an orange ball, yellow or white goalposts or white field lines.

An initial seed LUT is required to begin the feedback process. This LUT must be able to classify objects sufficiently for detection under the startup illumination environment. The quality of the initial LUT will be improved automatically by the proposed algorithm, due to the feedback process, allowing an initial LUT of lower quality to be used. This initial seed LUT can be provided by hand classification, or more expensive algorithms [1,5] that can detect and classify the salient regions of the colour space.

Upon each object detection, the pixel colours within the image region of the detected object are passed to the algorithm. For each of these pixels, the corresponding voxel in the current LUT is found. If any of the voxel's neighbours are of the same colour class as the detected object and the voxel is unclassified, the voxel is assigned the object's colour class, updating the LUT. Two voxels are considered to be neighbours if they share a face.

Once the LUT has been updated for each pixel in the detected object's image region, the *volume*, and *surface area* of the colour class are compared to preset maximum values. The volume of a colour class is defined as the number of voxels belonging to the class. The surface area of a colour class is defined as the number of *removable* voxels belonging to the class (see Sect. 2.1). If the volume or surface area of the class exceed their preset values, a *shedding* process is performed that removes a layer of voxels from the exterior of the colour class in the LUT. This works to remove old voxels from the LUT that no longer accurately represent their colour class due to illumination changes.

The shedding step of the algorithm can be performed through two different methods: layer based shedding and voting based shedding. These shedding methods are compared in Sect. 2.1. The process described earlier is the basic update process that is used with the layer based shedding method. Pseudocode for this update process is presented in Algorithm 1.

A different update process is required to support the voting shedding method. In the voting update process, an integral vote count is maintained for individual voxel in each colour class. Each voxel belonging to a colour class in the initial LUT has its vote count set to a preset maximum value. When an object of a given colour class is detected, the vote count of each voxel in that class is first decremented by a preset value. Then, as in the layer update method, colours from the detected object's image region are used to classify voxels in the LUT, with their votes also updated. When a new voxel is added to the colour class, its vote is set to zero. When a voxel of the current colour class is encountered, its vote count is incremented by a preset value (as presented in the pseudocode in Algorithm 3). Voxel vote counts are capped at a preset maximum value. If the volume or surface area of the class exceed their preset values, a shedding method is run that considers vote counts during shedding. Pseudocode for the voting based update process is described in Algorithm 1.

Algorithm 1. UPDATELUTLAYER(S, c)

Input: The set $S \in \mathbb{R}^3$ of pixel colours within the image region of a detected object

Input: c the colour class of the detected object

1 **foreach** $x \in S$ **do**
2 | $v \leftarrow$ GETLUTVOXEL(x)
3 | **if** CLASS(v) $= \varnothing \wedge c \in \{$CLASS($w$) $: w \in$ NEIGHBOURS(v)$\}$ **then**
4 | | CLASS(v) $\leftarrow c$
5 | **end**
6 **end**
7 **if**
   VOLUME(c) $>$ MAXVOLUME(c) **or** SURFACEAREA(c) $>$ MAXSURFACEAREA(c)
   then
8 | PERFORMSHEDDING(c)
9 **end**

In addition to these procedures and functions in Table 1, there are configuration parameters that influence the actions of the algorithm at runtime. Appropriate values for MAXSURFACEAREA(c), MAXVOLUME(c), MAXVOTES(c), VOTEGROWTHRATE(c), VOTESHRINKRATE(c), must be chosen by experimentation before the algorithm is run.

2.1 Shedding

Shedding is the process of reducing the volume of a colour class by removing its outermost voxels. To describe the shedding process, three different classifications of LUT voxel are defined:

Internal. A voxel that has six neighbours, all of the same class as in Fig. 1a.

Table 1. Procedures and functions that appear in the algorithms

CLASS(v)	Returns the current colour class assigned to the voxel v.
DECREMENTVOTES(c)	Decrements the vote count of all voxels with the colour class c by VOTESHRINKRATE(c).
GETLUTVOXEL(x)	Returns the voxel in the LUT indexed by the pixel x.
NEIGHBOURS(v)	Returns the six neighbours of the given voxel, adjacent to each of its six cube faces.
SURFACEAREA(c)	Returns the surface area of colour class c.
VOLUME(c)	Returns the number of voxels with the given colour class c in the LUT.
REMOVEABLESURFACE(c)	The voxels of a colour class that are removable (as defined by the method in Sect. 2.1)
VOTE(v)	Returns the vote count of the given voxel.

Algorithm 2. UPDATELUTVOTING(S, c)

Input: The set $S \in \mathbb{R}^3$ of pixel colours within the image region of a detected object
Input: c the colour class of the detected object
Input: The Lookup Table $\mathcal{L}$
1   DECREMENTVOTES(c)
2   **foreach** $x \in S$ **do**
3 | $v \leftarrow$ GETLUTVOXEL(x)
4 | **if** CLASS(v) = c **then**
5 | | INCREMENTVOTE(c, v)
6 | **end**
7 | **else if** CLASS(v) = $\varnothing \wedge$ CLASS(c) $\in \{$CLASS(w) : $w \in$ NEIGHBOURS(v)$\}$ **then**
8 | | CLASS(v) $\leftarrow c$
9 | | VOTE(v) $\leftarrow 0$
10 | **end**
11 **end**
12 **if**
    VOLUME(c) > MAXVOLUME(c) *or* SURFACEAREA(c) > MAXSURFACEAREA(c)
    then
13 | PERFORMSHEDDING(c)
14 **end**

Algorithm 3. INCREMENTVOTE(c, v)

Input: The colour class c
Input: The voxel v
1   $n \leftarrow$ VOTE(v) + VOTEGROWTHRATE(c)
2   VOTE(v) $\leftarrow \min(n,$ MAXVOTES(c))

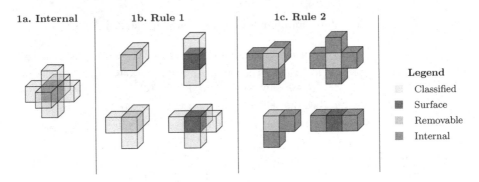

Fig. 1. Examples of the different classifications of voxels due to the rules outlined in Sect. 2.1

Surface. Any voxel that is not internal.

Removable. A surface voxel that satisfies at least one of the following conditions:

(a) For every neighbouring voxel of the same colour class, the opposite neighbouring voxel must be vacant (example Fig. 1b).

(b) Has at least two non-opposite internal voxels as neighbours. At least two neighbouring voxels must be internal voxels and those internal voxels must share a neighbour that is not the central voxel (example Fig. 1c).

These conditions are heuristics designed to preserve the topology of the colour class volume by avoiding the creation of new holes or splitting connected components of the colour class in the LUT. However, there are two notable cases where the topology could change. Firstly, when a thin section of volume develops a 'staircase' pattern. These staircases can be removed, resulting in two separate volumes. Secondly, when the volume extends into a torus shape, creating a hole in the resulting volume. The algorithm makes no attempt to address these problems, as avoiding them significantly increases the number of voxels that must be checked. Also, if these cases do arise, the system is able to recover as one half of the shape can disappear or the two halves will rejoin.

The two shedding algorithms that have been developed are:

Layer Shedding. Removes the class labels from all removable voxels of the given colour class in the LUT.

Voting Shedding. Zeroes all votes for and removes class labels of all voxels of the given colour class that have a vote count below the mean vote count among the surface voxels of that class.

Pseudocode for the layer shedding and voting shedding algorithms is included as Algorithms 4 and 5.

Algorithm 4. PERFORMLAYERSHEDDING(c)

Input: The colour class c to be shed
1 $S_r \leftarrow$ REMOVEABLESURFACE(c)
2 **foreach** $v \in S_r$ **do**
3   $\quad$| CLASS(v) $\leftarrow \varnothing$
4 **end**

Algorithm 5. PERFORMSHEDDINGVOTING(c)

Input: The colour class c to be shed
1 $S_r \leftarrow$ REMOVEABLESURFACE(c)
2 $m \leftarrow$ AVERAGE($\{$VOTE(w) $: w \in S_r\}$)
3 **foreach** $v \in S_r$ **do**
4   $\quad$| **if** VOTE(v) $< m$ **then**
5   $\quad$| $\quad$| CLASS(v) $\leftarrow \varnothing$
6   $\quad$| $\quad$| VOTE(v) $\leftarrow 0$
7   $\quad$| **end**
8 **end**

2.2   Implementation

During the implementation of the algorithm a number of optimisations were made. These optimisations were not included in the description of the algorithm as they obfuscate its intention.

The surface voxels for each colour class are stored in a hashset. The set is updated any time that a voxel is inserted or removed as this could change a voxel's classification to or from removable.

The voting algorithm stores a value to represent zero for each colour class. This value is incremented rather than decrementing the vote counts of each voxel when DECREMENTVOTES is called.

The worst case complexity of the algorithm is $\mathcal{O}(nr)$ where n is the number of pixels in the detected object and r is the number of removable voxels for the colour class. Shedding should only happen with regularity when the volume is moving. When the volume is stationary and shedding does not occur, the complexity of this algorithm is $\mathcal{O}(n)$.

In addition to these optimisations, the pixels sent to this algorithm from the object detectors can be subsampled in order to reduce the number of updates to the LUT at the cost of reduced adaptability.

3   Experiments

An experiment was conducted to evaluate the ability of the developed methods to adapt to changes in illumination colour. The robot was placed on a 2013 style kid size humanoid league soccer field [10] with the goals and ball entirely within its camera frame. A ceiling mounted digital projector was pointed at the goals and made to project continuously changing coloured light onto the robot's field of view. This was the only source of light in the room. The experimental setup is illustrated in Fig. 2.

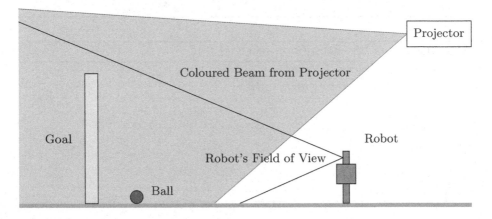

Fig. 2. The experimental setup. The robot views the ball and goals, which are illuminated by a ceiling mounted projector.

3.1   Colour Sequences

Twenty-one separate video sequences were captured for the experiment. Each video sequence was two minutes long with varying illumination colours $\mathcal{C}$ produced by the projector over time. Each of the videos set a different value of a parameter α that controlled the amount of colour variation produced by the projector. This is referred to as 'saturation' in the results. However, it is not equivalent to the saturation parameter of the HSV colour space. Each video was resampled to generate twelve videos of different durations ranging from 5–60 s long at 5 s intervals. This simulates the slower and faster colour changes that might occur during normal operation of the robot. As a result, a total of 252 videos were tested. The formula for the colour displayed by the projector at time t in a trial of duration d is given in Eq. (1), where the function HSV creates an HSV colour and the function RGB converts an HSV colour to its RGB representation.

$$\mathcal{C} = \frac{\alpha}{100}\,\text{RGB}(\text{HSV}(\frac{360t}{d}, 1, 1)) + (1 - \frac{\alpha}{100})\,\text{RGB}(\text{HSV}(0, 0, 1)) \qquad (1)$$

Each video was recorded at a resolution of 320 × 240, and a frame rate of 30 Hz. Figure 3 presents a sequence of selected frames from one of the videos.

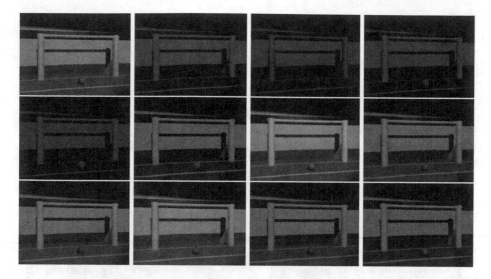

Fig. 3. Selected frames from one of the colour sequences from the experiment. The ball and goals are both visible and are lit only by coloured light from the projector. The colour throughout the sequence follows a full hue cycle while maintaining constant saturation parameter α of 60 % (Color figure online).

3.2   Trials

Three methods were run on each of the 252 videos: a regular static LUT, an adaptive LUT using the layer shedding scheme and an adaptive LUT using the voting shedding scheme. The colour space used for the LUT was YUV, with a resolution of 6 bits for y, 7 bits for u and 7 bits for v.

Two object detectors were used in the experiments: a ball detector that detects an orange ball and a goal post detector that detects yellow goal posts. Both of these detectors were based on the RANSAC algorithm [3]. The output of the detectors were used to track the orange and yellow colour classes.

During each trial, the positions of the detected objects in each frame were compared to the known positions of the objects in the frame. Detections that were within a threshold of the true object positions were recorded as successful.

Detection of the ball was considered successful if the x and y coordinate of the ball's centre and the ball's radius were each detected with an error of less than 2 pixels from their true values. The true radius of the ball was 9.3 pixels. A goalpost detection was considered successful if each of its corner points differed from its corresponding true corner point by less than 10 pixels in both dimensions. The left and right goalposts were 130 pixels and 110 pixels tall respectively. Both goalposts were approximately 20 pixels wide.

A detection rate for each sequence was calculated by dividing the total number of successful detections in a trial by the number of frames in the video and by the number of objects in each frame (a constant three for the goals and ball).

Each trial was run in realtime on a late 2013 Macbook Pro 15" 2.6 GHz Intel Core i7.

4   Results

Figure 4 summarises the performance of the static LUT, the adaptive LUT with layer based shedding and the adaptive LUT with voting based shedding during the experiment described in Sect. 3. Each figure presents a contour plot where the x and y axes are the saturation (defined in Sect. 3.1) and trial duration and the z axis is the detection rate as a percentage of possible detections of the trial.

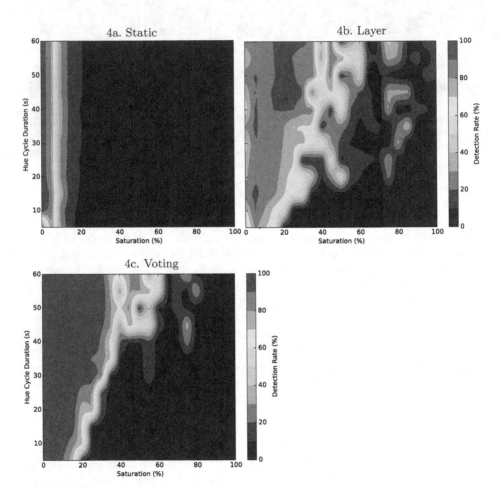

Fig. 4. Contour plot describing the performance of the three different LUT methods. The dark red area on the left is where the methods performed best (Color figure online).

As Fig. 4a shows, the static LUT achieves a detection rate above 90 % for trials with very low colour variation (i.e. saturation below 5 %) that falls to 0 %

before saturation reaches 20 %. The performance of the detector is independent of the duration of the trial.

The results for the adaptive LUT using the layer shedding method are presented in Fig. 4b. A detection rate of over 80 % is achieved for trials with 5 % saturation and duration ranging from 10 s to 50 s and trials with saturation between 10 % and 30 %, and durations above 40 s. Outside of these regions, the detection rate falls to zero in an irregular manner and approaches zero for most trials with a saturation value above 50 %. It falls to zero faster for lower durations.

The adaptive LUT using the voting shedding scheme produced the results in Fig. 4c. It achieved a detection rate above 90 % for all trials with a saturation value lower than 10 %, for all trials with a saturation parameter below 30 % and a duration longer than 30 s. The detection rate falls to 0 % for the shortest trials as the saturation parameter approaches 20 %, and falls to 0 % for most other trials as it increases beyond 40 %.

5   Discussion

The experiment showed that a static LUT performed very poorly when the illumination colour varied. Reasonable detection performance was only achieved for a very narrow range of hue values. This is in contrast with the two adaptive LUTs that performed well across a much wider range of illumination parameters. As a result, the voting shedding scheme displayed a clear advantage over the simpler, layer shedding scheme. It achieved high detection rates over a larger range of parameters, while exhibiting more predictable performance across the range of parameters tested.

In addition to the formal experiment, the performance of the algorithm was tested on a DARwIn-OP in a simulated RoboCup game environment. In this test, the robot's camera was running at 640×480 pixels at 30 Hz. The system was tested detecting goals and balls while the robot was walking. Other robots were also on the field occasionally occluding the ball and goals. During these tests, the detection and computational performance of the system was monitored with and without the adaptive LUT running.

It was discovered that the algorithm could cause the LUT to diverge from useful classification. This would occur whenever the volume in the LUT included black. Black acts as a singularity in the colour space as all hues tend to black with reduced brightness. Given enough examples of heavy shading or occlusion from dark objects, the LUT would include black ($y \rightarrow 0$) and the classification of false positives would increase. This exasperated the problem. Including minimum brightness constraints for pixels used by the algorithm resolved this issue as the colour classes could no longer reach this singularity.

To measure the computational performance of the system, the frame rate and CPU usage were analysed during testing both with and without the adaptive LUT. The frame rate of the system in both cases was 22 Hz. This is due to an IO bottleneck, rather then a processing bottleneck, which resulted in the total CPU usage being below 100 %. The additional computational overhead for the adaptive LUT was measured to be 10 percentile points.

6   Conclusion

This paper has presented a new technique for adapting lookup tables in realtime to achieve robust colour classification under varying illumination. The approach used information from object detections to add missing colour class information to the lookup table. It proposed the use of a shedding procedure to remove old colour class information that no longer accurately represented its colour class from the LUT. Together, these update and shedding procedures are intended to allow the LUT to adapt to changes in illumination. Two different shedding procedures were designed - a layer based method and a voting based method. They were compared based on their ability to adapt to illumination changes in a controlled experiment. The results show that the voting based shedding algorithm is both more accurate and consistent than the layer based method and that both adaptive LUT methods significantly outperformed a simple static LUT.

In contrast to most of the already existing studies that focus on illumination intensity changes, this method evaluated the hue changes of lighting.

Potential directions for future work could include testing the new method on a wider variety of real world scenarios and applying the algorithm to other problems requiring adaptive classification in a continuous space, in particular the classification of texture.

Acknowledgments. The authors would like to acknowledge Brendan Annable for early discussions and implementation ideas as well as Ellie-Mae Simpson, Jake Fountain and Amy Kendrick for their assistance in proofreading this document and their helpful suggestions.

References

1. Budden, D., Mendes, A.: Unsupervised recognition of salient colour for real-time image processing. In: Behnke, S., Veloso, M., Visser, A., Xiong, R. (eds.) RoboCup 2013. LNCS (LNAI), vol. 8371, pp. 373–384. Springer, Heidelberg (2014)
2. Serhan Daniş, F., Meriçli, T., Levent Akın, H.: Using saliency-based visual attention methods for achieving illumination invariance in robot soccer. In: Chen, X., Stone, P., Sucar, L.E., van der Zant, T. (eds.) RoboCup 2012. LNCS (LNAI), vol. 7500, pp. 273–285. Springer, Heidelberg (2013)
3. Flannery, M., Fenn, S., Budden, D.: RANSAC: identification of higher-order geometric features and applications in humanoid robot soccer. In: Proceedings of the 2014 Australasian Conference on Robotics and Automation (ACRA 2014). ARAA (on-line) (2014)
4. Heinemann, P., Sehnke, F., Streichert, F., Zell, A.: An automatic approach to online color training in RoboCup environments. In: 2006 IEEE/RSJ International Conference on Intelligent Robots and Systems, pp. 4880–4885 (2006)
5. Henderson, N., King, R., Chalup, S.K.: An automated colour calibration system using multivariate gaussian mixtures to segment hsi colour space. In: Kim, J., Mahony, R. (eds.) Proceedings of the 2008 Australasian Conference on Robotics and Automation (ACRA 2008), vol. 6. ARAA (on-line) (2008)

6. Iocchi, L.: Robust color segmentation through adaptive color distribution transformation. In: Lakemeyer, G., Sklar, E., Sorrenti, D.G., Takahashi, T. (eds.) RoboCup 2006. LNCS (LNAI), vol. 4434, pp. 287–295. Springer, Heidelberg (2007)
7. Luan, X., Qi, W., Song, D., Chen, M., Zhu, T., Wang, L.: Illumination invariant color model for object recognition in robot soccer. In: Tan, Y., Shi, Y., Tan, K.C. (eds.) ICSI 2010, Part II. LNCS, vol. 6146, pp. 680–687. Springer, Heidelberg (2010)
8. Mayer, G., Utz, H., Kraetzschmar, G.K.: Playing robot soccer under natural light: a case study. In: Polani, D., Browning, B., Bonarini, A., Yoshida, K. (eds.) RoboCup 2003. LNCS (LNAI), vol. 3020, pp. 238–249. Springer, Heidelberg (2004)
9. Neves, A.J.R., Trifan, A., Cunha, B.: Self-calibration of colormetric parameters in vision systems for autonomous soccer robots. In: Behnke, S., Veloso, M., Visser, A., Xiong, R. (eds.) RoboCup 2013. LNCS (LNAI), vol. 8371, pp. 183–194. Springer, Heidelberg (2014)
10. RoboCup Technical Committee: RoboCup soccer humanoid league rules and setup for the 2013 competition in Eindhoven (2013). http://www.robocuphumanoid.org/wp-content/uploads/HumanoidLeagueRules2013-05-28.pdf
11. Schulz, D., Fox, D.: Bayesian color estimation for adaptive vision-based robot localization. In: Proceedings of the 2004 IEEE/RSJ International Conference on Intelligent Robots and Systems (IROS 2004), vol. 2, pp. 1884–1889 (2004)
12. Sridharan, M., Stone, P.: Autonomous color learning on a mobile robot. In: Proceedings of the National Conference on Artificial Intelligence, vol. 20, pp. 1318–1323. AAAI Press/MIT Press, Menlo Park, Cambridge, London (1999, 2005)
13. Sridharan, M., Stone, P.: Structure-based color learning on a mobile robot under changing illumination. Auton. Robots **23**(3), 161–182 (2007)
14. Sridharan, M., Stone, P.: Color learning and illumination invariance on mobile robots: a survey. Robot. Auton. Syst. **57**(6), 629–644 (2009)

Fuzzy Logic Control of a Humanoid Robot on Unstable Terrain

Chris Iverach-Brereton[⊠], Jacky Baltes, Brittany Postnikoff,
Diana Carrier, and John Anderson

University of Manitoba, Winnipeg, MB R3T2N2, Canada
chrisib@cs.umanitoba.ca
http://aalab.cs.umanitoba.ca/

Abstract. This paper describes a novel system for enabling a humanoid robot to balance on highly dynamic terrain using fuzzy logic. We evaluate this system by programming Jimmy, a small, humanoid DARwIn-OP robot, to balance on a bongo board – a simple apparatus consisting of a deck resting on a free-rolling wheel – using our novel fuzzy logic system and a PID controller based on our previous work (Baltes et al. [1]). Both control algorithms are tested using two different control policies: "do the shake," wherein the robot attempts to keep the bongo board's deck level by CoM manipulation; and "let's sway," wherein the robot pumps its legs up and down at regular intervals in an attempt to induce a state of dynamic stability to the system. Our experiments show that fuzzy logic control is equally capable to PID control for controlling a bongo board system.

1 Introduction

In this paper we present a fuzzy logic control system for controlling a humanoid robot standing on a bongo board. This is a continuation of our previously-published research [1] into active balancing on highly-dynamic surfaces using Jimmy[1], a DARwIn-OP humanoid robot.

For humanoid robots to be useful in the broadest possible applications they must be able to traverse all manner of terrain without falling. While recent developments in hardware and software have seen humanoid robots improve dramatically in capabilities when traversing mostly-level ground with good traction – e.g. between 2009 and 2013 the world record for the HuroCup sprint event[2] has improved from 01:07.50 to 00:25.50 [2] – the ability to traverse arbitrary terrain with unknown traction remains beyond the state-of-the-art.

In order to analyse the problems of active balancing on unknown and unstable terrain we selected the bongo board, a simple apparatus consisting of a deck with

---

[1] Jimmy is named after Jimmy Ball of Dauphin, Manitoba, winner of the Silver Medal in the 400 m sprint at the 1928 Olympics, and the Bronze Medal in the 4 × 400 m relay at the 1928 and 1932 Olympics.

[2] In the HuroCup sprint event the robots must walk or run 3 m forward followed by 3 m backwards.

© Springer International Publishing Switzerland 2015
L. Almeida et al. (Eds.): RoboCup 2015, LNAI 9513, pp. 202–213, 2015.
DOI: 10.1007/978-3-319-29339-4_17

a single free-rolling wheel positioned below it, as a sample problem. A humanoid robot stands on top of the board and must control their CoM in such a way as to keep the bongo board's wheel centred and prevent the ends of the bongo board from striking the ground. Figure 1 shows a robot on the bongo board.

The remainder of this paper is organised as follows: Sect. 2 presents an analysis of the bongo board and how it relates to the well-known inverted pendulum problem. Here we also discuss our previous research using PID control to balance the bongo board. Section 3 describes a fuzzy logic system for controlling a humanoid robot on a bongo board directly inspired by a solution for the inverted pendulum problem. We summarise our experimental procedure and perform a quantitative analysis of the performance of the fuzzy logic controller compared to our previous PID-based solution in Sect. 4. Finally we discuss practical applications of this research and provide avenues of future research in Sect. 5.

Fig. 1. Jimmy standing on a bongo board. (Baltes et al. [1])

2 Analysis and Related Work

This section provides a brief analysis of the cart-and-rod problem and how it relates to the bongo board. The development of a Fuzzy Logic system for controlling a cart-and-rod inverted pendulum is summarised. Finally we discuss our earlier work in controlling a humanoid robot on a bongo board, including a Proportional-Integral-Derivative (PID) controller-based system by Baltes et al. [1].

2.1 Analysis of the Inverted Pendulum

The inverted pendulum problem is a well-known problem in control theory wherein a mass m is placed at the top of a rigid rod of length l. The other end of the rod is connected to a fulcrum inside a powered cart. The goal of the system is to control the forward velocity of the cart in such a way as to keep the mass and rod upright.

Figure 2 shows the classic cart-and-rod inverted pendulum problem and the forces acting on the system. Gravity, g, pulls down on the mass, applying torque τ at the fulcrum. The cart accelerates at $a(t)$, bringing the fulcrum towards the mass and applying torque to counteract τ.

Many solutions for the inverted pendulum problem have been demonstrated including PID controllers [3], reinforcement learning [4,5], and fuzzy logic [6,7].

2.2 Fuzzy Logic Control for Inverted Pendula

Yamakawa demonstrated that a simple set of fuzzy implications could be used to balance a cart-and-rod inverted pendulum [7]. In his implementation the cart

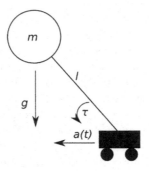

Fig. 2. The cart-and-rod inverted pendulum problem. A mass m is placed on a rod of length l anchored to a fulcrum inside a mobile cart. Gravity g acts on the mass, applying torque τ and pulling the mass down. The cart accelerates at a rate of $a(t)$ to keep the fulcrum positioned below the mass and keep the rod in an upright position.

is powered by a simple electric motor, with the speed $\dot{y}$ determined by the voltage supplied to the motor. His fuzzy rules take the inclination, θ, and angular velocity, $\dot{\theta}$, of the pendulum as inputs and outputs the horizontal velocity of the cart, $\dot{y}$, as shown in Algorithm 1. Yamakawa's implementation defines seven fuzzy input and output sets: Positive Large (PL), Positive Medium (PM), Positive Small (PS), Near-Zero (ZR), Negative Small (NS), Negative Medium (NM), and Negative Large (NL). Rules are not defined for states where the system is highly unstable (e.g. θ and $\dot{\theta}$ are both positive-large). Because the goal of the fuzzy control system is to keep the cart-and-rod in a relatively stable state with θ and $\dot{\theta}$ being small we can make the assumption that the system is working correctly when implementing these rules. If the system is in a relatively unstable state then the control system has already failed its stated purpose. Therefore Yamakawa's rules do not define any behaviour for situations when the rod is severely inclined (e.g. θ is NL) or when the rod is falling quicky (e.g. $\dot{\theta}$ is PL).

Algorithm 1. Yamakawa's fuzzy rules for controlling a cart-and-rod inverted pendulum. (Yamakawa [7])

```
if θ is PM and θ̇ is ZR then        if θ is NM and θ̇ is ZR then        if θ is NS and θ̇ is PS then
    ẏ is PM                              ẏ is NM                              ẏ is ZR
end if                              end if                              end if
if θ is PS and θ̇ is PS then        if θ is NS and θ̇ is NS then        if θ is ZR and θ̇ is ZR then
    ẏ is PS                              ẏ is NS                              ẏ is ZR
end if                              end if                              end if
if θ is PS and θ̇ is NS then
    ẏ is ZR
end if
```

2.3   Analysis of the Bongo Board

The bongo board can be seen as a variation of the cart-and-rod problem, only instead of the cart controlling the motion of the mass above the cart the rider controls the motion of the deck and wheel below them. A bongo board system is shown in Fig. 3, illustrating the major variables at work in the system.

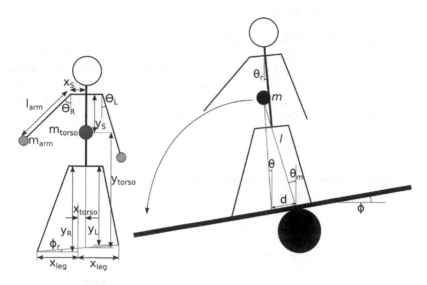

Fig. 3. An idealised bongo board in an unstable position with the rider. The rider's torso is offset horizontally by x_{torso} and vertically by y_{torso} with respect to the midpoint between the robot's feet. The legs have been adjusted to lengths y_R and y_L for the right and left legs respectively. ϕ_r gives the inclination of the rider due to the difference in heights of the legs. The arms have been set to angles θ_R and θ_L with respect to the right and left shoulders. The robot's CoM, m, is offset from the torso because of the arm positions. θ denotes the inclination of m with respect to the line drawn from the midpoint between the robot's feet. θ_m denotes the angle between m and the point of contact between the deck and the wheel. The distance from this point of contact and m is given by l. The deck is inclined by ϕ from the horizontal. (Baltes et al. [1])

The rider is assumed to have five degrees of freedom they can use to exert forces on the system: the angle of each shoulder in the frontal plane (θ_L and θ_R for the left and right shoulders respectively), the lateral and vertical offsets of the torso (x_{torso} and y_{torso} respectively), and the angle of inclination of the torso relative to the deck (ϕ_r). Humans observed balancing on a bongo board tend to rely on lower-spine and hip mobility to control ϕ_r. Many humanoid robots lack this level of torso flexibility, but may independently control the length of each leg by extending or contracting the knee to control ϕ_r [1].

Unlike the cart-and-rod inverted pendulum, the bongo board does not have a fixed fulcrum; the point of contact between the deck and the wheel translates along the deck as the wheel rolls from side-to-side and as the deck rotates around the periphery of the wheel. This means that, in the absence of forces exerted by the rider, two events will occur:

1. the deck will rotate under the effect of gravity, following the same principle as the falling mass in the cart-and-rod problem; and
2. the deck will roll downhill along the wheel, mimicking the effect of a mass sliding down an inclined plane.

Figure 4 shows a simplified bongo board in an unstable position. The passive rider (i.e. a rider exerting no forces on the system) has been replaced with a point-mass positioned at height l above the deck. As gravity pulls the mass down the deck rotates around the wheel with torque τ. This rotation causes lateral force F_α as the falling mass forces the pivot-point to translate. Finally, because the deck is inclined gravity will pull the entire deck-rider assembly downhill with force F_θ. The equations for τ, F_α, and F_θ are given below:

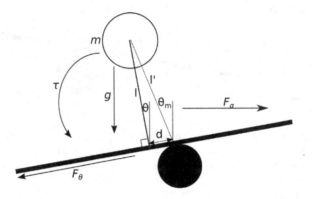

Fig. 4. An idealised bongo board in an unstable position. θ shows the rider's inclination relative to the deck, while θ_m shows the rider's inclination relative to the point of contact between the wheel and the deck. θ_r gives the rider's torso's absolute inclination. The mass, m lies at height l above the deck. As the bongo board falls the mass rotates downward applying force F_α to the system, while the entire mass-rod-deck assembly slides downhill, applying F_θ to the system. l' gives the distance from the point of contact between the wheel and the deck to the rider's CoM.

$$\tau = gml' \sin \theta_m \tag{1}$$
$$F_\alpha = ml(\ddot{\theta}_m \cos \theta_m - \dot{\theta}_m^2 \sin \theta_m) \tag{2}$$
$$F_\theta = mg \cos \theta_m \tag{3}$$

By rotating the arms in the frontal plane, shifting the torso's CoM, and inclining the torso relative to the deck the rider exerts several forces and moments on the system that counteract τ, F_α, and F_θ.

2.4   Control Systems for the Bongo Board

Because of the complex nature of the bongo board relatively few practical implementations capable of balancing the system exist. Anderson and Hodgins [8] demonstrated that it is possible for a small humanoid robot to balance on a deck attached to a fixed pivot point using an adaptive torque-based approach. McGrath et al. [9] showed that active balancing and correction for inclination with a small humanoid robot using gyroscope feedback was possible. Other work in the area of balancing in highly-dynamic environments are primarily theoretical [10,11]. While push-recovery strategies, such as Pratt et al.'s work [12], are useful for traversing unstable terrain, they rely on the ground providing a consistent normal force to arrest the robot's movement. Because the bongo board's deck is unsupported, except for the area in contact with the wheel, foot-placement strategies alone are insufficient for keeping the bongo board system stable.

Baltes et al. [1] demonstrated the first practical implementation of a humanoid robot capable of balancing on a bongo board. Their implementation uses PID controllers to adjust five degrees of freedom: the angle of each shoulder in the frontal plane, the vertical and horizontal positions of the torso's CoM relative to the deck, and the inclination of the robot's torso. Figure 5 shows how the robot's CoM is controlled; by extending or contracting the legs the CoM is moved vertically; by adjusting the lengths of the legs and rotating the hips and ankles in the frontal plane the CoM can be moved horizontally; and by contracting one leg while extending the other the robot can incline its entire torso to the left or right.

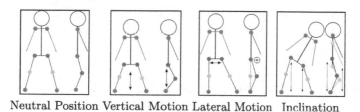

Neutral Position Vertical Motion Lateral Motion Inclination

Fig. 5. Diagrams showing how the rider in the neutral position (extreme left) and how it uses the joints in the legs and hips to raise/lower its CoM (centre-left), shift its CoM side-to-side (centre-right), and incline its torso (extreme right).

Baltes et al. [1] propose three different control policies, dubbed "stiff upper lip," "do the shake," and "let's sway," each using a slightly different set of control laws and motions. The "stiff upper lip" policy was determined to be ineffective, but "do the shake" and "let's sway" were both able to control the bongo board for short periods of time.

Baltes et al.'s control policies use PD controllers to control the robot's arm and torso rotation. In Eqs. 4 and 5 the terms K_{a_p} and K_{a_d} refer to the P- and D-gains used to control the arms, while K_{t_p} and K_{t_d} refer to the P- and D-gains used to control the inclination of the torso. All gains were manually tuned.

The "do the shake" policy consists of two PD controllers that independently control the robot's torso inclination and arm rotation. To compensate for sensor latency a linear predictive sensor model [13] to extrapolate the robot's current state based on previous sensor readings. If the predicted inclination and angular velocity are small then the torso inclination is left alone and the arms are spun to make small corrections. If the predicted inclination or angular velocity is large the arms and torso inclination are both used to correct the robot's motion. The control law used for the "do the shake" policy is given in Eq. 4.

$$
\begin{aligned}
\theta'_{torso} &= predicted(\theta_{torso}, \dot{\theta}_{torso}) \\
d_{arms} &= K_{a_p}\theta'_{torso} + K_{a_d}\dot{\theta}'_{torso} \\
\theta_{torso} &= K_{t_p}\theta'_{torso} + K_{t_d}\dot{\theta}'_{torso}
\end{aligned}
\tag{4}
$$

The "let's sway" policy was largely identical to "do the shake" save for the addition of a regular oscillation to the robot's motion; the legs continuously pump up and down in an attempt to induce a state of dynamic stability to the system. The control law for the "let's sway" policy is given in Eq. 5.

$$
\begin{aligned}
\theta'_{torso} &= predicted(\theta_{torso}, \dot{\theta}_{torso}) \\
\theta_{desired} &= \sin(\omega t) \\
d_{arms} &= K_{a_p}(\theta'_{torso} - \theta_{desired}) + K_{a_d}(\dot{\theta}'_{torso} - \dot{\theta}_{desired}) \\
\theta_{torso} &= K_{t_p}(\theta_{torso} - \theta_{desired})' + K_{t_d}(\dot{\theta}'_{torso} - \dot{\theta}_{desired})
\end{aligned}
\tag{5}
$$

Based on the angular velocity and inclination of the robot's torso as recorded during several trials using all three control policies and through qualitative observations Baltes et al. concluded that the introduction of a regular oscillation from the "let's sway" policy improved the robot's overall stability, albeit not in a statistically significant way.

3   Fuzzy Rules for the Bongo Board

Based on Yamakawa's rules for the cart-and-rod inverted pendulum and Baltes et al.'s PID-based system for controlling the motion of a bongo board we define a set of rules to control the five degrees of freedom that the bongo board's rider uses:

- θ, the angle from the deck to the rider's centre of mass;
- x_{torso}, the lateral offset of the rider's torso;
- y_{torso}, the vertical offset of the rider's torso;
- θ_L, the angle of the rider's left shoulder; and
- θ_R, the angle of the rider's right shoulder.

As with Yamakawa's rules, we define our rules such that the rate of change of each degree of freedom is output. For simplicity we control the angular velocity of both arms simultaneously. Therefore the fuzzy rules specify the angular velocity of both arms as a single output. The fuzzy rules defined for balancing the bongo board are shown in Algorithm 2, using the same abbreviations described in Sect. 2.2.

The fuzzy rules use the inclination and angular velocity of the bongo board's deck, ϕ and $\dot{\phi}$ respectively, as inputs. Given knowledge of the robot's torso's inclination and angular velocity and the current positions of the joints in robot's legs the inclination and angular velocity of the deck can be trivially calculated.

The fuzzy rules output the desired angular velocity of the arms ($\dot{\theta}_{arm}$) and the torso ($\dot{\theta}$) as well as the linear velocities of the rider's CoM along the x-axis (parallel to the deck of the bongo board) and y-axis (perpendicular to the deck of the bongo board).

We set the PS, NS, PM, and NM thresholds for ϕ and $\dot{\phi}$ based on the data recorded during our initial research with PID controllers [1]. The thresholds for the output variables ($\dot{\theta}_{arm}$, $\dot{x}$, $\dot{y}$, and $\dot{\theta}$) were determined experimentally by placing the robot on the bongo board and observing its behaviour, increasing or decreasing each threshold based on the robot's performance and the sensor data recorded. In all cases the ZR threshold was left at zero.

Algorithm 2. Linguistic rules for balancing a bongo board.

```
if φ is ZR and φ̇ is ZR then        if φ is PS and φ̇ is PS then        if φ is PM and φ̇ is ZR then
    θ_arm is ZR                         θ_arm is NS                         θ_arm is NM
    ẋ is ZR                             ẋ is ZR                             ẋ is NS
    ẏ is ZR                             y is ZR                             ẏ is NS
    θ̇ is ZR                             phi is NS                           θ̇ is NM
end if                              end if                              end if
if φ is NS and φ̇ is PS then         if φ is NS and φ̇ is NS then         if φ is NM and φ̇ is ZR then
    θ_arm is ZR                         θ_arm is PS                         θ_arm is PM
    ẋ is ZR                             ẋ is ZR                             ẋ is PS
    ẏ is ZR                             ẏ is ZR                             ẏ is PS
    θ̇ is ZR                             θ̇ is PS                             θ̇ is PM
end if                              end if                              end if
if φ is PS and φ̇ is NS then
    θ_arm is ZR
    ẋ is ZR
    ẏ is ZR
    θ̇ is ZR
end if
```

These rules specify that when the deck is inclined one direction, but the angular velocity is in the opposite direction the system should produce near-zero outputs; when the bongo board is either in a stable position, or is self-stabilizing (i.e. rotating in such a way that the deck becomes more level) the system should allow this process to continue uninterrupted. The vertical and horizontal torso offsets are only used when the system is highly unstable; these DOFs are used only as a last-resort to stabilise the system when arm motion and torso inclination prove insufficient.

4   Evaluation

To evaluate the performance of the fuzzy logic controller compared to Baltes et al.'s PID-based system we perform five 30-second trials using the PID and fuzzy logic controllers using both the "let's sway" and "do the shake" control policies. Due to the low success rate of the "stiff upper lip" policy reported by Baltes et al. [1] we omit this control policy from our experiment. These trials

Table 1. Minimum, maximum, and average deck inclination and difference in inclination from the previous tick. Measurements are in ° and °/tick (1 tick ≈ 8 ms).

Do the shake policy								
Control algorithm	Min ϕ	Max ϕ	Average ϕ	σ	Min $\dot\phi$	Max $\dot\phi$	Average $\dot\phi$	σ
PID	−41.7816	46.9645	−2.6833	9.1060	−22.7700	28.4904	1.917×10^{-3}	1.5709
Fuzzy logic	−34.2622	42.1050	−0.7879	7.4964	−24.0847	22.7207	1.629×10^{-15}	1.6288
	Min θ	Max θ	Average θ	σ	Min $\dot\theta$	Max $\dot\theta$	Average $\dot\theta$	σ
PID	−42.5811	46.9645	−2.6620	9.1182	−24.3399	28.4904	−0.0010	1.6141
Fuzzy logic	−34.2668	42.1050	−0.8253	7.6646	−23.7593	23.2903	−0.0022	1.5821
Let's sway policy								
Control algorithm	Min ϕ	Max ϕ	Average ϕ	σ	Min $\dot\phi$	Max $\dot\phi$	Average $\dot\phi$	σ
PID	−44.6387	41.6949	−2.8597	8.9129	−67.7595	68.3437	4.125×10^{-15}	2.1396
Fuzzy logic	−36.7589	43.0499	−1.2893	9.4946	−20.0433	21.6206	-3.332×10^{-14}	2.5342
	Min θ	Max θ	Average θ	σ	Min $\dot\theta$	Max $\dot\theta$	Average $\dot\theta$	σ
PID	−44.3100	42.4191	−2.8706	8.8555	−68.0614	68.0614	−0.0010	2.1837
Fuzzy logic	−36.9610	44.4448	−1.4046	10.3874	−21.8892	19.1989	−0.0004	2.4143

are performed using a physical robot standing on a wooden bongo board, as shown in Fig. 1. During the trials the robot operates on battery power with no external cables (e.g. power, ethernet) connected. A human operator is present to reset the apparatus should the bongo board fall and the robot is unable to autonomously right the board.

During each trial the robot records the angular velocity and inclination of the torso, as well as the positions of each joint. From these data we can calculate the angular velocity and inclination of the bongo board's deck.

Table 1 shows the angular velocity and inclination of the robot's torso ($\theta, \dot\theta$) and of the deck ($\phi, \dot\phi$) across all trials.

4.1 Comparison of Control Policies

Both the "do the shake" and "let's sway" policies exhibited similar performance regardless of the control algorithm used. The "do the shake" policy performed best when used with the fuzzy logic controller, while the "let's sway" policy performed better with the PID controller. The differences in performance were slight and are not statistically significant.

As observed by Baltes et al., the "let's Sway" policy did successfully maintain a lower average angular velocity in the robot's torso [1]. The difference was very small; the difference in average torso velocities 4.00597×10^{-5} deg./s. The "let's sway" policy does not appear to offer any significant benefits to balancing, but is not detrimental either.

The "do the shake" policy used with the fuzzy logic controller maintained the lowest average deck inclination with the smallest standard deviation of all experiments. Additionally, this combination of control algorithm and control policy had the lowest recorded average $\dot\phi$. These findings are mirrored in the average torso inclination and angular velocity, indicating that the "do the shake" policy may be preferable over the "let's sway" policy.

4.2   Comparison of Control Algorithms

Both the fuzzy logic and PID controllers were successfully able to control the motion of the bongo board and maintain a stable position for short periods of time. Qualitatively the fuzzy logic controller appeared to maintain a stable position for longer continuous periods of time, but was unable to automatically recover; when using the PID controller if either end of the deck struck the ground the robot would react very strongly, autonomously bouncing the deck back to a horizontal position. The fuzzy logic controller did not exhibit this self-recovery property.

The fuzzy logic controller's inability to self-recover after the deck struck the ground is primarily attributable to the fact that the rules are written with the assumption that the system is relatively stable and ϕ and $\dot{\phi}$ are small. When the deck strikes the ground the system undergoes extreme deceleration, sometimes exceeding $1g$. This large change in velocity requires a correspondingly large output, which is unaccounted for in the fuzzy rule-set. The PID controller in contrast has no strict upper bound on the magnitude of its output; the large deceleration due to the deck-strike is passed directly into the PID controller, which in turn produces a very strong response as its output. The introduction of additional rules to the fuzzy rule-set to specifically address the large changes in velocity experienced during a deck-strike may allow the fuzzy logic controller to self-recover in a similar fashion as the PID controller.

The fuzzy logic controller, when used with the "do the shake" policy maintained the lowest average deck inclination with the lowest standard deviation, indicating that overall the fuzzy logic controller was slightly more stable than the PID controller. This improvement is not statistically significant, but does indicate that, like the cart-and-rod inverted pendulum, the bongo board can be controlled by both PID control and fuzzy logic.

Fig. 6. A humanoid robot equipped with skis demonstrating alpine skiing. We use the "do the shake" policy to control the robot's pitch and roll while skiing. (Winnipeg Free Press [16])

5   Conclusions and Future Work

This research has shown that two well-known solutions to the inverted pendulum problem – PID control and fuzzy logic – are both well-suited to the bongo board. Furthermore we have shown that the introduction of a rhythmic oscillation to the bongo board system, while not detrimental, has little benefit to the stability of the bongo board.

This research has numerous possible applications to humanoid robotics. Active balancing on unstable terrain will be an essential skill for humanoid robots to be useful in arbitrary environments. Examples of such environments include loose rubble, which may suddenly slip or give-way underfoot; ice or wet linoleum, which offers minimal traction; and deep carpet or foam, which compresses underfoot and provides uneven support.

Iverach-Brereton et al. demonstrated a simple shuffling gait for ice skating [14,15], but could not sustain a glide phase due in part to the difficulty in balancing on a single skate blade. The use a PID or fuzzy logic controller to control the robot's lateral balance may allow the robot to balance for longer periods on a single skate, allowing for a more sustained glide phase.

The bongo board is fundamentally, like balancing on a skate blade, is a largely two-dimensional problem; the skate blade, like the bongo board, provides adequate support for the robot to remain stable along the front-back axis. Balancing along the left-right axis only requires control over movement in the frontal plane. The wobble board – an apparatus consisting of a circular deck and a free-rolling, spherical fulcrum below it – conversely requires control over translation and rotation in the frontal, sagittal, and transversal planes due to the spherical motion allowed by the fulcrum. Implementing a solution for the wobble board remains part of our ongoing research.

We have recently begun research into alpine skiing using a humanoid robot, shown in Fig. 6. Balancing on skis while going downhill requires control over rotation and translation in the frontal and sagittal planes, but does not require control over rotation in the transversal plane; the length of the skis prevents the robot from twisting in an uncontrolled fashion. To ensure that the robot's skis remain in contact with the hill regardless of inclination while simultaneously keeping the robot's torso vertical we implemented a controller using the "do the shake" policy to control the robot's motions in the frontal and sagittal planes. A video demonstrating this automatic correction for slope can be seen here: https://youtu.be/XU17sbItYxI.

Our work on alpine skiing demonstrates that the bongo board solutions presented here are applicable to more practical problems of balancing on varied terrain, as well as balancing in higher-dimensional problems than the more traditional bongo board.

References

1. Baltes, J., Iverach-Brereton, C., Anderson, J.: Human inspired control of a small humanoid robot in highly dynamic environments or Jimmy Darwin Rocks the bongo board. In: Bianchi, R.A.C., Akin, H., Ramamoorthy, S., Sugiura, K. (eds.) RoboCup 2014. LNCS, vol. 8992, pp. 466–477. Springer, Heidelberg (2015)
2. Baltes, J., Tu, K.Y., Lip, S.L.: HuroCup Competition. FIRA, September 2013
3. Wang, J.J.: Simulation studies of inverted pendulum based on PID controllers. Simul. Model. Pract. Theory **19**(1), 440–449 (2011). Modeling and Performance Analysis of Networking and Collaborative Systems
4. Harmon, M.E., Harmon, S.S.: Reinforcement Learning: A Tutorial. WL/AAFC, WPAFB Ohio, vol. 45433 (1996)
5. Hehn, M., D'Andrea, R.: A flying inverted pendulum. In: 2011 IEEE International Conference on Robotics and Automation (ICRA), pp. 763–770. IEEE (2011)
6. Wang, L.X.: Stable adaptive fuzzy controllers with application to inverted pendulum tracking. IEEE Trans. Syst. Man Cybern. Part B Cybern. **26**(5), 677–691 (1996)
7. Yamakawa, T.: Stabilization of an inverted pendulum by a high-speed fuzzy logic controller hardware system. Fuzzy Sets Syst. **32**(2), 161–180 (1989)
8. Anderson, S., Hodgins, J.: Adaptive torque-based control of a humanoid robot on an unstable platform. In: 2010 10th IEEE-RAS International Conference on Humanoid Robots (Humanoids), pp. 511–517 (2010)
9. McGrath, S., Anderson, J., Baltes, J.: Model-free active balancing for humanoid robots. In: Iocchi, L., Matsubara, H., Weitzenfeld, A., Zhou, C. (eds.) RoboCup 2008. LNCS, vol. 5399, pp. 544–555. Springer, Heidelberg (2009)
10. Park, J., Haan, J., Park, F.: Convex optimization algorithms for active balancing of humanoid robots. IEEE Trans. Robot. **23**(4), 817–822 (2007)
11. Hyon, S., Cheng, G.: Gravity compensation and full-body balancing for humanoid robots. In: 2006 6th IEEE-RAS International Conference on Humanoid Robots, pp. 214–221 (2006)
12. Pratt, J., Carff, J., Drakunov, S., Goswami, A.: Capture point: a step toward humanoid push recovery. In: 2006 6th IEEE-RAS International Conference on Humanoid Robots, pp. 200–207 (2006)
13. Baltes, J., Iverach-Brereton, C., Anderson, J.: Sensor filtering for balancing of humanoid robots in highly dynamic environments. In: 2013 CACS International Automatic Control Conference (CACS), pp. 170–173, December 2013
14. Iverach-Brereton, C., Winton, A., Baltes, J.: Ice skating humanoid robot. In: Herrmann, G., Studley, M., Pearson, M., Conn, A., Melhuish, C., Witkowski, M., Kim, J.-H., Vadakkepat, P. (eds.) TAROS-FIRA 2012. LNCS, vol. 7429, pp. 209–219. Springer, Heidelberg (2012)
15. Iverach-Brereton, C., Baltes, J., Anderson, J., Winton, A., Carrier, D.: Gait design for an ice skating humanoid robot. Robot. Auton. Syst. **62**(3), 306–318 (2012)
16. Martin, N.: Robot Goes From Hockey Skates to Skis. Winnipeg Free Press, Winnipeg (2015 A2)

Towards an Architecture Combining Grounding and Planning for Human-Robot Interaction

Dongcai Lu and Xiaoping Chen$^{(\boxtimes)}$

Multi-Agent Systems Lab, Department of Computer Science and Technology,
University of Science and Technology of China, Hefei 230027, China
ludc@mail.ustc.edu.cn, xpchen@ustc.edu.cn

Abstract. We consider here the problem of connecting natural language to the physical world for robotic object manipulation. This problem needs to be solved in robotic reasoning systems so that the robot can act in the real world. In this paper, we propose an architecture that combines grounding and planning to enable robots to solve such a problem. The grounding system of the architecture grounds the meaning of a natural language sentence in physical environment perceived by the robot's sensors and generates a knowledge base of the physical environment. Then the planning system utilizes the knowledge base to infer a plan for object manipulation, which can be effectively generated by an Answer Set Programming (ASP) planner. We evaluate the overall architecture on several datasets and a task of RoboCup2014@home (http://www.robocup2014.org/). The results show that the new architecture outperformed some other systems, and yielded acceptable performance in a real-world scenario.

1 Introduction

Natural language is an intuitive and flexible modality for human-robot interaction. Robots are expected to interact naturally with humans and automatically complete complex tasks in the real world. For instance, when a user tells a robot, "I want the drink which is to the left of a food", the robot needs to understand the meaning of such a natural language sentence or the query of "the drink to the left of a food", and fetch the target object to the user. The previous architecture proposed in [3] attempted to integrate natural language process and task planning for solving the user's task like "give me a coke", such a task had clearly pointed out that the action "give", the human "me", and the object "coke". The architecture didn't consider the problem of understanding natural language queries like "the drink to the left of a food". So, it cannot solve the task "give me the drink to the left of a food". To do this, it must combine the natural language process and the perception knowledge base to ground the meaning of a natural language sentence.

In this paper, we present a novel architecture that combines grounding and planning to enable robots to reason and act in the real world. We consider two requirements in this effort: (i) A robot allows humans to ask natural language

© Springer International Publishing Switzerland 2015
L. Almeida et al. (Eds.): RoboCup 2015, LNAI 9513, pp. 214–225, 2015.
DOI: 10.1007/978-3-319-29339-4_18

queries about what objects they want using spatial relations statement, we call this as *grounding* problem-mapping natural language sentence to their referents in a physical environment. (ii) A robot should be capable of getting access to the target object without knocking down other objects near the target object.

To meet these requirements, the main ideas are sketched below. Firstly, the grounding system is divided into two subsystems: Natural language processing and visual perception processing. Acquiring both kinds of knowledge is necessary to ground natural language sentence in the real world. Secondly, a planning system is proposed to infer the best sequence of actions for the robot to get access to the target object.

This work consists of two contributions. The first contribution is the grounding system, which combines the general-purpose semantic parsing technology and new visual perception processing. Compared to previous work [3], the grounding system is added to our architecture that can understand the meaning of a natural language query instead of structured command. The second contribution is that our architecture introduces the *manipulation planning* for operating objects, first combining linguistic and visual perception knowledge base to generate a plan for manipulating objects, this is more efficient than the work in [4] when the robot intends to get an object which is blocked by other objects.

Recently, the availability of robotic agents has opened new perspectives in language acquisition and grounding, and there are many works on grounding. Our work is similar to Sergio [4], but we step further by introducing a ASP language to fetch the objects that is blocked by other items and propose a different approach to producing grounding. Sergio [4] propose a system for human-robot interaction that learns both models for spatial prepositions and for object recognition. It allows the robot to understand complex commands that refer to multiple objects and relations. Cynthia [6] present an approach for joint learning of language and perception models for grounded attribute induction. The perception model includes classifiers for physical characteristics and a language model based on a probabilistic categorial grammar that enables the construction of compositional meaning representations. There also have some works on grounding natural language directions [5] and Commands [11]. Kollar [5] present a system that follows natural language directions by extracting a sequence of spatial description clauses (SDC) from the linguistic input and then infers the most probable path through the environment given only information about the environmental geometry and detected visible objects. Tellex [11] introduces a new model G^3 for understanding natural language commands given to autonomous systems that perform navigation and mobile manipulation in semi-structured environments.

In Sect. 2, the hardware system and overall software architecture are presented. The implementing techniques for two main modules of the robot, grounding and planning, are addressed in Sects. 3 and 4, respectively. Experimental results are reported in Sect. 5, and conclusions are given in Sect. 6.

2   Overall Architecture

System Overview. The hardware framework of KeJia is shown in Fig. 1(a). Its sensors include a laser range finder, a 1394 camera and a kinect. The robot has an arm for manipulating portable items. The computational resources consist of a laptop and an on-board PC. It is worthwhile emphasizing that neither additional computational resources off-board nor remote control is needed for the robot when it performs its tasks. This means all the computation is carried out on-board. Similar to RHINO [2] and STAIR [7], distributed and asynchronous processing are adopted, with no centralized clock or a centralized communication module in our robot's system.

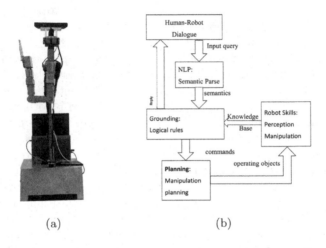

(a)   (b)

Fig. 1. In (a) is the hardware framework of KeJia. In (b) is overall architecture.

We have created an integrated architecture for our robot show in Fig. 1(b). The overall architecture in our robot is managed by human-robot dialogue, which contains speech recognition, dialogue management. The query from the dialog is extracted, classified and transferred into the NLP module. The NLP module learns a semantic parser used to map a natural language sentence to λ-calculus logical forms (Fig. 2(b)). In this part, natural language sentence contains nouns and preposition that correspond to the objects and spatial relations in perception knowledge base. The logical forms represent the semantics of a sentence. The grounding module then takes semantics and the perception knowledge base as inputs to ground the meaning of the input query, the perception knowledge base is extracted from visual perception processing in the robot perception skills module. A sample knowledge base is shown in Table 1. The grounding module uses some logical rules contain only conjunctions and existential quantifiers to infer a query's grounding, replies to the dialog or transfers the grounding into planning module depends on the number of this grounding.

Knowledge generated from the grounding module will be applied to the planning module for generating high-level plans. It is utilized to operate objects for getting the grounding object to the user without knocking down other objects in the scene.

3 Grounding

To solve the *grounding* problem, our grounding system is divided into two sub-problems: *semantic parsing*, learning the deep semantics of a natural language query; and *visual perception processing*, learning to recognize objects and extracting a knowledge base contains nouns and relations. Acquiring both kinds of knowledge is necessary to understand novel language in unexplored environments. For example, given an image (Fig. 2(a)) coming from the camera, and a natural language query, such as "the drink to the right of a food". Our grounding system first achieves a knowledge base (Table 1) from *visual perception processing*, then parses the natural language query into internal logical forms (Fig. 2(b)) from *semantic parsing* process. At last, the grounding system combines these results with logical rules which contain only conjunctions and existential quantifiers to get the query's grounding (Fig. 2(c)).

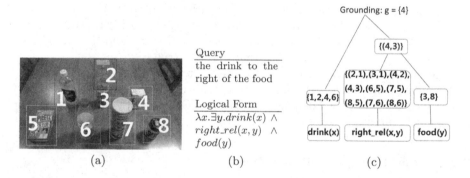

Fig. 2. Overview of Grounding system, (a): An environment containing 8 objects (image segments); (b): Semantic parsing for natural language query; (c): achieve grounding from combining knowledge base and semantics

3.1 Semantic Parser

The Semantic Parser is utilized to map natural language query to internal logical form that the robot can handle. In this paper, our Semantic Parser is based on CCG [10]. CCG is a linguistic formalism that provides a tight interface between natural language syntax and its semantics and has been used to model a wide range of language phenomena. The core of any CCG is the semantic lexicon. In

Table 1. Environment: example knowledge base

Object	Categories	Relations
1 (amsa)	drink, bottle	behind:{5,6}, left:{2,3,4}
2 (ice tea)	drink, box	behind:{6,7}, left:{4}, above:{3}, right:{1}
3 (porridge)	food, can	behind:{6,7}, left:{4}, below:{2}, right:{1}
4 (acid milk)	drink, box	behind:{7,8}, right:{2,3,1}
5 (pretz)	snack, box	in_front_of:{1}, left:{6,7,8}
6 (water)	drink, bottle	in_front_of:{1,2,3}, left:{7,8}, right:{5}
7 (barbecue chip)	snack, can	in_front_of:{2,3,4}, left:{8}, right:{5,6}
8 (mushroom sauce)	food, can	in_front_of:{4}, right:{5,6,7}

the lexicon, a word with its syntactic category and semantic form is defined. For example, the lexicon for our Semantic Parser would be as follows:

$$drink := N : \lambda x.drink(x)$$
$$food := N : \lambda x.food(x)$$
$$left := N/PP : \lambda f.\lambda x.\exists y.left\_rel(x,y) \wedge f(y)$$

The syntactic categories in CCG could be either *primitive* (e.g., NP or S), or *complex* (of the form A/B or $A\backslash B$). In the above lexicon *drink* has the syntactic category N which stands for the linguistic notions of noun, and the logical form denotes the set of all entities x such that *drink* is true.

A CCG also has a set of combinatory rules to combine syntactic categories and semantic forms which are adjacent in a string. Figure 2(b) shows a query z and its logical form l. For more details on CCG parsing, the reader should consult [12].

3.2   Visual Perception Processing

Object Recognition. Our object recognition module is applied to recognize what exactly the name of an object is. Sensors of our vision system consist of a Microsoft kinect and a high-resolution 1394 RGB camera from *PointGrey*. With the pre-calibrated intrinsic and extrinsic camera parameters, we obtain an aligned RGB-D image by combining the RGB image from 1394 camera with the depth image from Kinect. With such an aligned RGB-D image, our vision module is capable of detecting, tracking people and recognizing different kinds of objects.

We follow the approach as proposed in [9] to detect and localize table-top objects including bottles, cups, etc. The depth image is first transformed and segmented, then the largest horizontal plane is extracted using Point Cloud Library (PCL) [8], and point clouds above it are clustered into different pieces. After that the SURF feature matching against the stored features are applied to each piece [1]. The one with the highest match above a certain threshold is considered as a recognition. At last, to further enhance the detection performance and decrease FP (the number of the instances recognized as error) rate, we check each recognized cluster and filter out those vary too much in size.

Spatial Relations Module. Given a preposition and landmark object, the spatial relations module first outputs a distribution over the target objects and 3D points coming from Kinect Sensor that are located in the given preposition in relation to the given landmark object from the robot's point of view, then constructs a logical knowledge base τ given object instances and spatial relations.

Our system has used spatial relations: {above, below, in_front_of, behind, to_the_left_of, to_the_right_of}. The meaning of these spatial relations are modeled using a probabilistic distribution to predict the identity of a target object (or 3D point) g conditioned on a preposition w, and landmark object o. To obtain the spatial relations between a target object and a landmark object (except itself), the module computes the maximum probability on each relation: $\operatorname{argmax}_w P(g|w;o)$. Our system can get a 3D point from Kinect and compute its center pose as the pose of landmark object in the view of the robot, such as (x, y, z). So, the probabilistic of this spatial relation is:

$$\operatorname*{argmax}_{w} P(g|w; o) = \operatorname*{argmax}_{w} P((x, y, z)|w; (x', y', z')) \tag{1}$$

In the Eq. 1, (x, y, z) is the 3D pose of object w, (x', y', z') is the 3D pose of object o. We assume that v is a six-vector representing {in_front_of, behind, to_the_left_of, to_the_right_of, below, above}. For example, $v = (1, 0, 0, 0, 0, 0)$ is represented the $w = in\_front\_of$, we also assume that $\overrightarrow{diff} = ((x' - x), (x - x'), (y' - y), (y - y'), (z' - z), (z - z'))$. Now we can calculate the right side of the Eq. 1:

$$\operatorname*{argmax}_{w} P((x, y, z)|w; (x', y', z')) = \operatorname*{argmax}_{w} v * \overrightarrow{diff} \tag{2}$$

After the module obtains all spatial relations between landmark objects, then a logical knowledge base τ is constructed. The knowledge base produced by the perception module is a collection of ground predicate instances and spatial relations (see Table 1).

3.3   Evaluation

The evaluation determines groundings given a logical form l and a logical knowledge base τ. Intuitively, the evaluation simply evaluates the query l on the database τ to produce a grounding. We describe an evaluation by giving a recurrence for computing the grounding g of a logical form l on a logical knowledge base τ. This evaluation takes the form of a tree, as in Fig. 2(c). The basic logical rules are:

$$if \ \ l = \lambda x.c(x) \ \ then \ g = g^c$$
$$if \ \ l = \lambda x.\lambda y.r(x, y) \ \ then \ g = g^r$$

The groundings for more complex logical forms are computed recursively by decomposing l according to its logical structure. Our logical forms contain only conjunctions and existential quantifiers; the corresponding recursive computations are:

$$if \ l = \lambda x.l_1(x) \wedge l_2(x), \ then$$
$$g(e) = 1 \ iff \ g_1(e) = 1 \wedge g_2(e) = 1$$
$$if \ l = \lambda x.\exists y.l_1(x, y), \ then$$
$$g(e_1) = 1 \ iff \ \exists e_2.g_1(e_1, e(2)) = 1$$

4   Manipulation Planning

It is always possible to encounter an expected scenario which has not been covered by a certain theory for an action domain. In this work, we propose an approach to treating this problem. For example, how does the robot get the object 3 from the scene in Fig. 2(a). Here, We introduce the Answer Set Programming (ASP) language into the problem.

In this section, we will illustrate how the planning module automatically generates a sequence of grasping actions to get an object which is blocked by other objects using ASP rules. We focus on the robot's ability of *grasp* and the corresponding properties of the environment. The action name and fluent names used in the specification are following, where X and Y are variables ranging over possible objects in the environment:

- *grasp(X)*: the action of gripping the object X and picking it up
- *holding(X)*: the fluent that the object X is held in the grip of the robot
- *above(X, Y)*: the fluent that the object X is above the object Y
- *behind(X, Y)*: the fluent that the object X is behind the object Y

The effect of executing $grasp(X)$ is $holding(X)$ and described by the following ASP-rules:

$$h(holding(X), t + 1) \leftarrow occurs(grasp(X), t).$$
$$\neg h(above(X, Y), t + 1) \leftarrow occurs(grasp(X), t),$$
$$true(above(X, Y), t).$$

Also, there have three preconditions for action *grasp(X)* is described in ASP as a constraint. The first precondition of *grasp(X)* is *not holding(Y)* for any Y, The second precondition of *grasp(X)* is there have no object Y above object X for any Y, The third precondition of *grasp(X)* is there have no object Y in front of object X for any Y:

$$\leftarrow occurs(grasp(X), t), h(holding(Y), t - 1).$$
$$\leftarrow occurs(grasp(X), t), h(above(Y, X), t - 1).$$
$$\leftarrow occurs(grasp(X), t), h(behind(X, Y), t - 1).$$

The *frame problem* is resolved by "inertia laws" of the form:

$$h(above(X,Y),t) \leftarrow h(above(X,Y),t-1),$$
$$not\ occurs(grasp(X),t).$$
$$h(behind(X,Y),t) \leftarrow h(behind(X,Y),t-1),$$
$$not\ occurs(grasp(Y),t).$$

Now, with the knowledge extracted from the grounding module and ASP rules, the user's task can be solved by running them on an ASP solver *iclingo*. We use symbol G to denote the grounding of a query, Γ to denote the knowledge base, Π to denote the theory of the action domain of the robot. So, we could create a manipulation planning $\Delta = (\Pi, \Gamma, G)$.

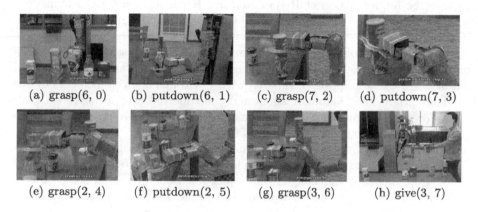

(a) grasp(6, 0)   (b) putdown(6, 1)   (c) grasp(7, 2)   (d) putdown(7, 3)

(e) grasp(2, 4)   (f) putdown(2, 5)   (g) grasp(3, 6)   (h) give(3, 7)

Fig. 3. Excute a plan for fetching the target object 3

For our robot, Π was described in this section. In scene 2(a), the grounding of the query "i want the food to the left of a drink" is object 3; it is called porridge and represented as $G = 3$. So, we give a manipulation planning task of $grasp(3)$, it can be converted to the ASP for solving the goal of grasp object 3:

$$\leftarrow not\ occurs(grasp(3),t).$$

The knowledge base Γ in the scene will be set as the fluent or initial state for the action domains of the robot. Some of the Γ are (see Table 1): $food(X)$, $can(X)$, $drink(Y)$, $bottle(Y)$, $left\_rel(X,Y)$, $right\_rel(Y,X)$, $above\_rel(X,Y)$. X and Y are the index of the object.

Times names of Π are 0,...,n. In practice, the last name n is computed by ASP solver *iclingo* automatically. In the example, we have obtained the Π, Γ, G, a computed plan of Δ is:

$$grasp(6,0), putdown(6,1), grasp(7,2), putdown(7,3),$$
$$grasp(2,4), putdown(2,5), grasp(3,6).$$

After all actions are executed to accomplish the task $grasp(3)$, the robot then gives the object to the user. The execution of the actions shown in Fig. 3.

5 Experiments

The experiments were designed to investigate the performance of our overall architecture for robot agents. In this section, we concentrate on the results of the groundings of objects and spatial relations. Moreover, we evaluate the overall architecture on three major datasets.

5.1 Data Sets

We evaluate our architecture on three datasets: RoboCup@home object sets (RCO), lab object sets (WEO), and grounding sets (WEG). Although there are many public data sets for object recognition, they are not suitable for our system. Because the inputs of Object Recognition are HR images and RGB-D images. So, we use data sets coming from RoboCup@home and our robot's platform (shown in Fig. 2(a)).

The RCO Sets contains 25 objects and 794 learning instances. WEO Sets contains 25 objects, 597 learning instances and 107 test instances. We collected spatial language descriptions compared with environment from members of our lab. The authors then manually annotated the collected descriptions with their logical forms and groundings. We call this datasets as WEG. Example environments and descriptions are shown in Table 2.

At last, three members of our laboratory tesed the overall architecture constructing of 30 dialogs in 30 scenes, each members 10 dialogs happened in 10 scenes. It means that we changed the object locations or moved some objects from a scene in each human-robot dialog. Example dialog is shown in Table 3.

Table 2. Example environments, language query, and grounding from WEG

Environment d	Query z and logical forms l	True grounding
	objects on the table $\lambda x.\exists y.object(x) \wedge on\_rel(x,y) \wedge table(y)$	{1,2,3,4,5,6,7,8}
	the drink to the left of the food $\lambda x.\exists y.drink(x) \wedge left\_rel(x,y) \wedge food(y)$	{1,6}
	the can below a box $\lambda x.\exists y.can(x) \wedge below\_rel(x,y) \wedge box(y)$	{3}
	the highest box $\lambda x.highest(x) \wedge box(x)$	{2}

5.2 Results

We use Leave One Out Cross validation (LOOCV) to evaluate our object recognition of visual perception processing, give precision and recall for the performance of the visual recognition system, defined as $Precision = TP/(TP + FP)$, $Recall = TP/P$. TP is the number of the instances recognized correctly, FP is the number of the instances recognized as error. P is the total number of objects.

Table 3. Example human-robot dialog from one user.

Example dialog	Grounding and planning
U: Tell me what objects are on the table?	Grounding: {3}. Manipulation
R: there are eight objects on the table. Which one do	Planning: {
you want?	$grasp(6, 0), putdown(6, 1),$
U: I want food	$grasp(7, 2), putdown(7, 3),$
R: Ok, there are two kinds of food on the table	$grasp(5, 4), putdown(5, 5),$
U: The food is below a box	$grasp(1, 6), give(1, 7).$}
R: Ok, you want the porridge. I will fetch it for you	

Table 4. The results of cross-validation for spatial relation module

Object recognition	TP	FP	Total	P	R	F1
RCO	757	31	794	0.96	0.95	0.95
WEO	570	5	597	0.99	0.95	0.97
Spatial relations	TP	FP	Total	P	R	F1
WEG	96	2	98	0.98	0.98	0.98

Table 5. The results of performance for overall architecture

Grounding	P	R	F1
WEG	0.95	0.81	0.87
Architecture	Total groups	Correct groups	Correct/total queries
User 1	10	8	23/26
User 2	10	9	26/28
User 3	10	8	21/25

For the relations evaluation, we use the annotated logical forms containing only one relation, of the form $\lambda x.\exists y.c_1(x) \wedge r(x, y) \wedge c_2(y)$. Using these logical forms, we measure the performance of r on the set of x, y pairs, then summarize these over all examples. Table 4 shows that two components of our visual perception processing system have a very high *Precision* on three data sets, it is important for our overall architecture to realize this work.

In this evaluation, we concentrate on the grounding system and the whole architecture. The experimental results are shown in Table 5. The performance on the grounding system has a precision of 0.95, due to the datasets are not large. However, it has a lower recall because of the sparse training data. On the Architecture test, three users of our lab use natural language statement containing spatial relations and object categories to communicate with the robot, each user makes 10 dialogues in a different environment. In Table 5, "Total Queries" means how many queries used by each user, "Correct Queries" means how many queries are produced correct grounding by the system. The result shows that the architecture displayed a rate of an average 0.83 correct performance for robotic object manipulation task when a person only uses spatial queries.

5.3   RoboCup@home Task

To compare our overall architecture (Fig. 1(b)) to others, this work has been demonstrated on the *Open Challenge* test in RoboCup2014@home, all of the teams are encouraged to demonstrate recent research results and the best of the robots' abilities, it focuses on the demonstration of new approaches/applications, human-robot interaction and scientific value. Table 6 shows the results of the top 5 teams, the first line in the table represents the name of a team; the second line represents the score that a team got on the *Open Challenge* test, the total score of the test is 2000. Also, our robot performed the demo described in this paper, showing it able to not only recognize objects, but also understand location relations and reason with such knowledge.

Table 6. The results of our overall architecture on Open Challenge test

Team	WrightEalge	NimbRo	TUe	ToBI	UChile
score	1507	1400	1340	1087	807

6   Discussion and Conclusions

As robots were moving toward more complex tasks and environments, it is necessary to create autonomous and intelligent systems that can reason and act in the real world. It will take a long time and massive efforts to achieve the goal. In this paper, we propose a different architecture from [3] that combines grounding and planning for grounding natural language sentence and acting in the real world. Our results in Table 5 and RoboCup@home task show that the overall architecture has proven general enough to yield acceptable performance in a real-world scenario, and the integration of grounding and planning is a fruitful avenue for constructing a robot cognitive architecture. Moreover, the architecture can be easily extended to solve various problems by robots. For example, following natural language instructions to do navigation tasks, finding a person in a crowd through natural language descriptions. In our future work, we are interested in extending this architecture to everyday household tasks.

Acknowledgments. This research is supported by the National Natural Science Foundation of China under grant 61175057 and the USTC Key-Direction Research Fund under grant WK0110000028.

References

1. Bay, H., Ess, A., Tuytelaars, T., Gool, L.V.: Speeded-up robust features. Comput. Vis. Image Underst. **110**(3), 346–359 (2008)
2. Burgard, W., Cremers, A., Fox, D., Hahnel, D., Lakemeyer, G., Schulz, D.: Experiences with an interactive museum tour-guide robot. Artif. Intell. **114**(1), 3–55 (1999)
3. Chen, X., Ji, J., Jiang, J., Jin, G., Wang, F., Xie, J.: Developing high-level cognitive functions for service robots. In: Proceedings of 9th International Conference on Autonomous Agents and Multi-agent Systems (2010)
4. Guadarrama, S., Riano, L., Golland, D., Gohring, D., Jia, Y., Klein, D., Abbeel, P., Darrell, T.: Grounding spatial relations for human-robot interaction. In: IEEE/RSJ International Conference on Intelligent Robots and Systems (2013)
5. Kollar, T., Tellex, S., Roy, D., Roy, N.: Toward understanding natural language directions. In: 5th ACM/IEEE International Conference on Human-Robot Interaction (2010)
6. Matuszek, C., FitzGerald, N., Zettlemoyer, L., Bo, L., Fox, D.: A joint model of language and perception for grounded attribute learning. In: International Conference on Machine Learning (2012)
7. Quigley, M., Ng, A.Y.: Stair: hardware and software architecture. In: AAAI 2007 Robotics Workshop (2007)
8. Rusu, R.B., Cousins, S.: 3d is here: point cloud library (pcl). In: Proceedings of the IEEE International Conference on Robotics and Automation (2011)
9. Rusu, R.B., Holzbach, A., Bradski, G., Beetz, M.: Detecting, segmenting objects for mobile manipulation. In: Proceedings of the 12th IEEE International Conference on Computer Vision: Workshop on Search in 3D and Video (2009)
10. Steedman, M.: Surface Structure and Interpretation. MIT Press, Cambridge (1996)
11. Tellex, S., Kollar, T., Dickerson, S., Walter, M., Banerjee, A., Teller, S., Roy, N.: Understanding natural language commands for robotic navigation and mobile manipulation. In: Proceedings of National Conference on Articial Intelligence (2011)
12. Zettlemoyer, L.S., Collins, M.: Learning to map sentences to logical form: structured classification with probabilistic categorial grammars. In: Proceedings of the 21st Conference in Uncertainty in Artificial Intelligence, pp. 658–666 (2005)

Poster Presentations

Robust Detection of White Goals

Pablo Cano[✉], Yoshiro Tsutsumi, Constanza Villegas,
and Javier Ruiz-del-Solar

Advanced Mining Technology Center and Department of Electrical Engineering,
Universidad de Chile, Santiago, Chile
{pcano,jruizd}@ing.uchile.cl

Abstract. The main goal of this paper is to present a simple, but robust
algorithm for detecting white goals in the context of the RoboCup SPL (Standard Platform League). White goals will be used for the first time in the SPL
competitions in 2015. The main features of the algorithm are a robust search
strategy for detecting the goal posts, and the use of the Y channel image, instead
of the color segments, for determining and characterizing the goal posts and the
horizontal crossbar. This last aspect is crucial for detecting a white goal placed
in a white background. The algorithm is validated in the real world (real robot in
a SPL field), showing its ability to detect the while goals even when they are
observed in a white background.

Keywords: Robot soccer · SPL league · Color vision systems · Object
detection

1 Introduction

As everybody knows in the soccer robotics community, the detection of the field's
objects (goals, lines, ball, players) is essential for playing soccer properly. Given that,
the robust detection of objects in dynamic and cluttered environments is a complex
task, and considering that legged soccer robots have normally low-computational
capabilities, at the beginning of the RoboCup soccer competitions the use of colored
objects and beacons was introduced, with the purpose of facilitating the use of
color-based vision systems.

Given that the final goal of the RoboCup is to build robots that will be able to
compete and defeat humans in real-world fields, the mentioned constraints are relaxed
from year to year. For instance, at the beginning of the former Four-Legged league,
which after moving from 4-legged robots to humanoid robots is now called Standard
Platform League (SPL), 6 colored beacons and 2 colored and solid goals were used.
The 6 colored beacons were first reduced to 4, then to 2, and finally they disappeared.
The colored and solid goals were first transformed into non-solid goals composed by 2
goal posts and 1 horizontal crossbar. Then the color of both non-solid goals was set to
yellow, and this year (2015), the goals will be white for the first time. That means that
they will be similar to the goals used in most human soccer matches.

© Springer International Publishing Switzerland 2015
L. Almeida et al. (Eds.): RoboCup 2015, LNAI 9513, pp. 229–238, 2015.
DOI: 10.1007/978-3-319-29339-4_19

The use of white goals is more challenging than the use of yellow ones, because in the image space they can be mixed up with the field lines, but also because in many human indoor environments the walls are white.

In this context, the main goal of this short paper is to present a simple, but robust algorithm for detecting white goals in the context of the SPL. The main features of the algorithm are a robust search strategy for detecting the goal posts, and the use of the Y channel image, instead of the color segments, for determining and characterizing the goal posts and the horizontal crossbar. This last aspect is crucial for detecting a white goal placed in a white background.

The proposed algorithm is validated in a real SPL field. The article is organized as follows: Sect. 2 describes relevant related work. Section 3 describes the proposed algorithm, and Sect. 4 the experimental results. Finally, conclusions are drawn in Sect. 5.

2   Related Work

In the RoboCup Soccer community the visual detection of field objects is a well know problem, which has been tackled by most of the teams. Some seminal works are the following. In 1999 Bandlow [1] developed a vision system based on color classification for object recognition and localization in a RoboCup scenario. Using the previous knowledge of the objects and their respective colors, the image is segmented in one of those possible colors, in order to find object candidates. Then, strictly defined constraints of color and shape are stored in a database and compared with the possible objects. Jamzad [2] introduced a novel color model for shape and object detection. They use "jump points" to take advantage of the perspective information of the image in order to obtain a fast object recognition process. Zagal [3] tries to automatize the object detection problem by using supervised learning approaches. Loncomilla and Ruiz del Solar [4] describe a complete different object recognition approach based on the use of interest points and descriptors. Hartl [5] adds robustness to the problem of objects detection using color similarities. Other works addressing important aspects such as color invariance, the use of natural light, robustness, and the use of context information are described in [6–12].

Specifically in the goal detection problem, Canas [13] use means of color based segmentation and geometrical image processing methods to detect goals and determine the positions of the robot according to it.

The here-proposed approach is based on the B-Human code release [14], which is based on [15]. In this approach, a new color segmentation method is proposed, where colors are mapped not only to unambiguous but also to ambiguous color classes. In order to find the goal post, the image is horizontally scanned at the projected horizon height of the robot, in order to find yellow pixels that represent a possible goal post (because until last year all goals were yellow). Then, after finding the edges of the goal post candidates, several filters are applied in order to discard the false positive detections. As mentioned, the proposed algorithm is based in [15], but the goal post is not searched at the projected horizon height of the robot, but at the field boundary. In addition, the goal is characterized using the original Y channel image, and not the segmented one.

3   White Goal Detection Algorithm

In order to find a white goal in the image under analysis, the following steps are made: *Detection of Posts Bases, Determination of Posts Heights, Determination of Posts Widths, Filtering* and *Characterization* (see Algorithm 1). These steps allow finding any possible post inside the image, to filter out the ones that do not represent a real goal, and to characterize the real ones considering the detected posts.

```
INPUT: the image
OUTPUT: a detected goal
find bases of every possible goal post
for each possible goal post base
    find upper and lower edges // determination of
                                  the goal post height
    calculate goal post width // determination of the goal
                                  post width
    filter goal post // filtering
end for
goal characterization
```

Algorithm 1. Pseudo code of the general algorithm that allows finding white goal posts in an image.

3.1   Detection of Posts Bases

In former approaches developed for solving this problem, a horizontal scan through the image is carried out, and every white-segmented pixel is analyzed as a possible goal-post-base candidate. The scanning needs to be done in only one row of the image and, as long as this row stays below the projected horizon of the robot, any existing goal post in the image should cross the scan line.

However, the challenge is to distinguish the white goal-post-base in a white background. This can be addressed by a proper selection of the scanning line, which does not need to be horizontal anymore. In the proposed algorithm a scan through the field boundary is performed. This boundary is a convex hull of the green pixels that represent the field. As it can be seen in Fig. 1, the goal post bases are always below the field boundary line, so this can be used as a scan line that will always cross the base of a goal post. Also, the background of this scan line will be always green, so it is still possible to use the segmented image to find the goal post's base.

The proposed algorithm is shown in Algorithm 2. The function *yBoundaryValue* (*x*) returns the row value of the field boundary for a given column x. The algorithm searches white segments by finding first the transition between a green and a white pixel, and then the transition between a white and a green pixel. Next, it uses the transition coordinates as the start and the end of a white segment (goal post candidate). The center of the base of the goal post candidate is calculated by averaging these two points.

Fig. 1. Field boundary detection examples.

```
INPUT: the image
OUTPUT: center of all posts candidates
x ← 0
while x < image.width do
    y ← yBoundaryValue(x)
    while imageColor(x,y) != white do
        y ← yBoundaryValue(x)
        x ← x + 1
    end while
    p_i ← (x,y)
    while imageColor(x,y) != green do
        y ← yBoundaryValue(x)
        x ← x + 1
    end while
    p_f ← (x,y)
    p_mid ← (p_i + p_f)/2
    savePostCandidateMid(p_mid)
end while
```

Algorithm 2. Pseudo code of the algorithm that scans the image searching for white segments that correspond to candidates of goal posts.

3.2 Determination of the Posts Heights

After detecting a possible goal post's base, a bi-directional vertical scan is performed, in order to find the edges of the goal post. Then, the upper edge of the goal post is scanned horizontally in order to determine if it is a left or a right goal post.

When scanning in the downward direction, no major problem exists, because the pixels inside the field are correctly segmented, then, finding the green base is an easy task. However, when going upwards, the white-background problem shows up, because a lot of non-goal post's pixels could be segmented as white pixels. Therefore, while the scan is performed, the image's real intensity values are considered, because small intensity gradients could indicate the end of the goal post. In Fig. 2 various shades of white are measured in the Y, Cb and Cr channels. It can be seen that the intensity gradients are more evident in the Y channel.

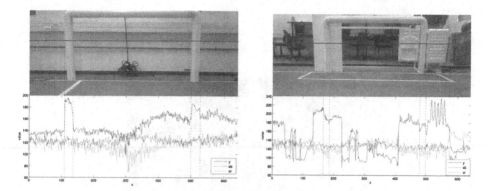

Fig. 2. Shades of white measurements in different channels. The blue line on the image indicates the position where the pixels values were evaluated. The graphs indicate the value of each channel of the YCbCr color space (Color figure online).

```
INPUT: initial point p = (x,y) and color
OUTPUT: final point  p_f = (x_f,y_f)
int ← imageIntensity(p)
p ← moveForward()
while f(int - imageIntensity(p)) < threshold and color ==
      imageColor(p)
    int ← imageIntensity(p)
    p ← moveForward()
end while
p_f ← p
return p_f
```

Algorithm 3. Generic pseudo code that scans the image in any direction until a color difference or an intensity gradient is found.

Algorithm 3 shows the pseudo code of the algorithm that scans the image in any direction using the pixel's intensity information. The function f represents the policy used to detect changes in the Y channel. It could be an absolute value between one pixel and the next one or an absolute value between the current pixel and all the previous ones low pass filtered by a Gaussian kernel. The function *moveForward* implements different ways of scanning the image, according to the current direction. It only moves inside the image and it handles the noise in the segmentation, using a configurable step size hysteresis. For the upward scan, it handles the possible tilt of the post, which could appear when the robot is in motion, or when the post is watched from a certain perspective. In Fig. 3 a tilted post is shown. As it can be observed, the upward scan is carried out satisfactorily.

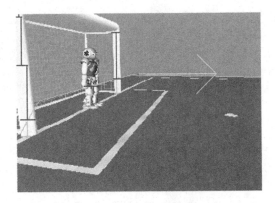

Fig. 3. Example of an upward scan in a tiled post. The function *moveForward*() recalculates the width and allows to continue the scan. It also clips the scan inside the image.

Then, using Algorithm 3 it is possible to determine the height of the post. Figure 4 shows an example where a traditional approach, which uses only the color-segmented image, and the proposed one, which uses the color-segmented image and the pixel values in the Y channel, are used for determining the goal post edges. It can be seen that only the proposed approach carried out this task properly, even when the background is white.

Fig. 4. Comparison between two different approaches used for finding the goal post's edges. Left, only the color-segmented image is used. Right, the color-segmented image and the pixel's intensity information are used (Color figure online).

3.3 Determination of the Posts Widths

After finding the upper and lower edges of the goal post, a line is drawn between these edges. Then, the goal post's width is analyzed in a discrete number of points inside the plotted line. For that purpose, in every chosen point of the initial line, a bi-directional horizontal scan is made. In this step, the same algorithm shown in Algorithm 3 is used.

Figure 5 shows the difference between scanning the image when using and not using the pixel's intensity information to find the goal post's width. Even though the algorithm did not work perfectly, the detection can be done, thanks to the filtering algorithm described in the next section. Also, the detection of the goal post's width in a white background environment cannot be done if the pixel's intensity information is not used.

Fig. 5. Goal post's width detection example. Left, only the color-segmented image is used. Right, the color-segmented image and the pixel's intensity information are used. The green and red lines represent good and bad goal-post width detections, respectively, according to the expected width of a goal post in the image. The green lines are obtained during the filtering step (Color figure online).

3.4 Filtering

Finally, all the possible posts are filtered out, in order to eliminate false positive detections. The principal filter uses the information of the lower edge and the width of the possible goal post to do the classification. First, the distance of the goal post to the robot is calculated using its lower edge position on the image, and the position of the projected horizon on the image. This allows calculating a scaled orthographic projection of the lower edge of the post that gives an estimated position of the post foot in local coordinates. Then, since the actual goal post proportions are known, the calculated goal post's width value can be compared with the expected one, which can be calculated using the goal post's distance. In Fig. 5, the width values that correspond to the expected width values are shown in green, while the wrong ones are shown in red. Thus, using the proportions between the correct and incorrect width values, the false positives are filtered out.

3.5 Characterization

The last step consists in characterizing a goal using all the post detections made before. To do this, a simple heuristic is applied, given that in one frame only one goal can be seen, so it is possible to watch only two posts at the same time. The distance between these two posts is also used to discard false positive detections, because the aspect ratio of a complete goal is known. Then, using the distance information of each post, the goal distance and angle to the robot is calculated. This steps are skipped if only one post is detected, and if more than two post are detected, all the detections are discarded.

4 Results

The described algorithm was tested using two real videos collected by the robot inside a SPL field. The first video considered only non-white backgrounds, and the second one white backgrounds. Please note that all the statistics shown below were calculated using the detections made by the robot itself.

The first video contained 1,873 frames, non-white backgrounds, and the goal was visible in 576 of the 1,873 frames. A person moved the robot to different positions on the field. Like so, the robot was able to look the goals and other parts of the field. During the video recording process other robots were placed on the field, making this situation closer to a real one. The second video consisted of 1,317 frames, white backgrounds, and the goal was visible in 354 of the 1,317 frames. As in the first case, a person moved the robot, and other robots were present in the field too.

The detection results are shown in the Table 1. As it can be observed, the detection rate is very high, $\sim 96\%$ when non-white background are used, and $\sim 86\%$ in the white background case. Also, in this last case a few false positive detections appear (4 in 1,317 frames). Nevertheless, both result are satisfactory, because one of the most important concerns was the possible confusion between the goals and others white objects present on the field, such as robots and lines, which did not happen during the experiments. Another concern was the difficulty in distinguishing the white goal with a white background, which almost did not happen in the experiments.

It is important to say that the false positive cases occurred when the difference between the goal post and the white background were minimal. In those cases, the upward scan came out of the goal post, and the algorithm detected a false goal post next to the real one. So in these cases, the result of this false detection was not that bad because a real goal post was close.

During the described tests, the computational time of the algorithm was measured in the robot. In average it takes 0.09 ms, with maximum and minimum picks of 0.15 ms and 0.05 ms respectively.

Table 1. Overview of the detection results.

	Non-white background	White background
True positive rate	0.96	0.86
Number of false positives	0	4

5   Conclusions

White goals will be used in the SPL league competitions, in 2015, for the first time. In order to address this new challenge, this paper proposes a simple, but robust algorithm for detecting white goals. The main features of the algorithm are a robust search strategy for detecting the goal posts, and the use of the Y channel image, instead of the color segments, for determining and characterizing the goal posts and the horizontal crossbar. This last aspect is crucial for detecting a white goal placed in a white background.

The proposed algorithm was validated in a real SPL field. The obtained results showed that the proposed algorithm is able to detect goals even when they are observed in white backgrounds, and that the goals are not mixed up with other white objects such as lines and robot's parts. Nevertheless, this algorithm strongly relays on a well-detected field boundary, so if this detection fails, the entire algorithm would fail, because it would be impossible to detect correctly the goal post bases. Although the boundary detection is usually good, this adds a new dependency to the proposed algorithm.

Acknowledgments. This work was partially funded by FONDECYT Project 1130153.

References

1. Bandlow, T., Klupsch, M., Hanek, R., Schmitt, T.: Fast image segmentation, object recognition, and localization in a RoboCup scenario. In: Veloso, M.M., Pagello, E., Kitano, H. (eds.) RoboCup 1999. LNCS (LNAI), vol. 1856, pp. 174–185. Springer, Heidelberg (2000)
2. Jamzad, M., Esfahani, E.C., Sadjad, S.B.: Object detection in changing environment of middle size RoboCup and some applications. In: Proceedings of the IEEE International Symposium Intelligence Control (2002)
3. Zagal, J.C., Ruiz-del-Solar, J., Guerrero, P., Palma, R.: Evolving visual object recognition for legged robots. In: Polani, D., Browning, B., Bonarini, A., Yoshida, K. (eds.) RoboCup 2003. LNCS (LNAI), vol. 3020, pp. 181–191. Springer, Heidelberg (2004)
4. Loncomilla, P., Ruiz-del-Solar, J.: Robust object recognition using wide baseline matching for RoboCup applications. In: Visser, U., Ribeiro, F., Ohashi, T., Dellaert, F. (eds.) RoboCup 2007: Robot Soccer World Cup XI. LNCS (LNAI), vol. 5001, pp. 441–448. Springer, Heidelberg (2008)
5. Härtl, A., Visser, U., Röfer, T.: Robust and efficient object recognition for a humanoid soccer robot. In: Behnke, S., Veloso, M., Visser, A., Xiong, R. (eds.) RoboCup 2013: Robot World Cup XVII. LNCS, vol. 8371, pp. 396–407. Springer, Heidelberg (2014)
6. Dahm, I., Deutsch, S., Hebbel, M., Osterhues, A.: Robust color classification for robot soccer. In: 7th International Workshop on RoboCup (2003)
7. Guerrero, P., Ruiz-del-Solar, J., Fredes, J., Palma-Amestoy, R.: Automatic on-line color calibration using class-relative color spaces. In: Visser, U., Ribeiro, F., Ohashi, T., Dellaert, F. (eds.) RoboCup 2007: Robot Soccer World Cup XI. LNCS (LNAI), vol. 5001, pp. 246–253. Springer, Heidelberg (2008)

8. Mayer, G., Utz, H., Kraetzschmar, G.K.: Playing robot soccer under natural light: a case study. In: Polani, D., Browning, B., Bonarini, A., Yoshida, K. (eds.) RoboCup 2003: Robot Soccer World Cup VII. LNCS (LNAI), vol. 3020, pp. 238–249. Springer, Heidelberg (2004)

9. Sridharan, M., Stone, P.: Towards illumination invariance in the legged league. In: Nardi, D., Riedmiller, M., Sammut, C., Santos-Victor, J. (eds.) RoboCup 2004: Robot Soccer World Cup VIII. LNCS (LNAI), vol. 3276, pp. 196–208. Springer, Heidelberg (2005)

10. Jüngel, M., Lötzsch, M., Hoffmann, J.: A real-time auto-adjusting vision system for robotic soccer. In: Polani, D., Browning, B., Bonarini, A., Yoshida, K. (eds.) RoboCup 2003: Robot Soccer World Cup VII. LNCS (LNAI), vol. 3020, pp. 214–225. Springer, Heidelberg (2004)

11. Iocchi, L.: Robust color segmentation through adaptive color distribution. Symp. A Q. J. Mod. Foreign Lit. 1(1), 1–11 (2006)

12. Palma-Amestoy, R., Guerrero, P., Ruiz-del-Solar, J., Garretón, C.: Bayesian spatiotemporal context integration sources in robot vision systems. In: Iocchi, L., Matsubara, H., Weitzenfeld, A., Zhou, C. (eds.) RoboCup 2008: Robot Soccer World Cup XII. LNCS (LNAI), vol. 5399, pp. 212–224. Springer, Heidelberg (2009)

13. Canas, J.M., Puig, D., Perdices, E., González, T.: Visual Goal Detection for the RoboCup Standard Platform League. In: X Workshop on Physical Agents, WAF, pp. 121–128 (2009)

14. Röfer, T., Laue, T., Müller, J., Bartsch, M., Batram, M.J., Böckmann, A., Lehmann, N., Maaß, F., Münder, T., Steinbeck, M.: B-Human team report and code release 2012 (2012). http://www.b-human.de/wp-content/uploads/2012/11.CodeRelease2012.pdf

15. Röfer, T.: Region-based segmentation with ambiguous color classes and 2-D motion compensation. In: Visser, U., Ribeiro, F., Ohashi, T., Dellaert, F. (eds.) RoboCup 2007: Robot Soccer World Cup XI. LNCS (LNAI), vol. 5001, pp. 369–376. Springer, Heidelberg (2008)

A Robust and Calibration-Free Vision System for Humanoid Soccer Robots

Ingmar Schwarz[✉], Matthias Hofmann, Oliver Urbann, and Stefan Tasse

Robotics Research Institute, Section Information Technology,
TU Dortmund University, 44221 Dortmund, Germany
ingmar.schwarz@tu-dortmund.de

Abstract. This paper presents a vision system which is designed to be used by the research community in the Standard Platform League (http://www.tzi.de/spl) (SPL) and potentially in the Humanoid League (http://www.robocuphumanoid.org) (HL) of the RoboCup. It is real-time capable, robust towards lighting changes and designed to minimize calibration. We describe the structure of the processor along with major ideas behind object recognition. Moreover, we prove the benefit of the proposed system by assessing recorded image data on the robot hardware. The vision system has already been successfully employed with the NAO robot by Aldebaran Robotics (http://www.aldebaran-robotics.com) in prior RoboCup competitions as well as several minor events.

1 Introduction

This paper presents a vision system for humanoid robots in the RoboCup. Nowadays, it is common to ensure stable lighting conditions during competition games played at RoboCup. This allows for easy color separation and object recognition. Therefore, colortable-based image processors which impose either manual or semi-automatic calibration on users, are still widely utilized in the area. Assuming perfect manual color segmentation allows very simple recognition algorithms. These, however, become error-prone once the manual calibration is imperfect or sometimes even when it is done by a different person with a different bias towards labeling intermediate areas. As a result, such systems are limited in their adaptability and are considered inflexible.

The RoboCup organization forces the development in the area of autonomous robotics by applying rule changes to its various disciplines. This way, RoboCup is evolving towards a more realistic game play while improving individual skills of the robots. To this end, future games will likely be conducted outdoors in natural environments. This stresses the need for robust image processing systems capable of dealing with such challenging tasks.

The remainder of the paper is structured as follows: While Sect. 1.1 describes the objectives of the system, Sect. 1.2 presents related work on the topic. Section 2 outlines the means being used in the image processor, including field color and line detection as well as robot, ball, and goal post recognition. Section 3 proves the benefit of the proposed module in various experiments, i.e. real-game conditions.

© Springer International Publishing Switzerland 2015
L. Almeida et al. (Eds.): RoboCup 2015, LNAI 9513, pp. 239–250, 2015.
DOI: 10.1007/978-3-319-29339-4_20

We analyze the run time performance along with detection rates. We conclude our work in Sect. 4, and describe future work.

1.1 Objectives

In the following, we list our objectives for the proposed vision system. Our vision system is real time capable on the NAO robot, meaning it takes less than 33 ms to process both images, which is the time needed to capture the next images on the NAO robot. To save run time for the other tasks such as motion, localization and behavioral, we set our limit at 15 ms. This allows tracking moving objects such as the ball sufficiently fast for its application in RoboCup. We postulate the vision system to deal with *changing environmental conditions* such as lighting, indoor as well as outdoor. To ensure compatibility to other robots or hardware upgrades, the system should not be limited to run on a single hardware setup. The last main objective for our vision system is to reduce the needed calibration as much as possible.

1.2 Related Work

As we avoid the usage of calibration intensive color tables, our work regarding field color detection follows up on the work of Reinhardt [1] which we extended (see Sect. 2.1). Reinhardt introduced the first colortable-less vision system in the SPL of the RoboCup. For obstacle detection, there are different approaches in the SPL: Metzler et al. [2] proposed an algorithm for obstacle detection which recognize the feet of the robot. An alternative method from Fabisch et al. [3] relies very much on the jersey. To improve robot detection, we use the jersey and the body of the robot in addition to the feet. Our work suggests various heuristics to detect each of the relevant objects by providing specialized modules and verification steps. In contrast to Härtl et al. [4], we use dynamic thresholds for segmentation. Moreover, our approach is independent of colorimetric shift, and we use different algorithms for object detection. There are more approaches to deal with lighting differences and self-calibration, e.g. Hanek et al. [5], Bruce et al. [6], and Jüngel [7], but none of these approaches were tested to play outside in natural lighting conditions.

2 Vision System

The vision system is supplied by pictures from the cameras, and the current camera matrices. The term camera matrix refers to a transformation matrix between the projected middle point of the feet of the robot and the position of the respective lower or upper camera. This facilitates the utilization of 3-dimensional information, and is used to determine the position on the field of the detected objects relative to the robot. This section is structured as follows: The first step of the vision system detects the field color (see Sect. 2.1). After that, the preprocessing step (see Sect. 2.2) extracts ball, line, and robot points, as well as goal segments. This information is required to compute and verify individual ball percepts (see Sect. 2.4), field lines with center circle (see Sect. 2.3), robots (see Sect. 2.6), and goal posts (see Sect. 2.5).

There is a trade-off between the robustness and reliability of the system, and the number of detected objects which might be false positives. The criteria when an object is deemed as detected have to be carefully determined. In general, we prefer to minimize the number of false positives. This mitigates the impairment of other vital functions of the system that depend on the results of the vision system, such as the localization and world modeling. Inaccuracies that influence the vision system have multiple causes: Due to the highly dynamic nature of a robot soccer game, robots are forced to look quickly at different points. This can result into motion blur, especially if a high exposure time is needed for low light environments. In addition, there are small differences in the cameras resulting in slightly different colors. Moreover, there is an interdependence between the vision system and the kinematics of the robot. Disturbances (e.g. caused by the walk of the robot) lead to inaccurate camera matrices. The vision system is therefore only executed when the camera matrix is plausible, i.e. the robot touches the ground with at least one sole.

2.1 Field Color Detection

Field color detection is the first vital step of the vision system as every object which has to be detected is located on the field, and thus the color is used to distinguish such objects from the field. Since light might change continuously, this step is repeated in each execution cycle. In many cases, the field color is not the most present color of the image as the robot might be located at the field border and looks outside the field. The SPL rules[1] define the field color as green-like. Our approach is similar to Reinhard [1]. We use a weighted color histogram on the YCbCr image from samples below the calculated horizon of the robot. Although the green part in the YCbCr colorspace is preferred by this weighting, other colors are accepted as well. The maxima of the weighted histograms of each channel are set as the current field color. We however do not use a fixed predefined color cube to cope with color changes induced by different lighting or cameras. The optimal color channel values from this step are used with a lighting- and color-dependent specific distance to classify a pixel as of field color throughout the vision system. These distances are taken from the histogram by using the width of the peak of the maximum. The field color is individually determined for each camera as those might differ significantly due to their different, automatically adjusting camera settings. To avoid field color changes when the robot looks out of the field, the field color is not allowed big changes once it is set.

---

[1] http://www.tzi.de/spl/pub/Website/Downloads/Rules2014.pdf.

2.2 Preprocessing

The main task of the preprocessor is to identify all possible object locations in the image by scanning the image along a fixed scan line grid. This allows the subsequent modules of the vision system to efficiently detect and verify the specific objects without scanning the image again. Inputs of the preprocessor are the previously computed field color (see Sect. 2.1), and the current camera matrix.

The process can be divided into three main steps: The first step uses *scan lines* which are then split into segments. To this end, only horizontal scan lines in the proximity of the horizon are used to find goal segments. The second steps *classifies* segments into field, line, ball, and unknown. The last step *filters and creates the data structures* needed by subsequent modules as the output.

The classification of field, ball and unknown segments is the core of the pre-processing and further described. The image is analyzed by means of horizontal and vertical scan lines that form a grid. The tightness of the grid can be configured by the user, significantly influencing the run time of the vision system (see Sect. 3.3). Moreover, the pixel tightness within the scan lines depends on the expected width of a line since a line is expected to be the smallest object to detect. Thus, the vision system is scalable and largely independent from the image resolution. The scan on each line stops when no field color is found and the segment exceeds the expected length of ball or line segments to save run time when multiple robots are in the image and to prevent detecting objects on these robots.

The scan process first splits the scan line at points of big color or brightness differences while counting pixels classified as of field color. The differences are measured at points where the direction of the color or brightness gradient changes to achieve robustness on blurred images. In the next step a segment is classified as of field color if at least 50 % is identified as field color. Afterwards, the segments are verified by their surrounding: Field segments are merged together and Line segments become verified line points when they are surrounded by field color, begin and end with leaps in brightness and if they do not exceed a specific width. Ball segments must begin and end with a color change using the cb- and cr-channel and are merged together after classification. Thus, it is noteworthy that no specific ball color is required.

All unidentified segments are then clustered to find the field border and points of interest for the robot detection. The upper image is only processed if the lower image does not detect a field border.

2.3 Line Detection

The detection of field lines plays a key role for the localization task as they are the most present feature on the field. The localization works best when the field lines and their widths are completely detected. The SPL rules state that field lines must be white and have a width of 5 cm. We distinguish between field lines and the center circle which has a radius of 75 cm. Hence, both types are processed differently in our vision system.

Edge detectors such as the Sobel and Roberts operators are commonly used for line recognition. The application of these algorithms is problematic in the SPL. Although Sobel is more robust, it is computationally expensive. Moreover, Sobel and Roberts are susceptible to motion blur and the quality of the detection decreases when lines are very thin. Hence, our approach connects the line points which have been computed in the preprocessing step instead scanning the image only in small parts to verify and extend field lines.

The line detection can be roughly divided into the following steps: First, we connect line points which are close to each other to form chains. Thus, we look for the point with the smallest Euclidean distance within the next two scan lines of the same type (horizontal or vertical). The line points from the preprocessing step are sorted by their scan line numbers for efficiency.

Second, *line segments* are built from such chains. Hence, we check the curvature of the chains in the image, and if it differs in one point too much from the other points, the segment is separated. The curvature as well as average and maximum errors (e_{avg}, and e_{max}) of a linear regression on the segment points are saved as meta information. We ensure that every point is assigned to exactly one line segment.

After the identification of line segments, we use their curvature to identify points on the center circle. All points on the center circle are projected on the field. We use the least squares fitting method [8] to find a circle by utilizing the set of the projected points (see Fig. 1). Moreover, we utilize an additional set of other line segments which are located in a suitable distance from the middle point to refine the circle.

Fig. 1. Center circle detection in a blurry image using line segments.

Field lines are formed from line segments which consist of a minimum number of points p_{min}, and are below a threshold regarding e_{avg}, and e_{max}. If segments have less than p_{min} points, we try to fuse the segment with an existing field line or other segment if it lies straight on and within a certain distance limit. Additionally, we check whether the field color is situated between the line segments. Lastly, we compute the width of a field line at its ends. The last step extends the recognized field lines if possible. This is done by short scan lines which are looking for differences in brightness and for the field color at the end of the line.

2.4   Ball Detection

The ball is a unique feature on the field, and thus a relevant landmark for localization, i.e. for the resolving of the field symmetry [9]. The ball perception influences tactical decisions and accurate approaching. The vision system uses the shape, the color, and the size of the ball as relevant features. We further assume that the ball always touches the ground, i.e. a 2D ball model. Hence, the ball is at least partly surrounded by the field. Reflections on the ball surface, shadows, and partial occlusions make ball detection a challenging task during the game.

The official SPL ball has a diameter of 65 mm, is orange-like, and is originally used for Hockey. The preprocessing module provides the center of the ball segment that has been computed in the preprocessing step, and the average Y, Cb, and Cr values of the possible ball to ensure color-independence of this component. For ball detection, we utilize only the Cb, and Cr channels due to significant value deviations in the Y channel, e.g. due reflections of the spotlights on the ball.

Further, ball detection can be divided into *shape classification* and *validation*. In the shape classification step, we employ radiating scan lines beginning from the center of the ball segment to determine the outer points of the ball. The ball segment including position, length and average color, has been computed in the preprocessing step. The scan lines run until the border of the image is reached, a repeating deviation of the ball color is recognized, or the theoretical ball radius (computed with the camera matrix), is exceeded by factor 3. It is important that small deviations of the ball color are tolerated. The shape of the ball is calculated with the least squares fitting method [8] as in center circle detection. The segment is rejected if no circle can be built. The verification step deletes ball perceptions when size and shape of the ball is not plausible.

2.5   Goal Detection

We primarily use shape and color information for goal detection with the idea of color independence (no preconfigured colors) as in previous parts of the system. Output of the module is the position of each visible goal post (2 at most) along with a tag indicating the side (right or left). A goal post has a diameter of 10 cm, and is white as defined in the SPL rules[2]. The base of the post is surrounded by field lines and the field color if not obstructed.

Input of the module is a set of extracted goal segments from the preprocessing. Goal segments consist of a start point and an end point as well as average values of the cb and cr channel of all pixels within the segment. Disturbances in

---

[2] However, the image processing system is evaluated on images from previous years which contain yellow-like goal posts.

the walk might cause distortions of the image's perspective. We therefore allow the goal post to be tilted. Consequently, as a first step, the angle of the goalpost in the image is determined. We connect the middle points of all goal segments to lines. All goal segments must have a consistent color and width, and we mainly use this kind of information to compute the upper and lower point of the posts. An additional set of horizontal scan lines is used to determine the width of the post which is vital for verification and is used to estimate the distance from the goal if the base of the goal post is not in the image. Moreover, if the upper point of the goal post is visible, the system extracts the crossbar to decide the side of the post.

2.6   Robot Detection

Reliable robot detection is very important in robot soccer as it allows for suitable path planning, and collision avoidance. Moreover, tactical decisions are often based on the positions of the robots on the field. Detected robots can be even used to support localization and world modeling [9].

In the SPL, all robots are of the same type. There are several possible approaches for robot detection like an exclusive search for team jerseys [3]. A major shortcoming of the approach is that it assumes the jersey or waistband always to be visible. In practice, jerseys might be occluded by robots, their body parts, and other objects.

Varying lighting conditions make it difficult to start with the jersey detection. Hence, we commence with feet detection, as the feet are usually surrounded by the field. Moreover, the feet of a visible robot should always be visible either in the upper or in the lower image. Input of the module is a set of obstacle points from the preprocessing step. Obstacle points are defined as points which could not have been assigned to any other object type.

We utilize scan lines to determine the width of the obstacle w_{scan} by looking for the field color in both horizontal directions for verification. The theoretical width of the object w_{opt} is computed with the camera matrix. We assume that the width of the obstacle at its feet is 21 cm which corresponds to the average width of the feet in any possible rotation. If the relation between w_{scan} and w_{opt} is too large, the obstacle is refused.

Lastly, color verification is performed. Since we assume the average walk height of the robot to be 53 cm, we look in suitable regions of the image for the jersey, and check for differences in the Cb and Cr channels to assess the team colors (which are red-like, and blue-like in the SPL). Moreover, we analyze whether the field color within the obstacle is more than 50 % by applying diagonal scan lines. The last step checks for average deviations in the parts of the robots which are not covered by the jersey whether they differ too much from the color of the robot (which is white and gray in the SPL). The detection is easily configurable to be compatible with other colors. For instance, robots from the humanoid kid size are recognized when the main color of the obstacle is changed from white to black.

3   Evaluation

The examination of the proposed algorithm is twofold. On the one hand, we present and discuss its object recognition performance under various conditions (see Sect. 3.2). On the other hand, a run time analysis is conducted as real-time capability is of importance in highly dynamic environments (see Sect. 3.3).

3.1   Hardware

We use the NAO robot developed by Aldebaran Robotics in our experiments. The NAO is about 55 cm tall, and used in the SPL. It comes with an upper and lower camera with opening angles of $47.64°$ in vertical direction, and $60.97°$ in horizontal direction with no overlap. This comparatively small opening angles impose fast and excessive movement of the head which increases motion blur. We use the auto exposure, and white balance that is provided by the camera driver for evaluation. Aldebaran Robotics provides an API to configure the camera. A complete list and a more detailed description can be found on the supplier's web page[3]. This image processor has been evaluated with the NAO V4 version which offers limited computational power (Intel Atom 1.5 Ghz single core processor, and 1 GB RAM).

3.2   Object Recognition

In this section, we assess the performance of the system on in a quantitative and qualitative way. First, we examine the detection rates by using recorded image data from the robot hardware. The image log has been manually checked to get an idea whether objects in the image should have been detected by the system. This means that a set of detection criteria has to be defined for each type of object (see Table 1).

Table 1. Detection criteria of each object.

Object	Detection criteria
Field lines	- Line parts with a minimum length of 30 cm
Center circle	- At least a quarter of the center circle is observable
Ball	- At least a quarter circle of the ball is visible
	- The ball is located on the playground within the field borders
Goal posts	- The goal post is not cut by the left or right image border
	- The base of the goal post lies in the image
Robots	- The feet of the robot can be seen

---

[3] http://www.aldebaran-robotics.com/en/Discover-NAO/datasheet.html.

The log file consists of 388 individual pictures that have been recorded at different places and conditions including the RoboCups and local Opens in the years 2008–2012 to check the performance of the algorithms in a realistic setting. This data is in the log file format of the B-Human framework [10], and available online.[4] At least a third of the images are noisy and blurry, and many pictures contain persons and objects that do not belong to the ordinary field.

The log file consists of 388 individual pictures that have been recorded at different places and conditions including the RoboCups and local Opens in the years 2008–2012 to check the performance of the algorithms in a realistic setting. At least a third of the images are noisy and blurry, and many pictures contain persons and objects that do not belong to the ordinary field.

Table 2 shows the detection rates, false positives, and the number of entities that should have been recognized (N). It can be seen that the detection rates vary significantly depending on the type of object. We observe that goal post detection is most accurate with 86 % of all objects that should have been detected. The low numbers of false positives are due to strict validation constraints in the heuristics of the algorithm. This leads to lower detection rates. As the vision system is operated at a maximum possible rate of 30 frames per second, we claim that the detection rates are sufficient for a successful application in a RoboCup game.

Table 2. Detection rates of the image processor and its components.

	Correct	False-positives	N
Field lines	61 %	8	601
Center circle	55 %	0	87
Ball	76 %	0	96
Goal posts	86 %	0	129
Robots	14 %	0	132

The system is able to detect the ball accurately even if less than half of the ball can be seen in the picture. Moreover, we detect balls that are approximately 4.5 m away from the robot (field size is 9×6 m) with the comparatively low resolution of 320*240. If a higher resolution (e.g. 640*480 pixels) is used, the ball can be recognized on the complete field.

Figure 3 depicts a reliable goal detection with side assignments. If the robot stands, goals are detected even from a distance of more than 9 m. Due to the shape and color of the goal posts, and the correction mechanisms integrated in the vision system, our approach is able to compensate motion blur, and disturbances. The detection of the center circle is dependent on the distance to it and less reliable than the field line detection, since the curvature of the center circle is then less visible due to the low viewpoint of the NAO robot. Our robot

---

[4] http://www.nao-devils.de/downloads/IPLogs.

detection does depend on the clustering of the preprocessor. If the other robot's distance is nearer, more scanlines intersect with that robot and consequently our detection rate increases. Robot's nearer than 2.5 m are detected with a success rate of more than 50 %.

The vision system has been successfully assessed in real-world conditions, e.g. the past three RoboCups as well as several events (see Fig. 2). To this end, it already worked in several outdoor environments, including for the shooting of the German crime series 'Tatort' without the need for calibration.

(a)   (b)

Fig. 2. Our robots playing at an exhibition in Zurich (a) and at the shooting for 'Tatort' (b).

3.3   Runtime Analysis

This experiment proves the real-time capability of the proposed system by recording the time needed to execute its various components. We reproduce situations that occur in regular RoboCup games to ensure the value of the assessment. To this end, a robot is placed on the field. Its task is to walk to the ball and kick it into the direction of the opponents' goal. Two robots are placed on the field, and the ball is rolled between the two robots. The experiment is conducted indoor in natural lighting conditions, i.e. no artificial light is turned on: This creates shadows and bright spots on the ground. The vision system is executed 1000 times in total which corresponds to 33 game seconds.

The run time is dictated by the preprocessing component which consumes more than 90 % of the total resources required as the majority of the scanning process takes place in this step. Thus, the number of used scan lines impose the overall run time. The algorithm can be easily scaled to fit users' requirements and platforms with different computational power. The current implementation allows the vision system to analyze both images in around 18 ms (17 ms for the preprocessing) on average using the highest camera resolution possible on the NAO robot, which is 1280*960 for the upper camera and 640*480 for the lower camera. The number of scan lines used is 92 for each image. If the resolution on both

(a)   (b)

Fig. 3. Goal posts with side assignments (a) and inaccurate detection of the left goal post due to motion blur (b).

cameras is halved, the run time decreases to around 14 ms total on average. The maximum run time in each case is only 2 ms higher while the minimum run time is at 5 ms if only the lower image is processed (i.e. the field border is detected in the lower image).

4   Conclusion and Future Work

This paper presents a real-time capable, robust, calibration-free, and easy configurable vision system that has been explicitly developed to meet the needs of the RoboCup Standard Platform League. To meet the run time requirements for the rest of the code, the number of scan lines is configurable which has been done throughout the past years. The concepts behind object recognition and the usage of scan lines can be slightly modified to allow their application in other domains. We have shown the benefits of the system in various experiments with the NAO robot, and stress that detection rates are sufficient for its application in RoboCup as the system is executed 30 times per second. At RoboCup 2014, we successfully performed a live game outside of the main hall with no calibration or adjustments to show the capabilities of our vision system. We further captured a video with experiments outside of our lab[5]. Throughout the last three years we adjusted the resolution for our cameras and the number of scan lines used and performed at several events (see Fig. 2) without the need to calibrate any further part of our vision system.

We deliberately avoid the employment of color tables due to their poor adaptability and robustness towards lighting changes. Instead, we use a variance of the field color detector proposed by Reinhardt [1]. The implementation is mostly independent from camera properties such as its resolution. This alleviates the

---

[5] https://www.youtube.com/watch?v=WXBEQF8k3b0.

migration to other hardware platforms than the NAO. We exploit domain knowledge to achieve lighting-independence, and ensure robustness against motion blur and colorimetric shift. Since our vision system partly relies on the camera matrix, calibration is required for the best performance. For next year, we plan to calibrate this part in our vision chain while playing to completely remove the need for calibration. Our robots use auto settings provided by the cameras and no calibration of those settings was needed in the past years.

For the future, we suspect rule changes towards a more realistic color scheme of the goals and the ball in the near future, i.e. the ball used in the SPL might become a black and white pattern. Our goal is to extend the vision system to work with such amendments. Additionally, we aim at improving the performance of the system by refining the robot detection component in future work.

References

1. Reinhardt, T.: Kalibrierungsfreie bildverarbeitungsalgorithmen zur echtzeitfähigen objekterkennung im roboterfußball. Master's thesis, Hochschule für Technik, Wirtschaft und Kultur Leipzig (2011)
2. Nieuwenhuisen, M., Behnke, S., Metzler, S.: Learning visual obstacle detection using color histogram features. In: Röfer, T., Mayer, N.M., Savage, J., Saranlı, U. (eds.) RoboCup 2011. LNCS, vol. 7416, pp. 149–161. Springer, Heidelberg (2012)
3. Fabisch, A., Laue, T., Röfer, T.: Robot recognition and modeling in the robocup standard platform league. In: Pagello, E., Zhou, C., Behnke, S., Menegatti, E., Röfer, T., Stone, P. (eds.) Proceedings of the Fifth Workshop on Humanoid Soccer Robots in Conjunction with the 2010 IEEE-RAS International Conference on Humanoid Robots, Nashville, TN, USA (2010)
4. Visser, U., Röfer, T., Härtl, A.: Robust and efficient object recognition for a humanoid soccer robot. In: Behnke, S., Veloso, M., Visser, A., Xiong, R. (eds.) RoboCup 2013. LNCS, vol. 8371, pp. 396–407. Springer, Heidelberg (2014)
5. Hanek, R., Schmitt, T., Buck, S., Beetz, M.: Towards RoboCup without color labeling. In: Kaminka, G.A., Lima, P.U., Rojas, R. (eds.) RoboCup 2002. LNCS (LNAI), vol. 2752, pp. 179–194. Springer, Heidelberg (2003)
6. Bruce, J., Balch, T., Veloso, M.: Fast and inexpensive color image segmentation for interactive robots. In: Proceedings. 2000 IEEE/RSJ International Conference on Intelligent Robots and Systems (IROS 2000), vol. 3, pp. 2061–2066. IEEE (2000)
7. Jüngel, M.: Using layered color precision for a self-calibrating vision system. In: Nardi, D., Riedmiller, M., Sammut, C., Santos-Victor, J. (eds.) RoboCup 2004. LNCS (LNAI), vol. 3276, pp. 209–220. Springer, Heidelberg (2005)
8. Chernov, N., Lesort, C.: Least squares fitting of circles. J. Math. Imaging Vis. **23**(3), 239–252 (2005)
9. Tasse, S., Urbann, O., Hofmann, M.: SLAM in the dynamic context of robot soccer games. In: Chen, X., Stone, P., Sucar, L.E., van der Zant, T. (eds.) RoboCup 2012. LNCS, vol. 7500, pp. 368–379. Springer, Heidelberg (2013)
10. Röfer, T., Laue, T., Müller, J., Bartsch, M., Batram, M.J., Böckmann, A., Böschen, M., Kroker, M., Maaß, F., Münder, T., Steinbeck, M., Stolpmann, A., Taddiken, S., Tsogias, A., Wenk, F.: B-human team report and code release 2013 (2013). http://www.b-human.de/downloads/publications/2013/CodeRelease2013.pdf

Power Usage Reduction of Humanoid Standing Process Using Q-Learning

Ercan Elibol[1], Juan Calderon[1,2], Martin Llofriu[1]([✉]), Carlos Quintero[2],
Wilfrido Moreno[1], and Alfredo Weitzenfeld[1]

[1] University of South Florida, Tampa, FL, USA
{ercan,juancalderon,mllofriu}@mail.usf.edu,
{wmoreno,aweitzenfeld}@usf.edu
[2] Universidad Santo Tomás, Bogotá, Colombia
{juancalderon,carlosquinterop}@usantotomas.edu.co
http://usf.edu/, http://www.usta.edu.co/

Abstract. An important area of research in humanoid robots is energy
consumption, as it limits autonomy, and can harm task performance.
This work focuses on power aware motion planning. Its principal aim is
to find joint trajectories to allow for a humanoid robot to go from crouch
to stand position while minimizing power consumption. Q-Learning (QL)
is used to search for optimal joint paths subject to angular position
and torque restrictions. A planar model of the humanoid is used, which
interacts with QL during a simulated offline learning phase. The best
joint trajectories found during learning are then executed by a physical
humanoid robot, the Aldebaran NAO. Position, velocity, acceleration,
and current of the humanoid system are measured to evaluate energy,
mechanical power, and Center of Mass (CoM) in order to estimate the
performance of the new trajectory which yield a considerable reduction
in power consumption.

Keywords: Humanoid · Dynamic modeling · Energy analysis · Optimization · Q-learning

1 Introduction

Energy efficiency is a significant challenge of humanoid robots, and mobile
robots in general. These robots contain many different components that consume energy, but a great portion is consumed by DC motors that transform
direct current to mechanical energy to drive them. While all components should
be analyzed for energy efficiency, DC motor activation and control consume most
of the energy required by many dynamic and static motion tasks.

E. Elibol—This work is funded by NSF IIS Robust Intelligence research collaboration
grant #1117303 at USF and U. Arizona entitled "Investigations of the Role of Dorsal
versus Ventral Place and Grid Cells during Multi-Scale Spatial Navigation in Rats
and Robots," and supported in part by the Agencia Nacional de Investigacion e
Innovación (ANII).

L. Almeida et al. (Eds.): RoboCup 2015, LNAI 9513, pp. 251–263, 2015.
DOI: 10.1007/978-3-319-29339-4_21

NAO robots are used in the Standard Platform League (SPL) of RoboCup. It is known that battery duration is one of the main constraints for longer humanoid robot autonomous performance. In order to use the SPL Robots for extended time, it is necessary to recharge batteries at least once during a single game.

Humanoids body weight, power needs and consumption of individual components play a significant role in energy utilization, balance and stability [1].

In terms of humanoid tasks, different approaches have been used to address the problem of stability. In [2], a humanoid robot stands up from sitting on a chair by using data previously collected from human demonstrations, where stable humanoid motion is accomplished by emulating human-like movements and speed. In [3], a three link simulated inverted pendulum learns to stand up using a tiered reinforcement learning method. A hierarchical architecture is applied on a three links two joints single legged robot during learning to stand up by trial and error. Our approach is dealing with a multiple goal settings; the robot has to learn the motion task while minimizing energy consumption. The study presented in [4] used a genetic algorithm fitness function to analyze the relationship between walking distance and energy consumption while keeping the knee joint on the supporting leg straight. In [5] a trajectory generation method for humanoid robots is proposed to achieve stable movement by using consumed energy as a condition, and generating a series of joint motions with a feedback technique to increase its stability. Reinforcement learning algorithms have been widely applied to other legged motions tasks [6]. Most of such work involves learning how to walk using biped robots [7–10]. Other related work includes learning to perform a penalty kick with a biped robot and learning to keep robot balance with an inverted pendulum model [11]. The work by Kuindersma et al. [12] had energy consumption optimization explicitly coded as a learning goal. Their work focused on moving the robot arms to compensate for balance disturbances and coded for the energy utilization of their movements in the cost function. Our model however, deals with the energy required to accelerate the whole robot body upwards, which requires a dynamic humanoid model of motions of the produced torque at each joint.

The motion of standing up from crouch position seems like a simple and common motion for humans, but it is quite complex, dynamic, and can become challenging for biped robots. Calderon et al. [13], present a joint stiffness control algorithm with the aim of reducing energy usage during the standing up procedure of a NAO robot. The goal of our new research is to optimize the standing up motion focusing on energy usage. In order to reduce energy consumption, a simulated kinematic and dynamic model uses Q-Learning to improve joint angular trajectories and implement an optimized route on a physical robot.

The rest of the paper is composed of Sect. 2 - Humanoid Modeling, Sect. 3 - Q-Learning Power Optimization, Sect. 4 - Energy and Power Performance Evaluation, Sect. 5 - Experimental Setup, Sect. 6 - Results, and Sect. 7 - Conclusions.

2   Humanoid Modeling

For humanoid modeling we used a NAO robot having 25 degrees of freedom (DoF), including two legs, two arms, a trunk, and a head, as shown in Fig. 1.

(a)   (b)

Fig. 1. Robot model and Humanoid Robot (NAO) in the sagittal plane.

A three DoF model in the x-z plane is used with the objective of reducing the complexity of the humanoid mathematical model (Fig. 1). The model has three joints and four links. The joints are ankle, knee, and hip pitch. The links represent foot, lower leg, thigh and trunk. The reduced dimensionality model is provided by the nature of the treated movement. This means that in a jumping movement both legs (left and right) are performing same action. The model is using the first two degrees of freedom (ankle and knee for both legs) and the last one corresponds to hip attached to the trunk. At the same time trunk is taken as a one mass, which includes arms, chest, and head.

Kinematic Model. The kinematical model is used to estimate the position of each joint and CoM for every link and the whole robot, see Fig. 1. L_i denotes the length of link i, θ_i is the absolute rotation of joint i, and L_{ic} shows the position of CoM for the corresponding link i.

Dynamic Model. The dynamic model is obtained using the Lagrangian formulation with the physical parameters of the NAO robot. The dynamic formulation has the following form in Eq. (1):

$$D(\theta)\ddot{\theta} + H(\theta, \dot{\theta})\dot{\theta} + G(\theta) = T_\theta \tag{1}$$

$D(\theta)$ is a 3×3 inertial matrix, $H(\theta, \dot{\theta})$ is a 3×1 vector of Coriolis and centrifugal forces, $G(\theta)$ is a 3×1 gravitational forces matrix. T_θ represents the torque vector and external applied forces at the joint. Finally the vectors θ, $\dot{\theta}$, $\ddot{\theta}$ represent rotational position, velocity, and acceleration of each joint.

Motor Model. The joint DC motor model used in simulations was estimated from collected actual motor responses to produce an approximation of each joint motor response during standup motions. A dynamic robot model used in simulations consists of three DoF. This simulation uses three different models, one for each joint. The transfer function model is shown in Eq. (2), parameters for the ankle are: $\omega = 0.0098$ rad/sec, $\zeta = 1.4$, for the knee are: $\omega = 0.0097$ rad/sec, $\zeta = 0.9$ and for the hip are $\omega = 0.0021$ rad/sec, $\zeta = 4.95$.

$$G_{(s)} = \frac{1}{\omega^2 s + 2\omega\zeta s + 1} \tag{2}$$

A second order system is used as the process model and the Predicting-Error Minimization (PEM) algorithm [14] is used as the estimation method. In order to increase the accuracy of the information given to the estimation algorithm, joint position and target position from the each joint motor was recorded previously while performing Aldebaran version of the stand-up process. The data was applied to PEM estimation method and the humanoid dynamic model to obtain an individual joint model as close to the real behavior of each motor as possible for the standing up motion.

3   Q-Learning Power Optimization

Q-Learning is a machine learning algorithm capable of learning a policy based solely in spurious feedback [15]. We used the canonical tabular version, in which the Q-value is updated according to Eq. (3), where α is a learning rate parameter.

$$Q(s, a) = Q(s, a) + \alpha\big(r + max_{a'}Q(s', a') - Q(s, a)\big) \tag{3}$$

A policy Π of which action to perform at each state can be derived from the Q table, as shown in Eq. (4).

$$\Pi : s \rightarrow a :: \Pi(s) = argmax_a Q(s, a) \tag{4}$$

The fact that this algorithm does not rely on a labeled training set, and does not rely on a model of action outcomes, make it suitable for its application to robotics. In fact, reinforcement learning algorithms including Q-Learning have been widely applied to robotics [6].

3.1   Power Aware Stand up Learning Algorithm

The proposed solution to the problem of learning to stand up with the minimum possible energy consumption is implemented as a QL algorithm. A planar humanoid robot is modeled with three joints: ankle, knee and heap, as explained in Sect. 2. The QL algorithm controls the ankle and knee joints only, whereas the hip joint position was set so the robot remains with the torso vertical to the ground.

The angular velocities and positions of the ankle and knee joints determine the state. Each joint state-space was discretized using a fixed length discretization step of $\pi/20$ rad. The same fixed length discretization was performed for velocities, with a discretization step of 0.1 rad/s.

The agent was allowed to perform one of three possible actions. Each of them performed a change on the ankle and knee velocities: decrement it, leave it unmodified or increment it. The decrements and increments were done by a fixed predefined value. Eq. (5) shows how the reward is computed. A negative reward is given whenever the humanoid performs a motion that leaves a joint in an invalid position (jointOutOfConstraints), according to the NAO robot limits. A negative reward is also given if the humanoid falls down (robotFell). It is considered to have fallen when the hip displacement along the sagittal plane is beyond a non-return point.

A positive reward is given if the humanoid reaches a target stand up position within some error tolerance (standingUp) and all joint angular velocities are below a threshold (notMoving). The position requirement is necessary for the humanoid to learn the task of standing up. The velocity constraints, on the other hand, ensures that the final inertia of the standup motion does not make the robot fall or force it to make a big energy effort to lower it. The average torque produced is subtracted from the positive reward value. This promotes solutions that minimize torque application, which in turn minimizes energy consumption. Only one non-zero reward is given in each episode, right at episode termination.

$$r = \begin{cases} -10, & \text{if jointOutOfConstraints or robotFell} \\ 3 - averageTorque, & \text{if standingUp and notMoving} \\ 0, & \text{otherwise} \end{cases} \quad (5)$$

Calibration. The ability of the learning algorithm to find a good solution depends on the value of a set of initial parameters. It was decided to perform a calibration process for the three parameters which we considered are the most important and they are: the learning rate α, the eligibility traces decay parameter γ and the exploration half-life decay parameter ϵ (ϵ-greedy with exponential decay). A coarse parameter sweep of 5 different values per parameter was performed. For each set of parameters, the algorithm was executed 5 times and the average reward was taken as a score. The set of parameters with the highest score was picked as the definitive set of parameter values.

4   Energy and Power Performance Evaluation

In order to evaluate power performance, we are assessing the average mechanical power, standard deviation and energy lost in every joint. Given joint j of the leg i, the mechanical power is the product of the motor torque τ and the angular velocity $\dot{\theta}$. The overall average power is obtained by averaging the mechanical absolute power delivered over a period T for all joints by Eq. (6):

$$P_{avm} = \frac{1}{T} \sum_{i,j} \int_0^T |\tau_{ij}\dot{\theta}_{ij}| dt \quad (6)$$

For some dynamic motions performed by humanoid robots, a sudden very high power demand can occur at the joints. Even though the average value of power usage can be small, the peak can actually be very high. The standard deviation measure is used to evaluate the distribution of power around the mean absolute power as seen in Eqs. (7) and (8):

$$P_{sd} = \sqrt{\frac{1}{T} \int_0^T \left(\sum_i \tau_{ij}\dot{\theta}_{ij} - P_{avm}\right)^2 dt} \tag{7}$$

For a humanoid robot, it is also necessary to consider the energy lost in the electric motors [16]. This can be defined as shown in Eq. (8):

$$E_{Lost} = \frac{1}{T} \int_0^T \tau^\intercal \tau dt \tag{8}$$

5   Experimental Setup

5.1   Learning Cycle

In order to be able to perform offline learning, a simulator was programmed using the motor model and the humanoids kinematic and dynamic models previously described in Sect. 3. Figure 2 shows the flow of events of a single iteration of Q-Learning episode. First, an action is selected by determining the state and querying the Q-Value table. Then, the motor models are used to compute the motor response to the required velocities. After that, kinematic models are applied to find joint positions, velocities and accelerations. This data is used by the dynamic model to compute the performed torques. Those torques, along with the kinematic information, are in turn used to compute the reward and update the Q-Value table.

A decision process is carried out to determine whether the episode has failed, succeeded, or it should continue. In the latter case, the cycle starts all over again.

Finally, the best standing routine was obtained by executing the calibrated algorithm 50 times. Then, the route with the highest reward was chosen.

5.2   Robot Execution

The obtained route was interpolated to a 4 second routine. This was done in order to be able to compare it with Aldebaran's stand-up routine, which was also set so as it would stand-up in 4 seconds. Figure 3 shows the robot at different points of the standing up routine. Then, the angleInterpolation function of the naoqi API was used. Twenty-five repetitions of the experiment were carried out for each routine. A custom made NAO local module *mllofriu/getSensorValues* (available in Github) was used to sample position and electric current values. The sample time was set to 10ms, which is the minimum loop latency allowed by the robot.

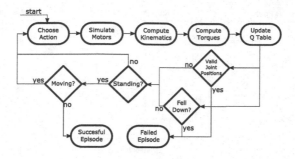

Fig. 2. The flow of events of a single iteration inside an episode.

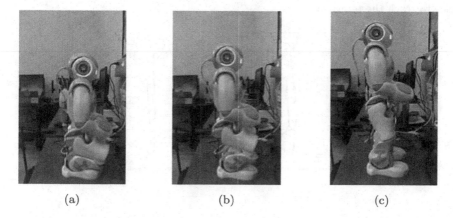

(a)   (b)   (c)

Fig. 3. The robot performing a stand up motion using the data obtained from Q-Learning.

6   Results

In this section, the results of both routines (Aldebaran routine and Q-Learning routine) are presented and discussed. The discussion is mainly based on the electric current consumption and position of each joint (ankle, knee, and hip), since the greater power consumption is located on these joints as discussed by Elibol et al. in [17]. The current consumption will be used to calculate electrical performance (motor input power) and the position will be used to calculate the location of Center of Mass, angular velocity, acceleration, produced torque, mechanical power and energy lost due to produced torque. The trajectories followed by joints in each routine are shown in Fig. 4. The Q-Learning trajectory is very different from Aldebaran's, which partially explains the difference in performance, as it will be discussed later in this section.

Figure 5 shows the humanoid robot executing the Aldebaran routine and the Q-Learning one. In this figure, the difference between the trajectories can be seen more clearly. Notice how Aldebaran trajectory tends to keep the body on top of the middle of the foot, while the Q-Learning trajectory leans all weight directly

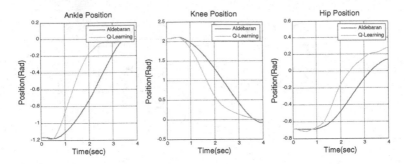

Fig. 4. Trajectories followed by every joint, Q-Learning trajectories follow different paths to reach the target positions.

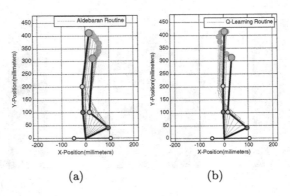

(a)   (b)

Fig. 5. Planar humanoid model shows Aldebaran (a) and Q-Learning (b) routines. Both routines show the initial and final position. Green circle is knee, blue circle is hip, and orange circle is head (Color figure online).

on top of the ankle joint. See also videos 1 (https://youtu.be/qsdiczXCBSQ) and 2 (https://youtu.be/YbnB6dx9cII) of the additional multimedia material for a sample learning iteration and a real robot testing iteration, respectively.

Because these trajectories are different, different current consumption are expected for each joint. Figure 6 shows the comparison of current profile between both routines for each joint.

The current required by ankle joint is far less for the Q-Learning routine than Aldebaran routine as shown by Fig. 6 and Table 1. This current saving is because of the decreased demand of produced torque at the ankle joint, which is effected by hip vertical position. A similar result is found when analyzing the hip current. Hip position is kept vertically during QL routine. The reduction of current of the hip and ankle joints responds to the Q-Learning reward schema, where lower torques are considered better. The knee, on the other hand, shows a slight increase in current consumption. Since hip position is moved to vertical position quicker in standing up motion (see Fig. 6b), this creates an increase torque consumption on the knee joints, which demands more current.

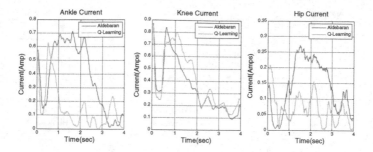

Fig. 6. Current consumption results averaged over 25 trials.

Table 1. Electrical current and power consumption comparison.

Joints	Average current Amp		Electrical input power Watt		
	QL	Aldebaran	QL	Aldebaran	Saved energy per link
Ankle	0.193	0.425	4.7864	10.54	54.59 %
Knee	0.401	0.348	9.9448	8.6304	−15.23 %
Hip	0.084	0.167	2.0832	4.1416	49.70 %
Total			16.814	23.312	27.87 %

Table 1 shows the average current consumption for each joint. A decrease of 54.59 % and 49.7 % in current consumption was achieved for the ankle and hip joints respectively. Also, it can be seen that the increase of the knee current consumption (15.23 %) is not as high as the amount saved by the other two joints together. This is also aligned with the Q-Learning reward schema, in which the overall torque is used, as opposed to minimizing each joint separately. By doing so, the algorithm was able to find a better tradeoff point between ankle, knee and hip joints applied torques minimizing current consumption. Other two important aspects of link movement are velocity and acceleration, since they directly affect the humanoids dynamics. This is shown in Eq. (3), where inertial matrix D and Coriolis and centrifugal matrix H depend directly on these variables.

Figure 7 shows the angular velocity and acceleration profile for each joint. The Q-Learning routine shows higher values of velocity and acceleration at the beginning of the movement. This increase is producing different effects on the performance of the routine. First, the trajectories of the joints were affected, which in turn affected the required torques needed to accomplish the routine. Secondly, the dynamics of the system were affected by quickly building up more inertia at the beginning and then reducing current and torque later. Figure 8 shows the required torque by each joint. The torque is reduced in the ankle and hip joints and slightly increased in the knee joint, which is in accordance with our analysis and supported by the experimental results.

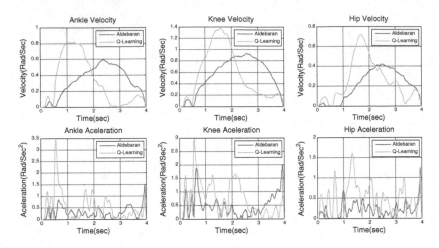

Fig. 7. Comparison of velocity and acceleration for: ankle(a), Knee(b), and Hip(c).

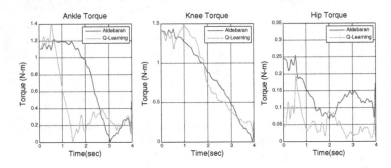

Fig. 8. Produced joint torque comparison between Aldebaran and Q-Learning routines.

The trajectories of Centers of Mass of both routines are shown in Fig. 9. The difference between them is highlighted. The Q-Learning location of CoM is causing a reduction of ankle and hip torques and an increase in knee torque as shown in Fig. 9. The mean and standard deviation of the CoM for both routines were calculated to have an idea about CoM Performance. The mean in the x axis for Aldebaran routine is 11.8 mm and 3.3 mm for QL. This means that the QL route kept the Center of Mass closer to the ankle joint than Aldebaran route. This is one of the reasons why this QL trajectory is saving energy. The standard deviation is 16.8 mm for Aldebaran and 11.8 mm for QL. This result suggests that QL is better at keeping balance.

The mechanical power performance of both routines was evaluated by calculating mechanical average power, standard deviation of mechanical power and lost energy due to required torque. Those parameters are shown in Table 2. The QL routine is producing less mechanical power and at the same time is losing

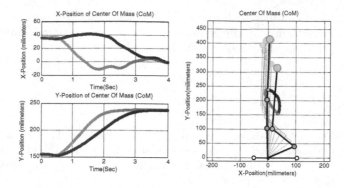

Fig. 9. CoM trajectories, x(t) &y(t), comparison between Aldebaran (blue), Q-L (green) routines (a), and spatial, y(t) vs. x(t), trajectory (b) (Color figure online).

Table 2. Produced mechanical power.

Joints	Mechanical power average $N.m.\frac{rad}{s}$		Mechanical power standard deviation $N.m.\frac{rad}{s}$		Lost energy Joule		Changes in lost energy
	QL	Aldebaran	QL	Aldebaran	QL	Aldebaran	
Ankle	0.133	0.204	0.233	0.195	0.614	0.86	-28.6%
Knee	0.555	0.421	1.26	0.888	0.963	0.955	$+0.8\%$
Hip	0.012	0.0237	0.012	0.05	0.065	0.147	-55.7%

less energy according with Eqs. (6, 7 and 8) previously presented. These results are consistent with the torque and velocity reduction shown above.

7   Conclusions

A dynamic model for a planar 4 link 3 joint robot is used with a Q-learning algorithm to learn how to stand up while reducing power consumption. Good quantitative and qualitative results are shown. The electrical power consumption was reduced for the ankle and hip joints, while the knee joint power consumption increased slightly. Mechanical energy loss is shown using different performance metrics as average mechanical power, standard deviation of mechanical power and energy lost due to required torque. By using new trajectories found by Q-learning for each joint, 28% of the electrical input power is saved for a single standing up routine. The Q-Learning strategy showed a better placement of the CoM over the ankle joint, greatly reducing the torque applied to it. In addition, a better management of inertia was observed, as the Q-Learning routine performed higher accelerations at the initial phases of the routine, lowering the torque required by both the hip and ankle joints later on. Additionally, a learning simulation platform was developed by integrating a motor model, dynamic and kinetic model of robot with a Q-Learning algorithm. Future work includes the use of this platform to optimize power consumption on more complex movements such as walking or jumping, which would increase the learning problem dimensionality.

References

1. Gonzalez-Fierro, M., Balaguer, C., Swann, N., Nanayakkara, T.: A humanoid robot standing up through learning from demonstration using a multimodal reward function. In: 2013 13th IEEE-RAS International Conference on Humanoid Robots (Humanoids), pp. 74–79. IEEE (2013)
2. Mistry, M., Murai, A., Yamane, K., Hodgins, J.: Sit-to-stand task on a humanoid robot from human demonstration. In: 2010 10th IEEE-RAS International Conference on Humanoid Robots (Humanoids), pp. 218–223. IEEE (2010)
3. Morimoto, J., Doya, K.: Acquisition of stand-up behavior by a real robot using hierarchical reinforcement learning. Robot. Auton. Syst. **36**(1), 37–51 (2001)
4. Yamasakitt, F., Endot, K., Kitanots, H., Asada, M.: Acquisition of humanoid walking motion using genetic algorithm - considering characteristics of servo modules. In: Proceedings of the 2002 IEEE, International Conference on Robotics 8 Automation, Washington, DC (2002)
5. Lei, X.-S., Pan, J., Su, J.-.B.: Humanoid Robot Locomotion. In: Proceedings of the Fourth International Conference on Machine Learning and Cybernetics, Guangzhou (2005)
6. Kober, J., Bagnell, J.A., Peters, J.: Reinforcement learning in robotics: a survey. Int. J. Robot. Res. **32**, 1238–1274 (2013)
7. Tedrake, R., Zhang, T.W., Seung, H.S.: Learning to walk in 20 minutes. In: Proceedings of the Fourteenth Yale Workshop on Adaptive and Learning Systems, vol. 95585 (2005)
8. Endo, G., Morimoto, J., Matsubara, T., Nakanishi, J., Cheng, G.: Learning CPG-based biped locomotion with a policy gradient method: application to a humanoid robot. Int. J. Robot. Res. **27**(2), 213–228 (2008)
9. Geng, T., Porr, B., Wrgtter, F.: Fast biped walking with a reflexive controller and real-time policy searching. Adv. Neural Inf. Process. Syst. **18**, 427–434 (2005)
10. Whitman, E.C., Atkeson, C.G.: Control of instantaneously coupled systems applied to humanoid walking. In: 2010 10th IEEE-RAS International Conference on Humanoid Robots (Humanoids), pp. 210–217 (2010)
11. Hester, T., Quinlan, M., Stone, P.: Generalized Model Learning for Reinforcement Learning on a Humanoid Robot. In: International Conference on Robotics and Automation (2010)
12. Kuindersma, S., Grupen, R., Barto, A.: Learning dynamic arm motions for postural recovery. In: 2011 11th IEEE-RAS International Conference on Humanoid Robots (Humanoids), pp. 7–12 (2011)
13. Calderon, J.M., Elibol, E., Moreno, W., Weitzenfeld, A.: Current usage reduction through stiffness control in humanoid robot. In: 8th Workshop on Humanoid Soccer Robots, IEEE-RAS International Conference on Humanoid Robots (2013)
14. Ljung, L.: System Identication - Theory for the User, 2nd edn. Prentice-Hall, Upper Saddle River (1999)
15. Sutton, R.S., Barto, A.G.: Reinforcement Learning: An Introduction. MIT Press, Cambridge (1998)

16. Silva, F.M., Machado, J.A.T.: Energy analysis during biped walking. In: Proceedings IEEE International Conference Robotics and Automation, vol. 1–4, pp. 59–64 (1999)
17. Calderon, J., Weitzenfeld, A., Elibol, E.: Optimizing energy usage through variable joint stiffness control during humanoid robot walking. In: Behnke, S., Veloso, M., Visser, A., Xiong, R. (eds.) RoboCup 2013. LNCS, vol. 8371, pp. 492–503. Springer, Heidelberg (2014)

An Episodic Long-Term Memory for Robots: The Bender Case

María-Loreto Sánchez[1,2](✉), Mauricio Correa[1,2], Luz Martínez[1,2], and Javier Ruiz-del-Solar[1,2]

[1] Advanced Mining Technology Center, Universidad de Chile, Santiago, Chile
`loreto.sanchez@ing.uchile.cl`,
`{mauricio.correa,luz.martinez,javier.ruizdelsolar}@amtc.cl`
[2] Department of Electrical Engineering, Universidad de Chile, Santiago, Chile

Abstract. The main goal of this paper is to propose a framework for providing an episodic long-term memory for a robot, which includes methods for acquiring, storing, updating, managing and using episodic information. This will give a robot the ability to incorporate past experiences when interacting with humans, so that the data that the robot learns transcends each session, and thus gives continuity to its activities and behaviors. As a proof of concept, the implementation of an episodic long-term memory for the Bender robot is described. This includes the implementation and evaluation of a behavior called *Conversation*, which allows Bender to interact with people using the information stored in the episodic memory.

Keywords: Service robots · Human-robot interaction · Episodic memory · RoboCup@Home

1 Introduction

A key aspect in achieving a long-term interaction and social relationship between humans and service robots is the requirement of a memory system for the latter. Memory constitutes an important part of a cognitive system implementation; in fact, it is the link of prior experiences with ongoing and future behavior. As stated by [1], a nontrivial level of social interaction requires that the robot should be able to use both semantic and episodic information. Both, semantic and episodic memories constitute the long-term memory. While semantic memory is a repository for facts, such as knowing that the capital of Chile is Santiago, episodic memory is the memory of past experiences, i.e., remembering names and places.

The amount of information to be processed in a lifetime is vast; therefore, efficient methods are required for acquiring, filtering, storing and updating a robot's episodic knowledge of its working environment. This information can be encoded in symbolic form and held in a storage module invoking the functionality of an episodic memory system. Episodic memory, first defined by Tulving [2],

© Springer International Publishing Switzerland 2015
L. Almeida et al. (Eds.): RoboCup 2015, LNAI 9513, pp. 264–275, 2015.
DOI: 10.1007/978-3-319-29339-4_22

refers to the memory of specific events occurring at a specific place and time and enables human beings to remember past experiences.

In this context, the main goal of this paper is to propose a framework for providing an episodic long-term memory for a robot, which includes methods for acquiring, storing, updating, managing and using episodic information. This will give robots the opportunity to incorporate knowledge of each working session within its memory, so that the data it learns transcends each session, and thus gives continuity to its activities and behaviors.

It must be stressed that the update of episodic information is a complex and time-consuming process that involves the consolidation of short-term memories into long-term ones [3]. In the case of humans this is carried out during the sleep process, and consumes a large amount of brain resources [4]. Therefore, the update of episodic information must be carefully designed and implemented in the robot case.

As a proof of concept, the implementation of the episodic long-term memory for the Bender robot is described.

The paper is structured as follows. Section 2 presents the related work. Section 3 describes the episodic long-term memory framework. Section 4 presents an implementation of the framework in the service robot Bender. This includes an evaluation of a behavior called *Conversation*, which allows the robot to interact with people using the information stored in the episodic memory. Finally, Sect. 5 presents conclusions and future works.

2   Related Work

During the past decade there have been several approaches in implementing episodic long-term memory in artificial systems. In 2004, Nuxoll & Laird [5] presented an implementation of a computational model of episodic memory in Soar [6], a cognitive architecture based on production rules. They developed a pacman-like domain moving around a grid searching for food-points as fast as it could, using its episodic memory to aid in selecting which direction it should move. In 2005 Ratanaswasd et al. [7] adapted human cognitive abilities to be used for task execution and control in their humanoid robot ISAC. With this, the robot acquired its knowledge through learning and past experience, previously stored within memory structures and retrieved during task executions. In the same line of work, Dodd & Gutierrez [8] connected a machine emotion system with ISAC's episodic memory system, this with the goal of producing more intelligent behaviors. They tested and analyzed the system through the explanation of a case that used a combination of emotional and statistical information to retrieve the "correct" episode. A cognitive control architecture for an artificial creature named RITY was presented in [9]. This architecture incorporates an episodic memory defined as a scalable structure that stores episodic perceptual snapshots as Rity's experience increases. This system also utilizes a temporally variant spatial map to store spatial information and a higher-level procedural memory using Finite State Machines. Jockel et al. [10] proposed

EPIROME, a framework to develop and investigate high-level episodic memory mechanisms applied in the domain of service robotics, enabling their service robot TASER to collect autobiographical memories to improve action planning based on past experiences. A model of episodic memory for autonomous agents and its implementation was presented by Deutsch et al. [11]. This model was derived from theories from psychology and neuro-psychoanalysis, and their simulation showed that agents with the episodic memory adapt faster to new situations than memory-less agents. Spexard et al. [12] presented an approach using a biologically inspired memory system consisting of short-term, scene, and long-term memory for autonomous mobile robots operating in natural, human centered environments, allowing robotic systems to store high level information as gestures or objects. In their work in [13], Ho et al. proposed a memory model for a long-term artificial companion, which migrates among virtual and robot platforms based on the context of interactions with the human user. The long-term memory contains episodic events that are chronologically sequenced and derived from the companion's interaction history, both with the environment and the user. In [14] Winkler et al. proposed CRAMm, a memory management system that can record comprehensive and informative memories without slowing down the operation of a PR2 robot. CRAMm also offers a query interface that allows the robot to retrieve the kinds of information stated above.

In this work we propose a simple and versatile episodic long-term memory framework, which uses two types of memory: short-term and long-term. For any interactive task, three simple actions are defined: acquisition, storage/update, and retrieval of information. The retrieval of episodes is accomplished through a simple access to the short-term and long-term memory databases. The simplicity of this framework allows it to be applied to different types of robots.

3   Episodic Long-Term Memory Framework

The proposed framework will enable robots to acquire, filter, store, update, manage and use episodic knowledge of its working environment. In particular, the robot will be able to store non-redundant information about people and objects with which it has interacted in a database, as well as places and dates where working sessions have been carried out, and behaviors that have been performed.

Having a long-term memory would allow a high-level behavioral system to be created, enabling robots to interact with people in richer forms, since robots could use previous memories of places, people, objects or performed behaviors, enhancing the interaction and establishing long-term relationships. This memory could be the link of prior experiences with ongoing and future behavior.

3.1   Long-Term Memory Implementation

The information that the episodic long-term memory will manage is (i) session's information: session's number (session's ID), date of start and end of session, time of start and end of session, country and city where the session is carried out, and first person seen in session; (ii) information about people with whom the robot has interacted: person's id, person's name and last name, facial image, age, city of origin and country of origin (in case this information is available); (iii) information about manipulated objects: object's id, object's name and object's picture; and (iv) information about the different behaviors performed: behavior's id, name of behavior, session in which the behavior is being performed and person performing it. This information to be stored in the memory of Bender was selected because it is the basic information needed to establish a long-term relationship with users.

The so-called *Memory Management* module manages all the mentioned information, and it decides which data is saved to the short-term memory and then to the long-term memory, and which one is saved directly to the long-term memory (Fig. 1). The session's number will be obtained based on the latest saved session in the long-term memory database. The city and the country where the session is carried out will be introduced by hand in text format before the session starts, i.e. when the session is called. Alternatively this information can be obtained from a GPS receiver. The date and the time of session will be obtained from the date and time of the computer. Session information will be transferred periodically to the long-term memory during the course of the session.

A so called *Interaction* module will be responsible to get the remaining episodic information to be passed to the memory management module, and finally transferred to the short-term memory. Person's name, last name, and facial image will be obtained by means of a face recognition behavior, which searches in the long-term and short-term memory databases for faces recognized from previous sessions and faces enrolled on the current session, respectively. Person's new ID will be obtained based on the latest saved new person in the long-term memory database, for first new person seen on session, or will be obtained based on the latest saved new person in the short-term memory database. Object's image will be obtained by an object recognition behavior, and passed to the memory management module through the interaction module. Object's ID will be obtained from the long-term memory database. Alternatively, this information can be obtained from Internet object's databases [15,16].

Information from the short-term memory will be selected and transferred to the long-term memory after each session. Figure 1 shows the general diagram for the process of long-term memory information saving.

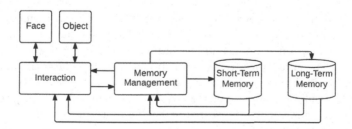

Fig. 1. General diagram for the process of long-term memory information saving.

3.2 Information Storage and Update

A database with 10 different tables was created to store the described information (Fig. 2). The *session* table stores the information about the session being carried out, the *person* and *imageper* tables store the information about people with whom the robot has interacted, the *object* and *imageobj* tables store the information about manipulated objects, and the *behavior* table stores the information about the different behaviors performed. The remaining 4 tables are intended to store the connections between the previously described tables: *session_person*, *session_object*, *session_behavior* and *behavior_person*.

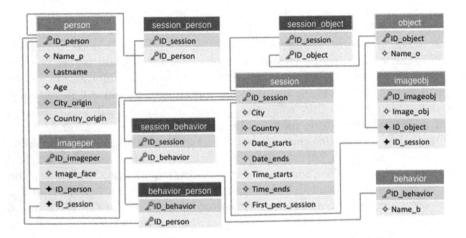

Fig. 2. Database diagram.

In order to describe the storage process, four different modules are implemented: Session Information, Face Information, Object Information and Behavior Information. Figure 3 shows the flow diagram of the storage information process.

- *Session Information*: When session starts, information of the session's place, the date of beginning and end, and the time of beginning and end of the session will be saved in session table in long-term memory. The date and time of the end of the session, as well as the first face enrolled or recognized in session will be transferred to session table in long-term memory.
- *Face Information*: Whenever the robot enrolls a new face, the module will store the person's first name and last name, face image and session's number in the short-term memory. When the robot recognizes a face, the module will store the person's face image and session's number in the short-term memory.
- *Object Information*: Whenever an object is recognized, the object's image and session's number will be stored at the short-term memory.
- *Behavior Information*: If a behavior has been performed during the session, the information of the session and person performing the behavior will be

saved in short-term memory. While performing a behavior that uses the information in long-term and short-term memory, the robot will check if further information of the person needs to be stored (age, city and country of origin); if so, it will ask and save the information in its short-term memory.

As already stated, it is necessary to do offline updates, where short-term memory information must be analyzed and selected to be transferred to the long-term memory. These offline updates will be carried on after every working session, were data will be reviewed and information redundancy will be deleted. Particularly, images of people or object can be saved several times per session, resulting in excessive memory use, thereby only two images of each person and two images of each object will transferred to the long-term memory.

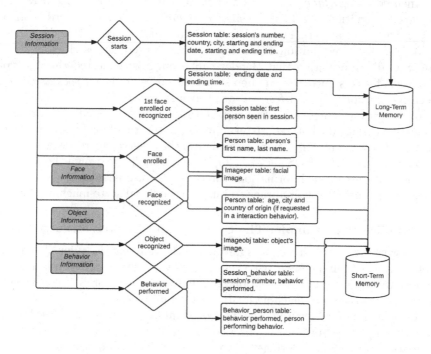

Fig. 3. Flow diagram for the storage process.

3.3   Use and Management of Information

Once there is data stored in the long-term memory, the robot will be able to retrieve it and use it in any working session. During sessions both, short-term and long-term memory will be accessed. Long-term memory for retrieving information from previous sessions and short-term memory for retrieving information saved in the current session.

Different behaviors intended for human-robot interaction can be implemented, in which the robot can retrieve information saved in the past and use it during the interaction with the human user, as well as store new information acquired.

4   Proof of Concept: The Bender Case

For the implementation of the proposed framework, an episodic long-term memory model has been developed and tested in the social robot Bender [17,18]. In this episodic long-term memory, data described in the previous section will be stored. As an example of a possible behavior, in order to illustrate the concept, a behavior called *Conversation* was created to retrieve and use the stored information in such way to enable Bender to interact with humans.

4.1   Computational Implementation

Bender is running the ROS communication middleware [19]. Face detection is implemented using *Haar features*, as explained in the work of Viola & Jones [20]; face recognition is achieved using *local binary patterns*, described in the work of Ahonen et al. [21]. The implemented face analysis module is described in [22,23]. For object recognition, SIFT descriptors are used, as described in [24,25]. Some episodic information that Bender can acquire is requested through voice commands, for this purpose two open source libraries are used: *pocketsphinx* for speech recognition [26], and *festival* for speech synthesis [27].

For the storage stage, a PostgreSQL database server is used for storing and managing the data for both short-term and long-term memory [28]. In order to achieve real-time communication with the database server to retrieve the information, the *sql_database* ROS package [29] is used. Finally, to design the behaviors that will make use of the information stored in the episodic memory, *smach* [30], a ROS-independent Python library to build hierarchical state machines, is used.

4.2   Application Example: The Starting and Conversation Behavior

When a working session starts, it is initialized the *Starting* behavior. The goal of this behavior is to find people, ask them if they want to perform the last carried out activity or just talk; it additionally saves the behavior to be performed. Figure 4 shows the state machine for the *Starting* behavior.

The behavior *Conversation* uses the stored information and enables the robot to recognize people from previous sessions, enroll new people and save further information from them. During the conversation, the robot can say if he has already been to the place where the session is occurring, he can also tell someone he recognizes, when and where they met for the first or the last time.

When starting the *Conversation* behavior, Bender will search for someone to talk with. If Bender does not find any human face, to enroll or recognize, he will ask for someone to come close and interact with him. If he detects a face and he is not able to recognize the person, he will ask for the name. Then, whether meeting for the first time or recognizing a person (previously met), Bender will announce that he is waiting for questions. He will be able to answer a series of different questions that the person in front of him will ask. If Bender does not hear anything from the person in front of him, he will require information he is missing from that person in order to save it in his database. Figure 5 shows the state machines for the *Conversation* behavior.

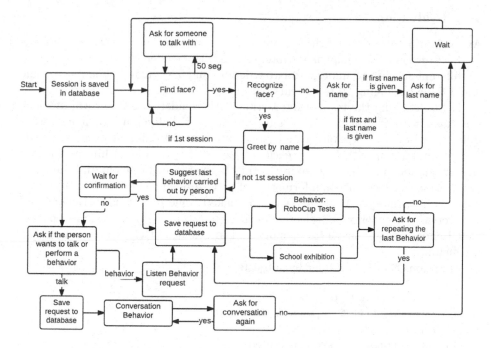

Fig. 4. *Starting* behavior to begin the working session.

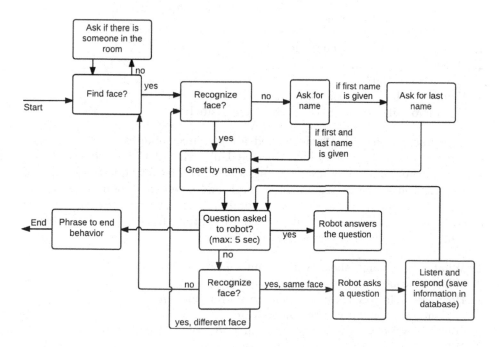

Fig. 5. State machine for the *Conversation* behavior.

4.3   Evaluation

An evaluation of the use of the episodic long-term memory framework by means of the *Conversation* behavior for interaction with Bender was carried out with 13 different subjects: 8 undergraduate students and 5 graduate students, 12 male and 1 female, ranging 20 to 32 years old.

Before starting the tests of interaction with Bender, the long-term memory was filled with information of 5 different persons, 4 different countries and 5 different cities were sessions were carried out, and 10 different objects. In the test of interaction with Bender, each subject stood in front of him and began the interaction. The first part of it was the stage of face enrolment and then face recognition (for recognition, information of just enrolled faces was recovered from short-term memory). After this, subjects asked some questions to Bender and vice versa. The criterion for selection of these questions is to present information that Bender already has in his memory and information that Bender can add to his memory.

Questions asked to Bender by subjects:

1. Do you know <name of person>?
2. Have you ever been to <place>?
3. Do you know someone from <place>?
4. How many countries have you visited?
5. How many cities have you visited?

Questions asked by Bender:

1. Can you tell me your name please?
2. Where are you from?
3. How old are you?
4. Would you like to know the countries I have visited?
5. Would you like to know the cities I have visited?

The interaction with Bender lasted about three minutes and was carried out in the Laboratory of Robotics of the University of Chile. After this, the subjects answered a survey with six questions. The first four questions required a numerical answer with values between 1 and 5, where 1 represents a very negative opinion and 5 a very positive opinion. The last two questions required a more open response, where subjects could express the opinion on the tested system. The questions are the following:

1. What do you think about waiting times in the interaction?
2. What do you think about the interaction?
3. How useful do you think is a memory system for a robot?
4. How natural was for you the interaction?
5. What relevant information do you think the robot should store?
6. What improvements to the system would you propose?

Table 1. Answers for questions 1 to 4 in survey.

Subject													Average
	1	2	3	4	5	6	7	8	9	10	11	12	13	
Q 1	4	3	3	4	3	2	3	4	4	4	3	3	3	3.3
Q 2	5	4	5	5	5	3	4	5	3	4	4	4	3	4.2
Q 3	5	5	5	3	5	5	5	4	5	5	5	5	5	4.8
Q 4	4	4	3	4	4	3	4	4	4	3	3	3	2	4.5

Table 1 shows the answer for questions 1 to 4. As can be seen, waiting time for questions and answers were the worst evaluated item (3.3 out of 5.0), and the perception of how useful memory is for robots, was of higher importance (4.8 out of 5.0). About interacting with the robot, subjects were comfortable and attracted by the opportunity to talk with Bender, which it was reflected in terms of what they thought of the interaction (4.5 out of 5.0), and naturalness of this in the survey performed (4.2 out of 5.0).

For question number 5, subjects found important to keep information of first name, last name, age, facial image, and place of origin of the person, as well as date and place of interaction, which is what is being stored at the moment. Likewise, subjects thought that occupation or job information, tastes and preferences of people, date of birth, relationship between people (friendship or relationship), type of interaction performed, locations of objects, location maps, mood of people and health problems, among others, would be useful to be stored.

For illustrative purposes, in [31] is shown a video of a user interacting with Bender using the behavior *Conversation*.

5 Conclusions and Future Work

Service robots need to interact with human beings to fulfill their purpose. An episodic memory is fundamental to achieving this objective, since long-term memory is composed of the sum of episodes over time, and from the retrieval of these episodes it can be built interesting and complex behaviors to achieve more realistic interactions.

In this work a framework for providing an episodic long-term memory for a robot is presented. This framework included methods for acquiring, storing, updating, managing and using episodic information. This framework was implemented and tested in the service robot Bender, using a behavior for human-robot interaction, called Conversation, in order to illustrate the concept. This behavior was created to retrieve and use the stored information in such a way to enable Bender to interact with humans in real-time, as well as to store new information learned during the interaction.

Interaction tests were performed with 13 different users. In these tests, the importance of long-term memory in conversational behavior in the relationship between humans and service robots was evaluated, finding that people who interacted with the robot could feel a greater closeness and empathy with him seeing that he could answer their questions satisfactorily. However, waiting time between questions and answer in the conversation was evaluated with regular rates (3.3 out of 5.0), and is an aspect that needs to be improved.

Test subjects found that information being currently stored in the episodic memory is very important and they suggested further information to be stored, such as occupation or job information, tastes and preferences of people, date of birth, relationship between people (friendship or relationship), type of interaction performed, locations of objects, location maps, mood of people and health problems.

In future works we are going to develop a more intuitive behavior, in order to increase the capability to maintain a more natural conversation with the users. In the same line, waiting times between questions and answers will be lowered to facilitate an even more natural interaction.

Acknowledgments. This work was partially funded by FONDECYT Project 1130153.

References

1. Wood, R., Baxter, P., Belpaeme, T.: A review of long-term memory in natural and synthetic systems. Adapt. Behav. **20**(2), 81–103 (2012)
2. Tulving, E.: Episodic and semantic memory. In: Donaldson, W. (ed.) Organization of Memory, pp. 381–403. Academic Press, New York (1972)
3. Bailey, C.H., Bartsch, D., Kandel, E.R.: Toward a molecular definition of long-term memory storage. Proc. Nat. Acad. Sci. USA **93**(24), 13445–13452 (1996)
4. Walker, M.P., Stickgold, R.: Sleep-dependent learning and memory consolidation. Neuron **44**, 121–133 (2004)
5. Nuxoll A., Laird, J. E.: A cognitive model of episodic memory integrated with a general cognitive architecture. In: Proceedings of 6th International Conference Cognition Modeling-ICCM 2004, Pittsburgh, PA, pp. 220–225 (2004)
6. Laird, J.E., Newell, A., Rosenbloom, P.S.: SOAR: an architecture for general intelligence. Artif. Intell. **33**(1), 1–64 (1987)
7. Ratanaswasd, P., Gordon, S., Dodd, W.: Cognitive control for robot task execution. In: IEEE International Workshop on Robot and Human Interactive Communication (RO-MAN), Nashville, Tennessee, 13–15 August (2005)
8. Dodd, W., Gutierrez, R.: The role of episodic memory and emotion in a cognitive robot. In: Proceedings of 14th Annual IEEE International Workshop on Robot and Human Interactive Communication (RO-MAN), pp. 692–697, Nashville, TN (2005)
9. Kuppuswamy, N.S., Cho, S., Kim, J.: A cognitive control architecture for an artificial creature using episodic memory. In: Proceedings of the 3rd SICE ICASE Internatioanl Joint Conference, pp. 3104–3110 (2006)
10. Jockel, S., Weser, M., Westhoff, D., Zhang, J.: Towards an episodic memory for cognitive robots. In: Proceedings of 6th Cognitive Robotics workshop at 18th European Conference on Artificial Intelligence (ECAI), pp. 68–74 (2008)

11. Deutsch, T., Gruber, A., Lang, R., Velik, R.: Episodic memory for autonomous agents. In: Proceedings of IEEE HSI Human System Interactions Conference, Krakow, Poland, 25–27 May (2008)
12. Spexard T.P., Siepmann F., Sagerer G.: Memory-based software integration for development in autonomous robotics. In: International Conference on Intelligent Autonomous Systems, pp. 49–53, Baden-Baden, Germany (2008)
13. Ho W.C., Lim M.Y., Vargas P.A., Enz S., Dautenhahn K., Aylett, R.: An initial memory model for virtual and robot companions supporting migration and long-term interaction. In: Proceedings of the 18th IEEE International Symposium on Robot and Human Interactive Communication, pp. 277–284. Toyama, Japan, September 27-October 2 (2009)
14. Winkler, J., Tenorth, M., Bozcuoglu, A., Beetz, M.: CRAMm - memories for robots performing everyday manipulation activities. In: Proceedings of the Second Annual Conference on Advances in Cognitive Systems, pp. 91–108 (2013)
15. 3D Warehouse. https://3dwarehouse.sketchup.com
16. Roboearth.: World Wide Web for robots. http://api.roboearth.org
17. Martínez, L., Pavez, M., Olave, G., Correa, M., Sánchez, M.L., Loncomilla, P., Ruiz-del-Solar, J.: UChile homebreakers team description paper (2015). http://bender.li2.uchile.cl/Files/TDP/UChileHomeBreakers_TDP2015.pdf
18. Bender.: a Personal Robot. http://bender.li2.uchile.cl/
19. Quigley, M., Conley, K., Gerkey, B., Faust, J., Foote, T., Leibs, J., Berger, E., Wheeler, R., Ng, A.: ROS: an open-source Robot Operating System. In: IEEE International Conference on Robotics and Automation (ICRA), Kobe, Japan (2009)
20. Viola, P., Jones, M.: Rapid object detection using a boosted cascade of simple features. In: Conference on Computer Vision and Pattern Recognition (CVPR), pp. 511–518 (2001)
21. Hadid, A., Ahonen, T., Pietikäinen, M.: Face recognition with local binary patterns. In: Pajdla, T., Matas, J.G. (eds.) ECCV 2004. LNCS, vol. 3021, pp. 469–481. Springer, Heidelberg (2004)
22. Ruiz-del-Solar, J., Verschae, R., Correa, M.: Recognition of faces in unconstrained environments: a comparative study. EURASIP Journal on Advances in Signal Processing. Recent Advances in Biometric Systems, A Signal Processing Perspective) (2009)
23. Verschae, R., Ruiz-del-Solar, J., Correa, M.: A unified learning framework for object detection and classification using nested cascades of boosted classifiers. Mach. Vision Appl. **19**(2), 85–103 (2008)
24. Ruiz-del-Solar, J., Loncomilla, P.: Robot head pose detection and gaze direction determination using local invariant features. Adv. Robot. **23**(3), 305–328 (2009)
25. Martínez, L., Loncomilla, P., Ruiz-del-Solar, J.: Object recognition for manipulation tasks in real domestic settings: a comparative study. In: Bianchi, R.A.C., Akin, H.L., Ramamoorthy, S., Sugiura, K. (eds.) RoboCup 2014. LNCS, vol. 8992, pp. 207–219. Springer, Heidelberg (2015)
26. Speech recognition system pocketsphinx. http://www.speech.cs.cmu.edu/pocketsphinx/
27. Speech synthesis system festival. http://www.cstr.ed.ac.uk/projects/festival/
28. Open Source Database PostgreSQL. http://www.postgresql.org/
29. ROS.org: sql_database. http://wiki.ros.org/sql_database
30. ROS.org: smach. http://wiki.ros.org/smach
31. Demonstrative video of the Conversation behavior. http://youtu.be/pg_UW02LuGQ

Context-Based Coordination for a Multi-Robot Soccer Team

Francesco Riccio$^{(\boxtimes)}$, Emanuele Borzi, Guglielmo Gemignani,
and Daniele Nardi

Department of Computer, Control, and Management Engineering,
Sapienza University of Rome, via Ariosto 25, 00185 Rome, Italy
{riccio,borzi,gemignani,nardi}@diag.uniroma1.it

Abstract. The key issue investigated in the field of Multi-Robot Systems (MRS) is the problem of coordinating multiple robots in a common environment. In tackling this issue, problems concerning the capabilities of multiple heterogeneous robots and their environmental constraints need to be faced. In this paper, we introduce a novel approach for coordinating a team of robots. The key contribution of the proposed method consists in exploiting the rules governing the scenario by identifying and using "contexts". The robots actions and perceptions are specialized to the current context to enhance both single and collective behaviors. The presented approach has been largely validated in a RoboCup scenario. In particular, we adopt a soccer environment as a testing ground for our algorithm. We evaluate our method in several testing sessions on a simulator representing a virtual model of a soccer field. The obtained results show a substantial improvement of the team adopting our algorithm.

Keywords: Multi-robot coordination · Context-awareness · RoboCup soccer

1 Introduction

In recent years, researchers managed to develop intelligent robots for a wide variety of environments outside research labs. Nowadays, robotic deployment reaches a broad range of fields, finding several applications such as in human-dangerous environment explorations, surveillance, or health care assistance. In such environments, often multiple robots are required to cooperate to achieve a higher effectiveness or carry out tasks that otherwise could not be completed.

In fact, Multi-Robot Systems (MRSs) present many advantages with respect to single robot systems. A MRS is to be considered more robust, with respect to system failures; scalable, depending on the environment specification and task requirements; and more efficient in performing a given task. Multi-Robot Systems have been widely studied in the framework of RoboCup competitions.

The purpose of this work is to present new approaches and methodologies to coordinate multiple autonomous robots and to guarantee their effective behaviors. Specifically, we develop our work in the soccer RoboCup scenario where a

© Springer International Publishing Switzerland 2015
L. Almeida et al. (Eds.): RoboCup 2015, LNAI 9513, pp. 276–289, 2015.
DOI: 10.1007/978-3-319-29339-4_23

team of robots is required to coordinate to effectively play a soccer game. In this scenario, we present an algorithm that exploits the high level information of occurring situations to obtain a specific behavior in response of multiple environmental stimuli. The aim of this work is to provide a high level of knowledge about the current state of the world, allowing a team of robots to have a more effective way of perceiving the environment and the entities in it. The key contribution of this paper is an approach for modeling the context features of a particular environment and an algorithm for integrating different coordination techniques for a team of robots. The approach has been deployed on several simulated and real robots, including a team of humanoid NAOs. On these robots, we carried out multiple experiments to evaluate the effectiveness of our contribution.

In the remainder of the paper, we first present an overview of related work, focusing on past research on multi-robot coordination. Next, we describe our approach to coordination highlighting all of our contributions thoroughly. Then, we present an application of the approach to the case of a team of humanoid robots in a soccer scenario. This setting is then used to quantitatively evaluate the proposed approach. Finally, we conclude with a discussion of our contribution and remarks on future work.

2 Related Work

The coordination of multiple robots has been broadly studied during the last few years, considering multiple scenarios and heterogeneous agents that need to operate in a specific environment. In particular this problem has been broadly studied in the RoboCup soccer community where a team of robots needs to autonomously play a soccer game against another robotic team.

One of the first attempts to coordinate a team of heterogeneous robots was proposed by a joint project of seven different Italian universities. The ART-Azzurra-99 Team [6], later extended in [3], developed a coordination system able to efficiently coordinate heterogeneous agents in a team. Their approach relied on a task assignment technique. The algorithm automatically distributes tasks that the team need to accomplish, based on auction techniques.

An alternative algorithm is given in [2] where an asynchronous distributed system for task allocation which either relies on the perception of each robot or on a token passing approach in order to allocate the robots within the team.

Lou et al. [4] propose an improved algorithm for task allocation based on an auction system. They divide the set of possible tasks in subgroups and assign a task to each robot without violating precedence constraints among tasks.

Wiegel et al. in [9] propose a task allocation for the soccer middle-size league[1] based on utility estimations. First, they define the set of preferred poses for the team depending on the current situation, and then compute the utility values with respect to generate set of reference poses.

---

[1] http://www.robocup.org/robocup-soccer/middle-size/.

More recently, MacAlpine *et al.* proposed a more advanced form of robot coordination [5]. In this article, the authors introduce a formation system algorithm that is exploited within the 3D simulation league. The algorithm computes a global world model that is shared between the agents and is locally evaluated. After each evaluation each robot broadcasts the obtained result. The team is split in an offensive and in a defensive group and the role of each member is assigned depending on the ball position and the distance from specific positions.

Finally, additional solutions to the heterogeneous robot coordination challenge were proposed in [1,7,8]. These solutions respectively rely on an estimation of the world-state, an estimation of a mapping function between robots and tasks or between robots and roles. In these works the authors employ generic world-state evaluations [1], specific world rules [7] or utility estimation functions and artificial potential fields to position the robots within the environment [8].

Considering the analyzed approaches, we notice that they focus either on sharing encoded information among the team (*local estimation*) or on reconstructing a suitable interpretation of the world with respect to each single robot (*distributed world knowledge*). Conversely, our method focuses on integrating such approaches according to the environment model. In fact, we propose a coordination algorithm, based on both a distributed world knowledge and task-role assignments, as described in the next section.

3   Approach

Our approach relies upon two well known methods for coordinating a team of robots: *distributed task assignment* and *distributed world modeling*. In order to coordinate a team, distributed task assignment relies on the exchange between robots of meaningful task-related values. Generally, such task-related values are utility estimations with respect to a given task. Conversely, distributed world modeling exploits the direct exchange between robots of their internal world representation. The proposed approach aims at combining the robustness of the two approaches. In the rest of this section, we describe the two main components on which our approach relies upon, namely the Coordination System and the Context System.

3.1   Coordination System

The coordination system is in charge of generating a suitable mapping function between the set of acting robots R and the set of tasks T. We conceptually separate the coordination system in two main steps. First, we update a *distributed world model* in accordance with the *events* occurring during the game. Then, we exploit the generated world model to compute utility estimations that are used to assign tasks among the robots.

Distributed World Modeling. The Distributed World Model (*DWM*) is defined as a dynamic global world knowledge about the current state of the environment and status of the task. The *DWM* is formalized by considering a set of partial models, each of which is a local representation of the world state for the i-*th* robot. Thus, given a team of robots $R = \{r_1, r_2, ..., r_n\}$, a *distributed world model DWM* is defined as the knowledge of the world reconstructed from a set of partial models $LM = \{LM_1, LM_2, ..., LM_n\}$. Formally, if for each robot r_i we define $LM_i(t)$ as the local model of the robot r_i at a particular time t, then we can define the distributed world model of the team as

$$DWM(t) = f(LM, t) \tag{1}$$

where f is a *reconstruction function* generating the distributed world model considering the partial models of each individual agent. The reconstruction function needs to be specified depending on the environment constraints and task specification. However, exchanging local models has an high computational cost and it is time consuming. Moreover, it assumes a reliable network condition which is hardly verified in real applications.

To overcome this issue, we design an event-based system which allows the robots to infer the local model for the i-*th* robot at time t by evaluating $LM_i(t-1)$ and the occurring events. Hence, we define the *model update function* $\psi(\cdot)$ which takes as input environment dependent events E and an estimation of the previous local models $\overline{LM}(t-1)$, returning the updated local models $\overline{LM}(t)$:

$$\overline{LM}(t) = \psi(E(t-1), \overline{LM}(t-1)) \tag{2}$$

where $E(t-1)$ are the events occurred at t-1.

Accordingly, we reformulate the *reconstruction function* as:

$$DWM(t) = f(\overline{LM}(t), t) \tag{3}$$

Task Allocation. Depending on the status of the global world model DWM, we adapt the utility function of the team to maximize the performance with respect to the common goal. The key idea is that a static, unique utility function cannot fulfill the requirements imposed by the game in every situation. To achieve this flexibility, we develop a task assignment routine based on utility estimations, and as we will explain in the next section, is *context-dependent*. This routine is an instance of a marked based technique which evaluates at any time the configuration of the robots within the environment, generating the best association between robots and tasks. More specifically, given a set of tasks $T = \{\tau_1, \tau_2, ..., \tau_m\}$ for a team of robots $R = \{r_1, r_2, ..., r_n\}$, a *utility estimation vector* (*UEV*) can be defined as vector containing a list of estimations of "how good" a particular robot is for each task τ_i at a certain time. In other words, if we define $b_{i,j}(t)$ the estimation that the robot r_i computes for the role τ_j at time t, the *UEV* for such a robot can be expressed as:

$$UEV_i(t) = \left[b_{(i,1)}(t), \quad \cdots \quad , b_{(i,m)}(t)\right] \tag{4}$$

Consequently, we can define *utility estimation matrix* (*UEM*) a matrix where each row i is the *UEV* for each robot r_i. This matrix is computed individually by each robot and it is built by gathering the *UEV*s coming from all the teammates. Formally, this matrix will have the following form:

$$UEM_i(t) = [UEV_1(t), \ ... \ , \ UEV_n(t)].$$ (5)

By considering the score of each robot and the current configuration of the distributed world model, given the score of each robot $b_{(i,j)}$, we can define a *coordination mapping function* Φ, which assigns to the task τ_j the robot with the highest score $b_{(i,j)}$, breaking ties randomly.

3.2   Context System

Our aim is to use contextual knowledge to increase the robot performance in accomplishing a given task. In our scenario, contextual knowledge is used to help the robots to evaluate the events and their effects: how they are triggered; how they modify the environment; how long their effects persist in time. In order to consider contextual information, we introduce in our approach a representation of contexts. Contexts, can be thought as specific configurations of the operational environment. For example, in a soccer game a context could be when the ball rolls out of the field and it needs to be put back into the game, or when the robots need to coordinate in environments with low-bandwidth for communication.

Our approach exploits the output of the contextual system to handle the events occurring during the game. More specifically, the context system outputs *contextual features*, which are used to weight events and their effects on the distributed world model. Formally, we characterize the context system as a function *CS* that takes in input sensory data D, internal robot states S, and external environment dependent information I. *CS* outputs contextual weights related to the notified events. Formally *CS* can be defined as:

$$CS : [\ D \ \times \ S \times \ I \ ] \rightarrow C$$

where the vector of context weights $C \in \mathbb{R}^k$, with $k = \|E(t)\|$.

In our formulation such contextual features C are used to influence the regular operation of the coordination system in order to improve the efficiency in executing a task. Accordingly, at any time t the set of events is weighted as:

$$E_w(t) = C \cdot E(t).$$ (6)

At this point, by considering Eq. 2, we can influence the coordination system using each weighted event E_w, resulting in:

$$\overline{LM}_i(t) = \psi(E_w(t), LM_i(t)).$$ (7)

The context system influences the robot behavior by specializing their actions according to the current contexts. The key insight is that a more specialized and informed agent improves its performance. We exploit this concept in a RoboCup scenario to enhance the capabilities of a soccer robot team, but it could be applied to any coordination system.

4   Coordinating in the RoboCup Soccer Scenario

Our approach to coordination has been developed in the RoboCup Standard Platform League scenario. The approach has been deployed on a team of NAOs, which are commercial, autonomous, 25-DOFs humanoid robots. Such robots are equipped with a wide variety of sensors and actuators, including two CMOS cameras, multiple proximity sensors, four micro-phones, and two speakers. In the chosen scenario, a team of robots needs to coordinate in a $9 \times 6\,\mathrm{m}$ soccer field of the RoboCup Standard Platform League.

In this section, we describe in detail our approach applied to the RoboCup Standard Platform League scenario. Accordingly, we first introduce the modeling of context information related to the soccer scenario. Next, we illustrate how these contexts can be recognized during a soccer game. Finally, we describe how these contexts can be used to improve the coordination of a team of robots.

4.1   Representing Contexts

Defining and representing contexts and context information is a non-trivial task even in a simplified scenario such as a soccer game. To overcome this issue we propose a hierarchical structure, used to recognize possible contexts occurring in the soccer scenario.

In this setting, we formalize two different layers for properly representing contextual information, namely the *task-related* and the *environmental* layer. In the task-related layer, we encode a set of three basic contexts called *task-related contexts* (C_T):

- **Playing:** the robots know the current location of the ball, and the robot coordinate according to the default task-space comprising the common task in a soccer scenario, i.e. striker, defender, supporter and second supporter;
- **Search for ball:** the robots do not know the ball position and cooperate in order to minimize the time in locating the ball;
- **Throw-in:** the robots are searching for the ball but can modify their search strategy by exploiting particular rules governing the soccer scenario.

In the environmental layer, we instead characterize the world depending on the network reliability that allows us to define another two contexts, called *environmental contexts* C_E:

- **Network up:** the robots are in a suitable network condition, i.e. the messages exchanged among the robots are received in a fixed amount of time;
- **Network delayed:** the current network condition does not allow a reliable communication among the robots.

The environmental contexts do not affect the task-space, but actively influence the coordination system. For instance in a *network-delayed* setting the robots modify coordination parameters such as the *role-persistence*. This parameter is used to control how roles are swapped among active robots. Specifically, each robot waits a given amount of time before releasing the role and assuming the new role. This is used to avoid too frequent role swappings and to allow for a more robust task allocation during a game. Since the role-persistence is defined as a time interval, it is crucially important that messages coming from other robots are evaluated depending on quality of the network in order to consider possible delays and, consequently, misleading information. Accordingly, when the network communication is limited, the robots can adaptively change the amount of messages exchanged among the active robots to limit the traffic in the communication.

It is worth remarking that contexts within the same layer are mutually exclusive. For example, if the ball has been seen (i.e. we are in the Playing context), the team can be either in a network-up or in a network-delayed contexts. Therefore, in this scenario we define *context information* as a tuple of two elements, one for each layer, namely

$$C = \langle c_T,\ c_E \rangle$$

where $c_T \in C_T$ and $c_E \in C_E$ represent the context for the task-related layer and the environmental layer, respectively. Figure 1 illustrates our multi-modal hierarchy for representing context information.

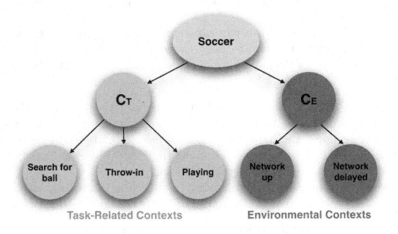

Fig. 1. Multi-modal context hierarchy in a soccer scenario.

4.2   Recognizing Contexts

As described in the previous section, during a soccer game we distinguish five different contexts. The three Task-Related Contexts (*Throw-In, Ball Lost,* and *Playing*) are recognized through the perceptions of the robots and the messages sent by the Game Controller to all the players. In particular, Throw-In considersthe scenario in which the ball has rolled out of the field and is flagged by the Game Controller. Ball Lost is recognized only through the perceptions of the robots which have lost the sight of the ball. Finally, Playing considers the scenario in which the game is normally played and is assumed as the default context.

In multi-robot coordination the robots share useful information via wireless communication. However, a reliable network support is not always guaranteed in real applications. RoboCup competitions are not an exception. To this end, the two Environmental Contexts (*Network up* and *Network delayed*) are recognized based on multiple measurements locally and periodically performed by each robot. Specifically, each robot evaluates the current state of the network at each time step by considering the Round Trip Time (RTT) of the sent packages. The RTT is defined as the time elapsed between the moment in which a package is sent from a source A to a source B and the moment in which B acknowledges to A that the package has been received.

During the game we periodically compute the RTT among the robots pairwise. This measurement allows us to have an understanding of the quality of the network for single channel of communication (i.e., between two robots of the team). Additionally, by considering all the communications within the team, we are able to understand the *global network level* (GNL_{RTT}) by averaging the single channel estimations. This percentage measures the quality of the network depending on the RTT. Such a measurement is set to 100 % if in average the RTT is smaller than a given threshold, or it is decremented if the acknowledge of a given package is never returned.

Finally, we define ranges of network reliabilities over the GNL_{RTT} in order to influence the robots' behavior. We manually set a discriminative value $\alpha = 50\%$ for determining environmental contexts (Fig. 1). We estimated the value of α on several testing sessions, noticing that a lost of packages associated to a $GNL_{RTT} \leq 50\%$ heavily affects the performance of the underneath coordination system. Accordingly, we define two ranges that result into the two different environmental contexts:

- Network up: the $GNL_{RTT} > \alpha$
- Network delayed: the $GNL_{RTT} \leq \alpha$

As a consequence, we influence the coordination system considering the contextual features generated by the two contexts. We identify a set of important parameters which are strongly network dependent, namely the rate of the packages sent, the weight of the events notified by a robot with a poor individual GNL_{RTT}, and the role-persistence used to switch roles among active robots. For instance, if the team realizes to be in the Network Delayed context, then we decrease the amount of exchanged data towards robots with a poor *RTT* and decrease the reliability of the information coming from them, as described in the following section.

4.3   Using Contexts

In this work we focus on the use of contextual information to more effectively coordinate a team of robots. Our main contribution is to demonstrate that by properly formalizing context and their influence within on the coordination system, we can decrease the number of variables to be optimized in a task allocation process, and most importantly, we decrease their operative numerical domain.

For instance, in our coordination system, we have a set of twelve roles that each robot can assume. Considering that each role cannot be selected by two robots simultaneously, without being able to recognize contexts we would have to search in a space composed by *12!* states. By using a contextual categorization of such roles, instead, we can split them in 3 different groups. Now, the search space for the task allocation within each context is composed by *4!* states. For large search spaces this can yield a big reduction in the complexity of the problem.

In the following, we describe how contextual information is used to shape the coordination system, both influencing the distributed world representation, and generating utility estimations, for the task assignment routine, that better satisfy the current context requirements.

Distributed World Modeling. For the considered scenario, we adopt an occupancy grid model to represent the environment and reason about it. This structure is suitable to encode meaningful information and synchronize the robots views. More precisely, each cell of the grid encodes information such as teammate and opponent positions, the estimated ball position, and an estimation of the wireless signal level (Fig. 2).

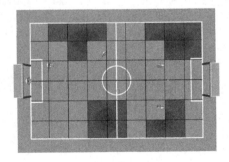

Fig. 2. Distributed world model in the soccer scenario.

For instance, in the *search context* the grid is used to locate the next-most-likely position of the ball and to minimize the time needed to recover it. In this setting, we divide the field into cells with a fixed granularity assigning to each cell a probability score, which represents the likelihood of finding the ball in that particular area. The robots are displaced within the field in accordance with the top scoring cells, and coordinate to minimize the time spent in reaching a given area to be explored. Such areas represent the targets of the coordination algorithm and are constantly updated, while the players are roaming.

More precisely, while searching, the robots lower the score of the near cells and rerank the most promising ones. This procedure automatically generates clusters of *explored* and *unexplored* cells. The former is used to synchronize the distributed world model, while the latter is used to define the targets of the coordination system. The robots share the centroid of explored clusters to exclude recently controlled areas from the search. In this context, the reconstruction function of Eq. 1 is defined as

$$\forall robot_i \quad cell_j = \arg\min_i \{cell_{ij}\} \tag{8}$$

where $cell_j$ is the score of the j-*th* cell in the DWM, which represents the minimum score for the j-*th* among all the partial local models of the robots LM_I. Such a score has a default value for each context and cell, which is accordingly modified during the execution of the searching task.

Task Allocation. The robots are coordinated using a DTA based on utility estimation. Therefore, we define a utility function, according to Eq. 5, that given a robot and a task, it returns a score representing how suitable the input robot is for the input task. Further, depending on the current context, the utility function changes to fulfill the requirement of the environment and the current task according to Eq. 7.

Let us assume to be in a playing context. In this case, the utility function takes into account the euclidean distance between the ball and the robot; the distance between the robot and a target position; the robot orientation; the elapsed time since the current role was assigned; and a bias term for each robot (such bias is used to solve ambiguous situations). This function is defined as:

```
uv(i,j) = penalty * playing_utility_function(i,j)
```

where $uv(i,j)$ is the utility score for the j-*th* role with respect to the i-*th* agent and *penalty* is a parameter used to prioritize the most important tasks. Each weight has been empirically calculated after several testing sessions and the *penalty* variable is used to assign negative rewards to fallen or penalized robots. The players share their utilities, and once their utility matrices are complete, they can decide the best role to assume. The robots check the matrix column-wise and assign to the j-*th* task the robot with the maximum score.

Similarly, in the searching contexts the robots use exactly the same criterion to allocate tasks with respect to the areas to look for the ball. The utility function takes the following form

```
uv(i,j) = searching_utility_function(i,j)
```

where $uv(i,j)$ is the utility score for the j-*th* centroid with respect to the i-*th* agent. Due to the new utility function, we can minimize the time for exploring most promising areas.

In order to guarantee more robustness to the task allocation, a role-time-persistence has been introduced. However, this choice imposes the usual trade-off between stability and reactivity. To efficiently adapt the role-persistence, we exploit the environmental context. In particular, we adapt the role-persistence relying upon the current configuration of the network.

5   Experimental Results

The algorithm has been extensively tested in a virtual RoboCup-dedicated environment developed by the B-human Team[2] and it has been deployed on a team of Nao robots. In order to show the improvements of our coordination algorithm, we set up several experiments with different configurations. Our aim is to highlight specific features of the algorithm, and to test the thesis behind our coordination approach (i.e. the fact that a properly informed coordination can considerably enhance the performances).

In the first setting, we demonstrate the improvement of our system with respect to a *non-context-aware* approach. In this configuration, we deploy two teams that share the same high-level behaviors, one featuring the context-based coordination (*blue team*) and the other modeling a unique utility estimation for the whole testing session (*red team*). Specifically, the red team does not modify its utility estimations, if the ball is not seen or if the ball rolled out of the field.

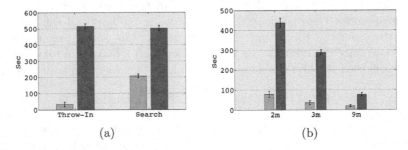

            (a)   (b)

Fig. 3. The blue team features the context-coordination, while the red team is not context aware. On the y-axis, the averaged cumulative sum of the time interval in which the robots do not see the ball is shown, while on the x-axis, we report the results for two different contexts (a), and three depth values for the robots field of view (b).

In this setting, we measured the cumulative time during which the ball was not seen by the team in 10 min of the game. In Fig. 3(a) we can notice that the recovering time is considerably reduced when deploying the proposed approach. Further, it is worth highlighting the importance of different levels of information in the context system. In a Throw-In context, in fact, the ball is subject to specific rules and the performance can be further enhanced.

---

[2] http://www.b-human.de/.

Instead, Fig. 3(b) highlights the robustness of our approach with respect to the robots' field of view. In this setting, we again measured the cumulative time during which the ball was not seen by the team during a 10 min game. In this experiment, we varied the robots' field of views to 2, 3 and 9 m. It is worth noticing that both teams improve their performance. However, the context-coordination preserves a better profile in all the configurations.

Finally, in order to show the effectiveness of our approach in managing network *contexts* we report the stability of the role switching with respect to the quality of the global network level GNL_{RTT} introduced in the previous section. To this end, we traced the network GNL_{RTT} in a real gaming session, and then we run several experiments by varying the simulated network stability with respect to the logged GNL_{RTT}. This, allows us to have more reliable simulated testing session and have a realistic profile of the GNL_{RTT} during a game. In such testing sessions, we report the error in meters that each robot has with respect to the assigned role. In this setting, both the blue and the red team features the context-coordination, however, the blue team role-persistence that controls the role switching is changed depending on the GNL_{RTT}, while the red team switches roles according to a fixed role-persistence threshold. Our goal is to generate a more robust behavior when the network has a poor reliability, and simultaneously, a more reactive role mapping when the team is experiencing a good communication setting. As in the previous experiments, the tests have been carried out in multiple sessions of 10 min of an SPL game. Figure 4 shows the results obtained in a testing session. We report, for each role, the error in meters by computing the average of the error that each robot has during the simulation, with respect the default position of the assigned role.

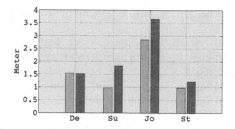

Fig. 4. The y-axis shows the error between the robots and the default position for the assigned role, while the x-axis reports the set of roles: Defender (De), Supporter (Su), Jolly (Jo) and Striker (St). The blue team adapts its role-persistence depending on the GNL_{RTT}, while the red team adopts a static role-persistence time threshold.

In this setting the two team feature the same coordination system and the performance is only conditioned by the role-persistence policy. Also in this configuration, we notice an improvement in formalizing environmental contexts as the network status. In our opinion, such context-related information is a starting point for handling network issues, and it needs to be better investigated in order to further improve team coordination.

6   Conclusion

In this paper we have presented a novel method for coordinating a team of robots. Starting from the idea that a more informed team will eventually show improved performance during task execution, we presented an approach that integrates utility estimations and a distributed world knowledge to come up with a mapping of robots to roles. The proposed method presented extracts context information to influence and modify the coordination rules and select the most suitable configuration in accordance to the current situation. Given the previously described experimental results, we are able to state that the Context-based coordination provides the expected improvements. Indeed, the specialization and the adaptation of the coordination algorithm significantly increases the performances of a team of robots.

Considering the results obtained with this approach, our intent is to generalize the method for addressing the problem of *"Multi-Robot coordinated search and target localization"*. The main idea is to deploy a coordinated team of robots to localize multiple targets (e.g. lost objects, control malfunction infrastructures, victim assessments) in an arbitrary environment and to improve the execution of the current robots' tasks by exploiting any kind of information that can help the robots to specialize their search and to improve their performances.

References

1. Abeyruwan, S., Seekircher, A., Visser, U.: Dynamic role assignment using general value functions
2. Farinelli, A., Iocchi, L., Nardi, D., Ziparo, V.A.: Task assignment with dynamic perception and constrained tasks in a multi-robot system. In: ICRA, Barcelona, Spain, pp. 1535–1540 (2005)
3. Iocchi, L., Nardi, D., Piaggio, M., Sgorbissa, A.: Distributed coordination in heterogeneous multi-robot systems. Auton. Rob. 15(2), 155–168 (2003)
4. Luo, L., Chakraborty, N., Sycara, K.: Multi-robot assignment algorithm for tasks with set precedence constraints. In: 2011 IEEE International Conference on Robotics and Automation (ICRA), pp. 2526–2533. IEEE (2011)
5. Stone, P., MacAlpine, P., Barrera, F.: Positioning to win: a dynamic role assignment and formation positioning system. In: Chen, X., Stone, P., Sucar, L.E., van der Zant, T. (eds.) RoboCup 2012. LNCS, vol. 7500, pp. 190–201. Springer, Heidelberg (2013)
6. Nardi, D., Adorni, G., Bonarini, A., Chella, A., Clemente, Giorgio, Pagello, Enrico, Piaggio, Maurizio: ART99 - azzurra robot team. In: Veloso, M.M., Pagello, E., Kitano, H. (eds.) RoboCup 1999. LNCS (LNAI), vol. 1856, pp. 695–698. Springer, Heidelberg (2000)
7. Stone, P., Veloso, M.: Task decomposition, dynamic role assignment, and low-bandwidth communication for real-time strategic teamwork. Artif. Intell. 110, 241–273 (1999)

8. Vail, D., Veloso, M.: Multi-robot dynamic role assignment and coordination through shared potential fields. In: Schultz, A., Parker, L., Schneider, F. (eds.) Multi-Robot Systems. Kluwer (2003)
9. Weigel, T., Gutmann, J.S., Dietl, M., Kleiner, A., Nebel, B.: CS Freiburg: coordinating robots for successful soccer playing. IEEE Trans. Rob. Autom. **18**(5), 685–699 (2002)

A Study of Layered Learning Strategies Applied to Individual Behaviors in Robot Soccer

David L. Leottau[1(✉)], Javier Ruiz-del-Solar[1], Patrick MacAlpine[2], and Peter Stone[2]

[1] Advanced Mining Technology Center, Department of Electrical Engineering,
Universidad de Chile, Santiago, Chile
{dleottau, jruizd}@ing.uchile.cl
[2] Department of Computer Science, The University of Texas at Austin,
Austin, TX 78712, USA
{patmac, pstone}@cs.utexas.edu

Abstract. Hierarchical task decomposition strategies allow robots and agents in general to address complex decision-making tasks. Layered learning is a hierarchical machine learning paradigm where a complex behavior is learned from a series of incrementally trained sub-tasks. This paper describes how layered learning can be applied to design individual behaviors in the context of soccer robotics. Three different layered learning strategies are implemented and analyzed using a ball-dribbling behavior as a case study. Performance indices for evaluating dribbling speed and ball-control are defined and measured. Experimental results validate the usefulness of the implemented layered learning strategies showing a trade-off between performance and learning speed.

Keywords: Reinforcement learning · Layered learning · Machine learning · Soccer robotics · Biped robot · NAO · Behavior · Dribbling · Fuzzy logic

1 Introduction

The use of computational/machine learning (ML) techniques such as Reinforcement Learning (RL) allows robots, and agents in general, to address complex decision-making tasks. However, one of the main limitations of the use of learning approaches in real-world problems is the large number of learning trials required to learn complex behaviors. In addition, many times the learning of abilities associated with a given behavior cannot be directly used, i.e. combined or transferred to other behaviors. These drawbacks can be addressed by transfer learning [1] or hierarchical task decomposition strategies [2].

Layered Learning (LL) [3] is a hierarchical learning paradigm that enables learning complex behaviors by incrementally learning a series of sub-behaviors. LL considers bottom-up hierarchical learning, where low-level behaviors (those closer to the environmental inputs) are trained prior to high-level behaviors [4].

© Springer International Publishing Switzerland 2015
L. Almeida et al. (Eds.): RoboCup 2015, LNAI 9513, pp. 290–302, 2015.
DOI: 10.1007/978-3-319-29339-4_24

The main contribution of this paper is describing and analyzing how LL can be applied to design individual behaviors in the context of soccer robotics. Three different layered learning strategies are implemented and analyzed using the ball-dribbling behavior as a case study [5]. Ball-dribbling is a complex behavior where a robot player attempts to maneuver the ball in a very controlled way while moving towards a desired target. Very few works have addressed ball dribbling behavior with humanoid biped robots [5–9]. Furthermore, few details are mentioned in these works concerning specific dribbling modeling [10, 11], performance evaluations for ball-control, or obtained accuracy to the desired path.

After modeling ball-dribbling behavior, some conditions needed to learn ball-dribbling under the LL paradigm are described. Afterwards, sequential, concurrent, and partial concurrent LL strategies are applied to the dribbling task and analyzed. Results from these experiments show a trade-off between performance and learning time, as well as between autonomous learning versus previous designer knowledge.

The paper is organized as follows: In Sect. 2 the Layered Learning paradigm and different LL strategies are detailed. Section 3 describes the ball-dribbling behavior, and Sect. 4 presents the application of the LL paradigm to the modeling and learning of ball-dribbling behavior. Experimental results are presented in Sect. 5, and conclusions are given in Sect. 6.

2 Layered Learning

Layered learning (LL) [3] is a hierarchical learning paradigm that enables learning complex behaviors by incrementally learning a series of sub-behaviors (each learned sub-behavior is a layer in the learning progression) [12]. LL considers bottom-up hierarchical learning, where high-level behaviors depend on behaviors in lower layers (those closer to the environmental inputs) for learning. From LL literature, three general strategies can be identified:

- **Sequential Layered Learning (SLL):** In the original formulation of the LL paradigm [3], layers are learned in a sequential bottom-up fashion. Lower layers are trained and then frozen (their behaviors are held constant) before advancing to learning of the next layer. While a higher layer is trained, lower layers are not allowed to change, which reduces the search space. However, it can also be restrictive because it limits the space of possible solutions that agents could search combining behaviors.
- **Concurrent Layered Learning (CLL):** CLL [4] allows lower layers to keep learning concurrently during the learning of subsequent layers. The agent may explore a behavior's joint search space combining all layers. Since CLL does not restrict the search space, its dimensionality increases, which can make the learning process more difficult.
- **Overlapping Layered Learning (OLL):** OLL [12] seeks to find a trade-off between freezing each layer once learning is complete (SLL) and leaving previously learned layers open (CLL). This extension of LL allows some, but not necessarily all, parts of newly learned layers to be kept open during the training of subsequent

layers. In the context of learning parameterized behaviors this means that a subset of a learned behavior's parameters are left open and allowed to be modified during learning of the proceeding layer. The parts of previously learned layers left open "overlap" with the next layer being learned. Three general scenarios for overlapping layered learning are distinguished in [12]: Combining Independently Learned Behaviors (CILB), Partial Concurrent Layered Learning (PCLL), and Previous Learned Layer Refinement (PLLR). This work considers the implementation of Partial Concurrent Layered Learning, where only part, but not all, of a previously learned layer's behavior parameters are left open when learning a subsequent layer with new parameters. The part of the previously learned layer's parameters left open is the "seam" or overlap between the layers [12].

3   Case Study: Soccer Dribbling Behavior

Soccer dribbling behavior with humanoid biped robot players is used as a case study [5]. Figure 1 at left shows the RoboCup SPL soccer environment where the NAO humanoid robot [13] is used. The proposed modeling of dribbling behavior will use the following control actions: $[v_x, v_y, v_\theta]'$, the velocity vector; and the following state variables: ρ, the robot-ball distance; γ, the robot-ball angle; and, φ, the robot-ball-target complementary angle. These variables are shown in Fig. 1 at right, where the desired target ($\oplus$) is located in the middle of the opponent's goal, and a robot's egocentric reference system is considered with the x axis pointing forwards. A more detailed description of the proposed modeling can be found in [5, 14].

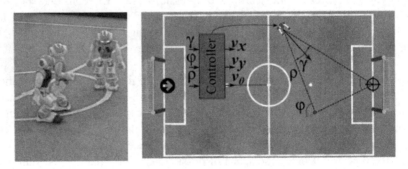

Fig. 1. A picture of the NAO robot dribbling during a RoboCup SPL game (left) and definition of variables for ball-dribbling modeling (right).

Ball-dribbling behavior can be split into three sub-tasks which must be executed in parallel: *ball-turning*, which keeps the robot tracking the ball-angle ($\gamma = 0$), *target-aligning*, which keeps the robot aligned to the ball-target line ($\varphi = 0$); and *ball-pushing*, whose objective is that the robot walks as fast as possible and hits the ball in order to push the ball towards a desired target, but without losing possession of the ball. So, the proposed control actions are the requested speed to each axis of the biped walk engine, where $[v_x, v_y, v_\theta]'$ are respectively involved with *ball-pushing*, *target-aligning*, and *ball-turning* [15].

From a behavioral perspective, ball-dribbling can also be split in two more general tasks, *alignment* and *ball-pushing*. This division into two behaviors has been proposed in [5], based on the idea that *alignment* can be designed off-line, unlike *ball-pushing*, which needs interaction with its dynamic environment in order to learn a proper policy. In this way, *alignment* is composed of *ball-turning* and *target-aligning*. A behavior scheme of ball-dribbling is depicted in Fig. 2(a).

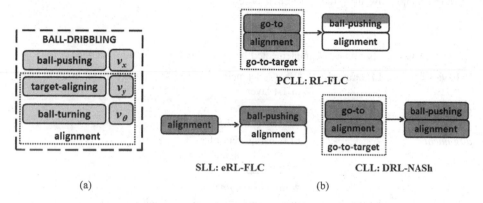

Fig. 2. (a) Behavioral scheme of the ball-dribbling problem. (b) Different layered learning strategies implemented; open behaviors are colored meanwhile frozen behaviors are white.

With respect to *ball-pushing*, the modeling of the robot's feet–ball–floor dynamics is complex and inaccurate because kicking the ball could generate several unexpected transitions, due to uncertainty of foot-ball interaction and speed when the robot kicks the ball (note that the robot's foot's shape is rounded and the foot's speed is different from the robot's speed v_x). Moreover, an omnidirectional biped walk intrinsically has a delayed response, which varies depending on the requested velocity $[v_x, v_y, v_\theta]'$. To learn when and how much the robot must slow down or accelerate is a complex problem, hardly solvable in an effective way with methods based on identification of system dynamics and/or kinematics and mathematical models [14]. To solve this problem as a Markov Decision Process (MDP) with an RL scheme for learning simultaneously, ball-dribbling dynamics have been successfully applied previously in the same domain [5, 14]. Thus, all the learning methods presented in this paper use an RL scheme for tackling the *ball-pushing* task.

4   Layered Learning of Dribbling Behavior

This section presents how three different strategies of the Layered Learning paradigm can be applied to the ball-dribbling task: PCLL, SLL, and CLL. These strategies are implemented by using a behavior in the first layer called *go-to-target*, where the robot goes to a desired target pose on the field. *Go-to-target* is composed in a very similar way to the ball-dribbling behavior depicted in Fig. 2(a); it also uses *alignment* but uses *go-to* instead of *ball-pushing* as depicted at the top of Fig. 2(b). *Go-to* behavior (see Table 1) is similar to *ball-pushing* as it also modifies v_x, but instead of directing the forward motion of the robot toward a ball it moves the robot forward toward a specific target location on the field. *Go-to-target* behavior is designed based on a Takagi-Sugeno-Kang Fuzzy Logic Controller (TSK-FLC) [16] which acts over the walk engine velocity vector. This behavior is currently part of the control architecture of the UChile Robotics Team [5, 17]. See Table 1 for descriptions of the behaviors' parameters and how they relate to each other.

Table 1. Summary of implemented behaviors and their learning methods

Behavior	LL strategy	What is learned in 1st layer	What is learned in 2nd layer
go-to	–	FLC parameters of v_x by using CMA-ES	–
alignment	–	FLC parameters of v_y and v_θ by using CMA-ES	–
go-to-target	–	*go-to* and *alignment*	–
Dribbling with RL-FLC	Partial concurrent (PCLL)	*go-to-target*	*ball-pushing*: A partial policy for v_x observing ρ, by using RL
Dribbling with eRL-FLC	Sequential (SLL)	*alignment*	*ball-pushing*: A policy for v_x observing $[\rho, \gamma, \varphi]'$, by using RL
Dribbling with DRL-NASh	Concurrent (CLL)	*go-to-target*	Three policies, for v_x, v_y, v_θ, which are learned in parallel observing the joint state $[\rho, \gamma, \varphi]'$, by using RL
Dribbling with DRL	–	*ball-pushing* (v_x), *target-aligning* (v_y), and *ball-turning* (v_θ) by using Decentralized-RL [14]	–

For this work, the *go-to-target* controller parameters have been learned by using the RoboCup 3D simulation optimization framework of the LARG lab within the Computer Science Department at the University of Texas at Austin. This optimization framework uses the Covariance Matrix Adaptation Strategy (CMA-ES) [18], performed on a Condor [19] distributed computing cluster.

4.1 Partial Concurrent Layered Learning

The RL-FLC work reported in [5] proposes a methodology for modeling dribbling behavior by splitting it into two sub-problems: *alignment*, which is achieved by using a Fuzzy Logic Controller (FLC), and *ball-pushing*, which is learned by using a RL based controller. This methodology has been successfully used during RoboCup 2014 in the SPL robot soccer competitions by *UChile Robotics Team* [17] and it is currently the base of their dribbling engine.

The PCLL strategy is applied as follows: The *go-to-target* behavior is learned in the first layer for tuning FLC's parameters. During learning of the second behavior layer the entire *alignment* behavior is frozen while the *ball-pushing* behavior is partially re-learned. That means, only the parameter for how ρ affects v_x is opened to the RL agent, meanwhile parameters for how γ and φ influence v_x are kept frozen. So, γ and φ are not considered in the state space. Thus, ball-pushing parameters are partially refined in the context of the fixed *alignment* behavior. Please see top Fig. 2(b) and Table 1.

Table 2. Description of states and actions for the RL-FLC scheme

		States space: s = [ρ]		
		Min	*Max*	*# bins*
Feature	ρ	*0mm*	*600mm*	*13*
		Actions space: a = [v_x]		
		Min	*Max*	*# discrete actions*
Action	v_x	*0mm/s*	*150mm/s*	*16*

Desired characteristics for a learned ball-dribbling policy are to have the robot walk fast while keeping the ball in its possession. That means ρ must be minimized (to keep possession of the ball), while at the same time maximizing v_x, which is the control action. Proposed RL modeling for learning the speed v_x depending on the observed state of ρ is detailed in Table 2. The proposed reward function is expressed in Eq. (1). This reward function reinforces walking forward at maximum speed ($v_{x.max'}$) without losing the ball possession ($\rho < \rho_{th}$).

$$r_x = \begin{cases} 1, & \rho < \rho_{th} \wedge v_x \geq v_{x.max'} \\ -1, & otherwise \end{cases} \tag{1}$$

4.2 Sequential Layered Learning

An enhanced version of the RL-FLC method is implemented using a SLL strategy. This enhanced approach (eRL-FLC) learns the *ball-pushing* behavior mapping the whole state space [ρ, γ, φ] by using a RL scheme. The modeling description is presented in [14]; it is designed to improve ball control because the former RL-FLC approach assumes the ideal case where target, ball, and robot are always aligned ignoring γ and φ angles, which is not the case during a real game situation.

The SLL strategy is applied as follows: The *alignment* behavior is learned in the first layer; then, during learning of the second layer, *alignment* is frozen and the whole *ball-pushing* behavior is learned by performing the ball-dribbling task in the context of the fixed *alignment* behavior. This is depicted at the bottom-left of Fig. 2(b) and summarized in Table 1.

The proposed RL modeling is depicted in Table 3, where only *ball-pushing* is learned. The proposed reward function is expressed in Eq. (2).

Table 3. Description of States and Actions for eRL-FLC and DRL schemes

Joint state space: $s = [\rho, \gamma, \varphi]^T$				
		Min	*Max*	*# bins*
Feature$_1$	ρ	0mm	600mm	13
Feature$_2$	γ	-50°	50°	11
Feature$_3$	φ	-50°	50°	11
Actions space: $a = [v_x, v_y, v_\theta]$				
		Min	Max	# discrete actions
ball-pushing	v_x	0 mm/s	150 mm/s	21
target-aligning	v_y	-50 mm/s	50 mm/s	21
ball-turning	v_θ	-45 °/s	45 °/s	21

4.3   Concurrent Layered Learning

A Decentralized Reinforcement Learning (D-RL) strategy is proposed in [14], where each component of the omnidirectional biped walk $[v_x, v_y, v_\theta]'$ [20] is learned in parallel with single-agents working in a multi-agent task. Furthermore, this D-RL scheme is accelerated by using the Nearby Action Sharing (NASh) approach [15], which is introduced for transferring knowledge from continuous action spaces, when no information different to the suggested action in an observed state is available from the source of knowledge. In the early training episodes, NASh transfers actions suggested by the source of knowledge (former layer) but progressively explores its surroundings looking for better nearby actions for the next layer.

In order to learn dribbling behavior with the DRL-NASh approach, the CLL strategy is applied as follows: The *go-to-target* behavior is learned in the first layer. During learning of the second layer *go-to* and *alignment* behaviors parameters are left opened and relearned to generate *ball-pushing* and *alignment* behaviors, thereby transferring knowledge from *go-to-target* through use of the NASh method. This is depicted at the bottom-right of Fig. 2(b) and summarized in Table 1.

Again, the expected policy is to walk fast towards the desired target while keeping the ball in the robot's possession. That means: maintaining $\rho < \rho_{th}$; minimizing γ, φ, v_y, v_θ; and maximizing v_x. The proposed RL modeling is detailed in Table 3. The corresponding reward functions per agent are expressed in Eqs. (2–4).

$$r_x = \begin{cases} 1, & \rho < \rho_{th} \wedge |\gamma| < \gamma_{th} \wedge |\varphi| < \varphi_{th} \wedge v_x \geq v_{x.max'} \\ -1, & otherwise \end{cases} \tag{2}$$

$$r_y = \begin{cases} 1, & |\gamma| < Ang_{th} \\ -1, & otherwise \end{cases} \tag{3}$$

$$r_\theta = \begin{cases} 1, & |\gamma| < Ang_{th} \wedge |\varphi| < Ang_{th} \\ -1, & otherwise \end{cases} \tag{4}$$

where $\rho_{th}, \gamma_{th}, \varphi_{th}$ are desired thresholds where the ball is considered controlled, meanwhile $v_{x.max'}$ reinforces walking forward at maximum speed.

5 Experimental Results and Analysis

5.1 Experimental Setup

As mentioned in the previous section, proposed LL schemes are implemented using the *go-to*-target behavior in the first layer, which is learned using CMA-ES. The second layer of all these schemes are performed by using a RL (SARSA (λ)) episodic procedure. After a reset, the robot is set in the center of its own goal (black right arrow in Fig. 1), the ball is placed in front of the robot, and the desired target is defined in the center of the opponent's goal ($\oplus$). The terminal state is reached if the robot loses the ball, or, the robot leaves the field, or, the robot crosses the goal line and reaches the target, which is the expected terminal state. Due to the comparative study purposes of this work, all the experiments are carried out in simulation. The training field is 6×4 meters. $Ang_{th} = 5°$, $v_{x.max'} = 0.9 \cdot v_{x.max}$, and fault-state constraints are set as: $[\rho_{th}, \gamma_{th}, \varphi_{th}] = [500\,mm, 15°, 15°]$.

Four different learning schemes are presented in this paper: RL-FLC implemented with PCLL; eRL-FLC implemented with SLL; DRL-NASh implemented with CLL; and Decentralized RL scheme (DRL) as a base of comparison. The DRL scheme is proposed in [14] and briefly introduced in Table 1, it learns from scratch without any type of transfer learning or LL strategy.

The evolution of the learning process of each proposed scheme is evaluated by measuring and averaging ten runs. In this way, the following performance indices are considered to measure dribbling-speed and ball-control respectively:

- *% of maximum forward speed* ($\%S_{Fmax}$): given S_{Favg}, the average dribbling forward speed of the robot, and S_{Fmax}, the maximum forward speed: $\% S_{Fmax} = S_{Favg}/S_{Fmax}$.
- *% of time in fault-state* ($\%T_{FS}$): the accumulated time in fault-state t_{FS} during the whole episode time t_{DP}. The fault-state is defined as the state when the robot loses possession of the ball, i.e., $\rho > \rho_{th} \vee |\gamma| > \gamma_{th} \vee |\varphi| > \varphi_{th}$, then:
- $\%T_{FS} = t_{FS}/t_{DP}$.

- *Global fitness* (*F*): introduced for the sole purpose of evaluating and comparing both performance indices together. It is computed as follows: $F = 1/2 \cdot [(100 - \%S_{Fmax}) + TFS]$, where $F = 0$ is the optimal policy.

5.2 Results and Analysis

Figure 3 shows the learning evolution of the four proposed schemes. Additionally, the policy of the run with the best performance from each scheme is tested and measured separately using 100 runs; average and standard error of those performances are presented in Table 4. The time to threshold index in Table 4 (learning speed) is calculated with a threshold of F = 27 %, according to global fitness plots in Fig. 3.

Table 4. Performance indices

Method	$\%S_{Fmax}$		$\%T_{FS}$		F	Time to Th. (Episodes)
	Avg.	Std. Err	Avg.	Std. Err	Avg.	
DRL-NASh (CLL)	74.83	0.049	14.69	0.080	19.92	1391
eRL-FLC (SLL)	61.49	0.032	16.84	0.061	27.67	66
RL-FLC (PCLL)	57.50	0.04	26.32	0.069	34.4	53
DRL	64.35	0.12	13.87	0.19	24.76	1594

The time to threshold of the DRL scheme is the longest between all the tested schemes; this is the expected result, taking into account that no LL or transfer knowledge strategies have been implemented for this scheme. However, DRL learns from scratch exploring the whole state-action space, allowing each sub-behavior (*ball-pushing*, *target-aligning*, and *ball-turning*) to learn about actions of the other two sub-behaviors. Even so, although DRL shows the lowest percentage of faults, it does not show the best global performance. The best performance is shown by the DRL-NASh scheme using CLL, which evidences the usefulness of CLL for this problem.

The DRL-NASh using CLL scheme shows the best global performance, the highest dribbling speed and the second best percentage of faults; however it takes on average around 1390 learning episodes before achieving asymptotic convergence, just around 13 % faster than the DRL scheme. It validates the fact that by using concurrent layered learning it is possible to find better performance; the drawback is that increasing the search space dimensionality makes learning slower. Discussion about the NASh strategy and how the performance of first-layer-behavior influences the learning time and final performance is presented in [15]. Exploring this subject is a potential alternative to speed-up learning times when Concurrent LL is used with RL agents.

The RL-FLC using PCLL approach shows the fastest asymptotic convergence and the lowest accuracy. This is expected because RL-FLC is the least complex learning agent, which has frozen the major part of its search space, decreasing its performance but accelerating its learning.

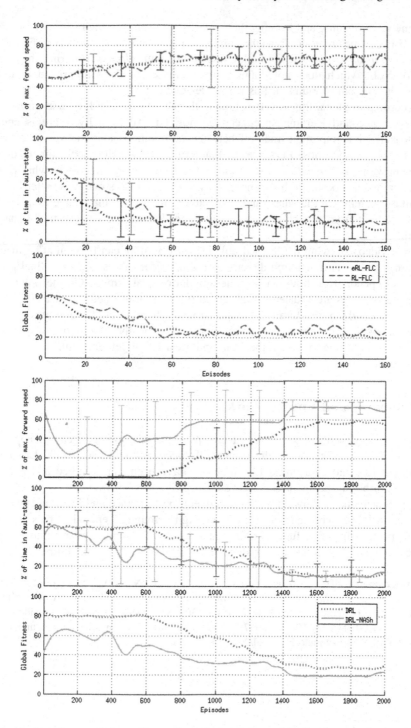

Fig. 3. Learning evolution with standard deviation bars of the four proposed schemes.

Benefits of opening and learning the whole *ball-pushing* behavior for the eRL-FLC using SLL scheme are noticeable when observing standard deviation bars in Fig. 3. For this case, *ball-pushing* learns its policy interacting with *alignment* during the second layer of SLL, which does not dramatically increase the dribbling speed though it reduces the amount of faults, just as it was designed.

According to global fitness versus time to threshold in Table 4, a trade-off in terms of performance and learning speed can be noticed. Additionally, there is another non-measured but important trade-off between autonomous learning versus previous designer knowledge. Those LL strategies that reduce the search space's dimensionality require previous knowledge of the problem for determining effectively what part of former learned layers should be opened, and what type of LL strategy is better for each particular problem. On the other hand, more autonomous learning strategies as CLL or merely learning from scratch require less designer knowledge but can make learning difficult.

Some videos showing the learned policies for dribbling can be seen at[1]. Currently the learned policy is transferred directly to the physical robots, thus, the final performance is dependent on how realistic the simulation platform is. On the other hand, since state variables are updated and observed frame by frame acting like a closed loop control action, which tries to minimize the error, a different initialization of robot, ball, and target positions does not affect performance dramatically. The robot always tries to follow a straight-line between the ball and desired target emulating the training environment.

6 Summary and Future Work

This paper has described how different Layered Learning strategies can be applied to design individual behaviors in the context of soccer robotics. Sequential LL, Partial Concurrent LL, and Concurrent LL strategies have been implemented and analyzed using ball-dribbling behavior as a case study.

Experiments have shown a trade-off between performance and learning speed. For instance, the PCLL scheme is capable of learning in around 53 episodes. This opens the door to make achievable future implementations for learning similar behaviors with physical robots. This is one of our short term goals and part of our future work.

Acknowledgments. This work was partially funded by FONDECYT under Project Number 1130153. Research within the Learning Agents Research Group (LARG) at UT Austin is supported in part by NSF (CNS-1330072, CNS-1305287), ONR (21C184-01), and AFOSR (FA8750-14-1-0070, FA9550-14-1-0087). David Leonardo Leottau was funded under grant CONICYT-PCHA/Doctorado Nacional/2013-63130183.

---

[1] https://www.youtube.com/watch?v=HP8pRh4ic8w
https://www.youtube.com/watch?v=_i8aNYSd6Iw&feature=youtu.be.

References

1. Taylor, M., Stone, P.: Transfer learning for reinforcement learning domains: a survey. J. Mach. Learn. Res. **10**, 1633–1685 (2009)
2. Takahashi, Y., Asada, M.: Multi-controller fusion in multi-layered reinforcement learning. In: 2001 International Conference on Multisensor Fusion and Integration for Intelligent Systems, MFI 2001, pp. 7–12 (2001)
3. Stone, P.: Layered Learning in Multiagent Systems: A Winning Approach to Robotic Soccer. MIT Press, Cambridge (2000)
4. Whiteson, S., Stone, P.: Concurrent Layered Learning. In: Second International Joint Conference on Autonomous Agents and Multiagent Systems, pp. 193–200. ACM Press, New York (2003)
5. Leottau, L., Celemin, C., Ruiz-del-Solar, J.: Ball dribbling for humanoid biped robots: a reinforcement learning and fuzzy control approach. In: Bianchi, R.A., Akin, H., Ramamoorthy, S., Sugiura, K. (eds.) RoboCup 2014. LNCS, vol. 8992, pp. 549–561. Springer, Heidelberg (2015)
6. MacAlpine, P., Barrett, S., Urieli, D., Vu, V., Stone, P.: Design and optimization of an omnidirectional humanoid walk: a winning approach at the RoboCup 2011 3D simulation competition. In: Twenty-Sixth AAAI Conference on Artificial Intelligence (AAAI 2012), Toronto, Ontario, Canada (2012)
7. Alcaraz, J., Herrero, D., Mart, H.: A closed-loop dribbling gait for the standard platform league. In: Workshop on Humanoid Soccer Robots of the IEEE-RAS International Conference on Humanoid Robots (Humanoids), Bled, Slovenia (2011)
8. Meriçli, Ç., Veloso, M., Akin, H.: Task refinement for autonomous robots using complementary corrective human feedback. Int. J. Adv. Robot. Syst. **8**, 68–79 (2011)
9. Bennewitz, M., Behnke, S., Latzke, T.: Imitative Reinforcement Learning for Soccer Playing Robots. In: Lakemeyer, G., Sklar, E., Sorrenti, D.G., Takahashi, T. (eds.) RoboCup 2006: Robot Soccer World Cup X. LNCS (LNAI), vol. 4434, pp. 47–58. Springer, Heidelberg (2007)
10. Tilgner, R., Reinhardt, T., Kalbitz, T., Seering, S., Fritzsche, R., Eckermann, S., Müller, H., Engel, M., Wünsch, M., Mende, J., Freick, P., Stadler, L., Schließer, J., Hinerasky, H.: Nao-Team HTWK Team Description Paper 2013. In: RoboCup 2013: Robot Soccer World Cup XVII Preproceedings. RoboCup Federation, Eindhoven, The Netherlands (2013)
11. Röfer, T., Laue, T., Judith, M., Bartsch, M., Jenett, D., Kastner, T., Klose, V., Maaß, F., Maier, E., Meißner, P., Sch, D.: B-human team description for RoboCup 2014. In: RoboCup 2014: Robot Soccer World Cup XVIII Preproceedings, Joao Pessoa, Brazil (2014)
12. MacAlpine, P., Depinet, M., Stone, P.: UT Austin Villa 2014: RoboCup 3D Simulation League Champion via Overlapping Layered Learning. In: 29th AAAI Conference on Artificial Intelligence AAAI 2015, Austin, Texas, USA (2015)
13. Gouaillier, D., Hugel, V., Blazevic, P., Kilner, C., Monceaux, J., Lafourcade, P., Marnier, B., Serre, J., Maisonnier, B.: Mechatronic design of NAO humanoid. In: 2009 IEEE International Conference on Robotics and Automation, pp. 769–774. IEEE, Kobe, Japan (2009)
14. Leottau, D.L., Ruiz-del-solar, J.: An accelerated approach to decentralized reinforcement learning of the ball-dribbling behavior. In: AAAI 2015, Workshop on Knowledge, Skill, and Behavior Transfer in Autonomous Robots, pp. 23–29, Austin, Texas, USA (2015)
15. Leottau, D.L., Ruiz-del-Solar, J.: An accelerated approach to decentralized reinforcement learning: a humanoid soccer robots validation. In: IEEE/RSJ International Conference on Intelligent Robots and Systems IROS 2015, Hamburg, Germany (2015, submitted)

16. Takagi, T., Sugeno, M.: Fuzzy identification of systems and its application to modeling and control. IEEE Trans. Syst. Man Cybern. **15**, 116–132 (1985)
17. Yanez, J.M., Cano, P., Mattamala, M., Saavedra, P., Leottau, D.L., Celemin, C., Tsutsumi, Y., Miranda, P., Ruiz-del-solar, J.: UChile robotics team. Team description for RoboCup 2014. In: RoboCup 2014: Robot Soccer World Cup XVIII Preproceedings, Joao Pessoa, Brazil, July 2014
18. Hansen, N.: The CMA Evolution Strategy: A Tutorial. https://www.lri.fr/~hansen/cmatutorial.pdf
19. Thain, D., Tannenbaum, T., Livny, M.: Distributed computing in practice: the Condor experience: Research Articles. Concurr. Comput. Pract. Exp. **17**, 323–356 (2005)
20. Forero, L.L., Yáñez, J.M., Ruiz-del-Solar, J.: Integration of the ROS framework in soccer robotics: the NAO case. In: Behnke, S., Veloso, M., Visser, A., Xiong, R. (eds.) RoboCup 2013. LNCS, vol. 8371, pp. 664–671. Springer, Heidelberg (2014)

Development of Humanoid Robot Locomotion Based on Biological Approach in EEPIS Robot Soccer (EROS)

Azhar Aulia Saputra[1,2]([envelope]), Achmad Subhan Khalilullah[2],
and Naoyuki Kubota[1]

[1] Graduate School of System Design, Tokyo Metropolitan University,
6-6 Asahigaoka, Hino, Tokyo 191-0065, Japan
{azhar,kubota}@tmu.ac.jp
[2] EEPIS Robotic Research Center (ER2C), Politeknik Elektronika Negeri Surabaya,
Jalan Raya ITS, Sukolilo, Surabaya, East Java 60111, Indonesia
subhankh@pens.ac.id
http://er2c.pens.ac.id/index.php

Abstract. In this paper we propose the development of EROS locomotion by using neural oscillator. We investigated muscular structure of human body for designing the neuron structure. Two motoric neurons, extensor neuron and flexor neuron, represent one structure of joint that generating the angle of joint. Sensoric neuron connection also designed for adapting the environment. Three kinds of sensor such as ground reaction sensor, tilt sensor, and angular velocity sensor are utilized for validate the proposed method. Evolutionary algorithm was used for optimizing synapse weight among motoric neuron, while recurrent neural network was used for the dynamical condition learning. The locomotion system of this research was shown using Open Dynamic Engine (ODE). The proposed method can generate locomotion pattern and its stability learning system improves the stability of locomotion. The proposed approach formed the walking locomotion that potentially can be developed to become adaptive locomotion.

Keywords: Neural oscillator · Locomotion · Recurrent neural network · EROS

1 Introduction

Many researchers develop the locomotion of humanoid robot soccer. Various methods have been used to support their research. In RoboCup humanoid soccer league, researchers compete to develop humanoid robot soccer capable to play with human soccer player plan to be held in 2050. The rules and the structure of robot are made to be similar with soccer rules and human structure. For humanoid robot structure, as the height of robot became higher, soles of feet size become smaller than before until similar as human size. Robot should be

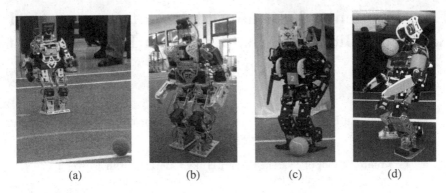

Fig. 1. EROS family. (a) E1205. (b) EROS-I. (c) EROS-II. (d). EROS-III

capable to walk, run, kick, and move during playing soccer on an unstable surface of grass. Many methods were used for locomotion in humanoid robot soccer. The conventional methods based on ZMP approach were discussed for locomotion approach [1–3]. Kajita et al. introduced a new method of a biped walking pattern generation by using a preview control of the zero-moment point (ZMP) [1]. Kim et al. also used ZMP as the feedback response to realize the dynamic locomotion HUBO robot [3]. In the previous works, the pattern equation based on ZMP approach was used to form the trajectory pattern for robot movement [4,5]. This approach was implemented in E-1205, EROS-I, EROS-II, and EROS-III as depicted in Fig. 1.

In this paper, the locomotion system was realized using biological structure approach. We adapted the human mechanism to generate the locomotion. We investigate the coupled muscle in human to acquire the inter-connection networking in neurons structure. We create the model neuron structure that composed from motoric and sensoric neuron. In order to acquire the effect value of synapse weight among motoric neuron, we used multi objective evolutionary algorithm. Recurrent neural network was used for learning the synapse weight between sensoric and joint neuron. To implement this locomotion system, we used computer simulation ODE. This paper is organized as follows. Section 2 explains the related work in this research. Section 3 talks about motion generator based on biological approach. Section 4 explains the stability system, Sect. 5 shows several experimental results and Sect. 6 concludes the paper.

2   Related Work

Locomotion model based on neural oscillator is a kind of biological system characterized by their behavior pattern with complexity of large degrees of freedom that can be stable and also flexible depending on the environment condition. Few researchers used central pattern generation with has basis in neurophysiological studies. Neural oscillator is a type of neural network for the generation of rhythmic motion [7–14]. Taga et al. have designed the coupled neural oscillator implemented for human locomotion. Global limit cycle was used where it was generated by a global entrainment between rhythmic activities of a nervous system to realize the stable and flexible locomotion [7]. Moreover, the mechanical formula was also created in order to acquire the feedback calculation. Vitor Matos et al. presented Central Pattern Generation (CPG) approach based on phase oscillators to bipedal locomotion [8]. This method can realize the transition from walking to running that can be adept for humanoid robot soccer that has high mobility.

Baydin also implemented CPG in 2012, in order to control humanoid biped walking mechanism. In his research evolutionary algorithm were implemented in order to determine the weight value and oscillatory parameter of CPG network [9]. Shan et al. used multi-objective GA for optimizing weight value [10]. While other researchers implemented reinforcement learning method for controlling CPG based locomotion [11].

In 2014, John Nassour presented an extended mathematical model of CPG for the locomotion [13]. His idea was to design the multi layers neuron connection in order to control the locomotion in various model of walking and to adapt the environmental condition. In order to support the stabilization, He et al. combine ZMP to CPG system [14]. In the proposed method the combination between the evolutionary algorithm for optimizing the pattern locomotion and recurrent neural network for increase the stabilization were used to build the locomotion system.

3   Locomotion Model

In this system, we used six joints in each leg. We adapted the mechanical structure of human as shown in Fig. 2. We investigated the musculoskeletal system that composed of the connection among rigid link. In this system, robot was installed with the ground reaction sensor with four point sensors in each sole of feet. Moreover, the robot was also equipped with tilt, accelerometer, and gyroscope sensor.

3.1   Neural Oscillator Based Locomotion Generator

In biological approach, neural oscillator was used as the basic element of the locomotion generator. We used neural oscillator generated by mutual inhibition between certain neurons with adaptation signal input as shown in Fig. 3. The neural oscillator model generated oscillatory signal activity, which consisted of two tonically excited neurons with self-inhibition effect linked reciprocally via inhibitory connections. We used neural oscillator model proposed by Matsuoka [15,16].

In Eqs. (1), (2), and (3), x_i, y_i, v_i, are the inner state, the output of the ith neuron, and a variable representing the degree of self-inhibition effect of the ith neuron,

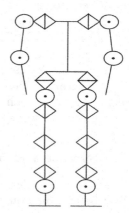

Fig. 2. Joint structure

respectively. S_i is an external input with a constant rate; Time constant of the inner state and the adaptation effect are notated by τ and τ'. w_{ij} represents the strength of the inhibitory connection between the neurons. $\sum_{j=1}^{n} w_{ij} y_j$ represents the total signal input from the other neurons that have mutual connection.

$$\tau \dot{x}_i + x_i = x_0 - \sum_{j=1}^{n} w_{ij} y_j - b v_i + S_i \tag{1}$$

$$\tau' \dot{v}_i + v_i = y_i \tag{2}$$

$$y_i = \max(x_i, 0) \tag{3}$$

$$\Theta_n = y_{2n} - y_{(2n+1)} \tag{4}$$

Neuron motoric system received two different types of sensory information: proprioceptive and exteroceptive information. In Eqs. (1) and (2), we compute by using Runge-Kutta gill method. In neuron oscillator based locomotion, there are two neurons (flexor and extensor) represented as the union of joint system are computed using Eq. (4). Θ_n represents nth joint angle that resulted by coupled neuron process.

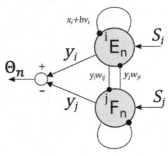

Fig. 3. Coupled neuròn

The robot has 18 degree of freedom; with 2 legs and 2 hands, each leg has 6 joints and each hand has 3 joints. There are 3 joints in hip: hip-x joint which has rotate axis in x-axis; hip-y joint which has rotate in y-axis; hip-z joint which has rotate in x-axis, 1 joint in knee, and 2 joints in ankle: ankle-x joint which has rotate in x-axis and ankle-y joint which has rotate in y-axis. The neuron structure for this locomotion system is shown in Fig. 4. We make limitation angle in each joint. Hip-x and ankle-x joints limitation are represented by $-\pi/2 < \theta_n < \pi/2$ hip-y and ankle-y limitation are represented by $-\pi/3 < \theta_n < \pi/3$, and knee joint limitation is represented by $-\pi/2 < \theta_n < 0$. The neuron structure is divided into server neuron and client neuron: neuron E_1, E_2, E_3, and E_4 act as the server neurons, which generate main signal to client neuron. These neurons are located in hip-x and knee joint. E_n implies an extensor neuron in nth joint. In order to optimize and summarize the weight of synapse, we separated the weight into seven parts: S1–S7. The weight values in each part are the same, while the detail of inter-connection among neuron can be shown in Fig. 4.

3.2   Synapse Weight Optimization

In this research, the optimization process was conducted to find the value of effect between neurons. There are many methods for parameters optimization. In this research, multi-objective optimization is used since we have three objective optimization problems. For solving multi-objective optimization problems, there

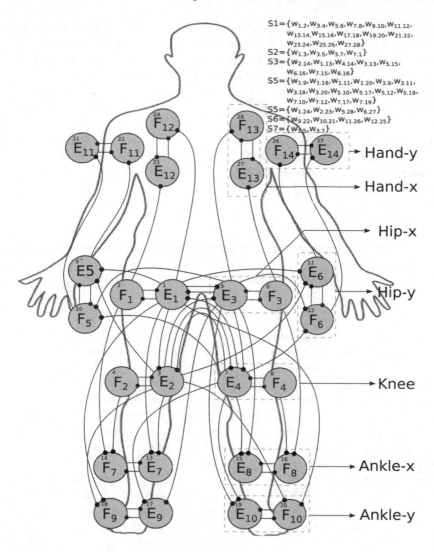

$$S1 = \{w_{1.2}, w_{3.4}, w_{5.6}, w_{7.8}, w_{9.10}, w_{11.12},$$
$$w_{13.14}, w_{15.16}, w_{17.18}, w_{19.20}, w_{21.22},$$
$$w_{23.24}, w_{25.26}, w_{27.28}\}$$
$$S2 = \{w_{1.3}, w_{3.5}, w_{5.7}, w_{7.1}\}$$
$$S3 = \{w_{2.14}, w_{1.13}, w_{4.14}, w_{3.13}, w_{5.15},$$
$$w_{6.16}, w_{7.15}, w_{8.16}\}$$
$$S5 = \{w_{1.9}, w_{1.18}, w_{1.11}, w_{1.20}, w_{3.9}, w_{3.11},$$
$$w_{3.18}, w_{3.20}, w_{5.10}, w_{5.17}, w_{5.12}, w_{5.19},$$
$$w_{7.10}, w_{7.12}, w_{7.17}, w_{7.19}\}$$
$$S5 = \{w_{1.24}, w_{2.23}, w_{5.28}, w_{6.27}\}$$
$$S6 = \{w_{9.22}, w_{10.21}, w_{11.26}, w_{12.25}\}$$
$$S7 = \{w_{1.5}, w_{3.7}\}$$

Fig. 4. Interconnection neuron diagram

are two main methods: Pareto front and weighting factor. One of the main drawbacks of these methods is choosing the most appropriate value of weighting factors. The development of locomotion based on neural oscillator implies the optimization of the oscillation of body tilt $\bar{\sigma}$ (minimization) in pitch and roll direction, the change rate of walking direction (minimization) for maintaining the go straight movement, and the velocity of movement $\bar{v}$ (maximization). In order to acquire the minimization value represent by $\bar{v}$, we used the remaining parameter of velocity, which can not be reached by robot walking in certain time. In the optimization case, we calculated the absolute average value of body tilt in pitch and roll direction, which was computed using Eqs. (7) and (8) and the velocity of robot walking was computed using Eq. (9). The weight among neuron should be determined to acquire the desired velocity with minimum oscillation of body tilt. In order to acquire the fitness evaluation computed by Eq. (10), we run the robot in ODE and analyze the output of sensor.

$$\sigma(t, w_{ij}) : |R| \tag{5}$$

$$L(t, w_{ij}), x(t), y(t) : R \tag{6}$$

In Eq. (5), the body tilt (σ) is normalized in absolute value. And in Eq. (6), L, x, y are the real number represented the length of movement in both x-axis, and y-axis respectively. These parameters were acquired in a time sampling during simulation process in ODE.

$$\bar{\sigma}_{pitch}(w_{ij}) = \left(^1/_T\right) \sum_{t=0}^{T} \sigma_{pitch}(t, w_{ij}) \tag{7}$$

$$\bar{\sigma}_{roll}(w_{ij}) = \left(^1/_T\right) \sum_{t=0}^{T} \sigma_{roll}(t, w_{ij}) \tag{8}$$

$$\bar{v}(w_{ij}) = \left(^1/_T\right) \sum_{t=0}^{T} \frac{d}{dt} L\left(t, w_{ij}, x(t), y(t)\right) \tag{9}$$

$$f = \arg \min_{w_{ij}} (\bar{\sigma}_{pitch}\varepsilon_1 + \bar{\sigma}_{roll}\varepsilon_2 + \bar{v}\varepsilon_3) \tag{10}$$

We implemented steady state genetic algorithm (SSGA) for optimizing the value of weight among neuron (w_{ij}). In SSGA, there is one individual that inserted to the new population in one generation. We used weighting factor in order to optimize multi-objective problems. ε_1, ε_2, ε_3 are the weight coefficients for tilt oscillation in pitch direction, roll direction, and for walking velocity respectively. In SSGA process, first we select the parent from initial generation; second we create the new individual by using mutation and crossover process. After that we evaluate the new individual by using fitness calculation computed in Eq. (10).

g_1	g_2	g_3	g_4	g_5	g_6	g_7

Fig. 5. Chromosome of an individual

In SSGA, the chromosome of the individual is composed of the weight parameters among neuron. Each parameter in represented by one gene. Since we have seven parts of weights, thus one individual has seven genes represented in Fig. 5. Each gene has minimum (h_{min}) and maximum value (h_{max}). The detail of mutation and crossover process was explained in our previous paper [17].

4   Stability System

The stability system is built to have responsive ability to the disturber comes from environmental condition such as normal force resulted depending on surface condition. We designed the connection between sensoric neuron and joint neuron system that explained in Table 1. Local effect approach in this model was used to summarize the structure neuron. In these cases, we have three balancing systems: ground reaction, pose control, and hand reaction. Each of them has different connection structure.

4.1   Synapse Structure

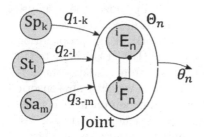

Fig. 6. Sensor connection structure in joint angle level

In the proposed model, We implemented three different kind of sensors as mentioned in previous section. The synapse connection structure is illustrated in Fig. 6 and explained in Table 1. In ground reaction, there are four sensoric neurons located in the four corners of foot, Sp_1–Sp_4 for left foot and Sp_5–Sp_8 for right foot. These sensoric neurons are connected to ankle joint neurons: ankle-x joint and ankle-y joint. When front sensoric neurons (Sp_1, Sp_2, Sp_5, Sp_6) detect the ground, they send the positive impulse to ankle-x joint. When the right side sensoric neurons (Sp_2, Sp_3, Sp_6, Sp_7) detect the ground, positive impulse will be sent to joint ankle-y. These sensoric neurons (ground sensor) also influence knee joint and hip-x joint. When sensoric neuron detects the ground, it sends impulse to knee and hip-x joint. In tilt and angular velocity sensors, there are two degrees of freedom: pitch (St_2 and Sa_2) and roll (St_1 and Sa_1). St_1 and Sa_1 are connected to hand-x joint, while St_2 and Sa_2 are connected to ankle-y, hip-y, hand-y joint. The output sensors are resulted during the simulation process.

$$\theta_n = \Theta_n + \sum_{i=1}^{m} Sa_i q_{3,i}\alpha_{3,i}^n + \sum_{i=1}^{l} St_i q_{2,i}\alpha_{2,i}^n + \sum_{i=1}^{k} Sp_i q_{1,i}\alpha_{1,i}^n \qquad (11)$$

In Eq. (11), m, l, k are number of tilt, angular velocity, and ground sensor, respectively. The values of synapse $q_{1,i}$, $q_{2,i}$, and $q_{3,i}$ are controlled by using recurrent neural network (RNN) for acquiring the best stabilization parameter. This system will be detailed explained in the next section. θ_n represents the angle joint in nth joint. $\alpha_{3,i}^n$, $\alpha_{2,i}^n$, $\alpha_{1,i}^n$ represent the impuls effect of angular velocity sensor, tilt sensor, and ground sensor in n-th joint explained in Table 1, Since 0 implies there is no effect, 1 and −1 imply positive and negative effect.

Table 1. Connection structure

							nth Joint							
n	1	2	3	4	5	6	7	8	9	10	11	12	13	14
St_1	1	0	1	0	0	0	0	0	0	0	0	1	1	0
St_2	0	0	0	0	1	1	0	0	0	0	1	0	0	1
Sa_1	1	0	1	0	0	0	1	1	0	0	0	1	1	0
Sa_2	0	0	0	0	1	1	0	0	1	1	1	0	0	1
Sp_1	0	0	1	-1	0	0	0	1	0	-1	0	0	0	0
Sp_2	0	0	1	-1	0	0	0	1	0	1	0	0	0	0
Sp_3	0	0	1	-1	0	0	0	-1	0	1	0	0	0	0
Sp_4	0	0	1	-1	0	0	0	-1	0	-1	0	0	0	0
Sp_5	1	-1	0	0	0	0	1	0	-1	0	0	0	0	0
Sp_6	1	-1	0	0	0	0	1	0	1	0	0	0	0	0
Sp_7	1	-1	0	0	0	0	-1	0	1	0	0	0	0	0
Sp_8	1	-1	0	0	0	0	-1	0	-1	0	0	0	0	0

4.2   Recurrent Neural Network

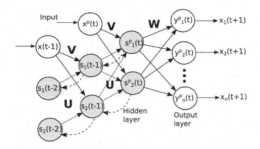

Fig. 7. Recurrent neural network structure

In order to optimize the synapse weight, we utilized the recurrent neural network back propagation through time (BPTT) as depicted in Fig. 9. This system updates the correction synapse weight from sensoric neuron to joint neuron in real time. In our recurrent neural network system p represents the state number of current condition of Robot. In this system we assumed the number of state condition is one. In Eq. (12), $x_i^p(t)$ is the input value in t time sampling; $s_h^p(t-1)$ is output value of hidden neuron in $t-1$ time sampling; b_j is bias input for hidden layer; $x_y^p(t)$ is output value of hidden neuron in t time sampling (Fig. 7).

$$s_y^p(t) = f(ns_y^p(t)) = f\left(\sum_i^l x_i^p(t)V_{ij}^p(t) + \sum_h^m s_h^p(t-1)U_h^p + b_j\right) \qquad (12)$$

$$y_k^p(t) = g(O_k^p(t)) = g\left(\sum_j^m s_j^p(t)W_{kj}^p(t) + b_k\right) \qquad (13)$$

After forward process is done and resulted the output neuron value y_k^p computed in Eq. (13), waiting time is required to acquire the sensor value $S(k, y_k^p(t))$ and analyze the weight value. In Eqs. (14) and (15), δ_k^p is error propagation in output layer and is error propagation in hidden layer. We used hand actuator as the response output $y_k^p(t)$. The number of neurons in the input layer, in the hidden layer, and in the output layer are denoted by l, m, and n, respectively.

$$\delta_k^p = (d_k^p - S(k, y_k^p(t)))g'(O_k^p(t)) \qquad (14)$$

$$\delta_j^p = \sum_k^n \delta_k^p W_{kj}^p(t)f'(ns_j^p(t)) \qquad (15)$$

$$\mathbf{W}^p(t+1) = \mathbf{W}^p(t) + \eta_w \mathbf{s}^p(t)\delta_k^p \qquad (16)$$

$$\mathbf{V}^p(t+1) = \mathbf{V}^p(t) + \eta_v \mathbf{x}^p(t)\delta_j^p \qquad (17)$$

$$\mathbf{U}^p(t+1) = \mathbf{U}^p(t) + \eta_u \mathbf{s}^p(t-1)\delta_j^p \qquad (18)$$

The activation function for hidden layer $f(x)$ used sigmoid function. d_k^p is the desire output and the output function for are sigmoid function. In this case, the desire output is the condition where robot maintains the angular velocity of robot becomes zero. We have 12 neurons in the input layer resulted from sensors (Sp_k, St_l, Sa_m), four neurons in the hidden layer, and 12 neurons in the output layer. The configuration of input neuron is explained in Table 2. We need to train the parameters $\mathbf{V}$, $\mathbf{U}$, and $\mathbf{W}$ as the weight among neuron. η_v, η_u, and η_w are the learning rate for $\mathbf{V}$, $\mathbf{U}$, $\mathbf{W}$ parameter recursively. In BPTT, the error propagation is done recursively. As depicted in Eqs. (16), (17), and (18), the weight of synapse between sensoric and joint neuron can be regenerated.

Table 2. Configuration of input and output neuron

x_i^p	x_1^p	x_2^p	x_3^p	x_4^p	x_5^p	x_6^p	x_7^p	x_8^p	x_9^p	x_{10}^p	x_{11}^p	x_{12}^p
$S(k, y_k^p)$	$S(y_1^p)$	$S(y_2^p)$	$S(y_3^p)$	$S(y_4^p)$	$S(y_5^p)$	$S(y_6^p)$	$S(y_7^p)$	$S(y_8^p)$	$S(y_9^p)$	$S(y_{10}^p)$	$S(y_{11}^p)$	$S(y_{12}^p)$
Input	St_1	St_2	Sa_1	Sa_2	Sp_1	Sp_2	Sp_3	Sp_4	Sp_5	Sp_6	Sp_7	Sp_8
Output	$q_{2,1}$	$q_{2,2}$	$q_{3,1}$	$q_{3,2}$	$q_{1,1}$	$q_{1,2}$	$q_{1,3}$	$q_{1,4}$	$q_{1,5}$	$q_{1,6}$	$q_{1,7}$	$q_{1,8}$

5 Experimental Result

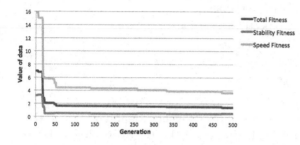

Fig. 8. Fitness evolution

We designed the robot by using computer simulation ODE adapting the rules of RoboCup 2015 [18]. Parameter properties from EROS robot were inserted in simulation robot as shown in Table 3. We conducted the experiment gradually as follow; first, optimizing the weight among neuron (w_{ij}) to acquire the walking locomotion pattern; second, optimization for increasing the stability in locomotion. In order to find the pattern of trajectory, we observed the signals generated from many combinations of interconnection in server neuron introduced by Matsuoka [16]. After the best pattern of locomotion was determined, we optimize the synapse weight among neurons. SSGA algorithm was applied that use speed of walking ($\bar{v}$) and tilt sensor ($\bar{\sigma}$) as the objective function. The parameters used for weight among neuron optimization was tabulated in Table 4.

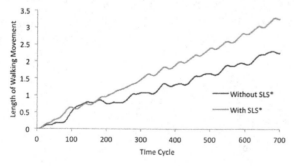

Fig. 9. Length of walking comparison

In this experiment, we acquire the fitness diagram from SSGA process that shown in Fig. 8 resulted from two objective values (Speed and Stabilization). The fitness total decreased significantly until 50-th generation. Stability fitness and speed fitness also decreased, which implied the oscillation data in pitch and roll direction to become smaller and the speed become higher. The good pattern locomotion were reach in 50-th

Table 3. Robot specification

Parameter	Description
Degree of freedom	18 degree of freedom
Height, weight	600 mm, 3000 g
Sensors	4 ground detection sensors in each leg
	Tilt sensor and angular velocity sensor

Table 4. Parameters of SSGA

Parameter	Value
Population size	16 individuals
Num. of generations	500 generations
Num. of objectives	2 objectives: speed and stabilization
Chromosome	7 real number
Evaluation time	9 s
$\varepsilon_1, \varepsilon_2, \varepsilon_3$	0.3, 0.3, 0.7

generation. We finished the generation until 500-th generation. The locomotion system takes one individual (S1–S7 = 1.546, 1.230, 1.901, 1,676, 2.158, 1.390, 1.120) as the best individual in SSGA, and then become the parameter in robot. Next, we analyzed the signal resulted from coupled motoric neuron.

By using evolutionary algorithm, walking pattern based on neural oscillator can be formed. However, since the stability level resulted from evolutionary computation was not strong enough to cover outside disturbances, the stability system is required to solve this problem.

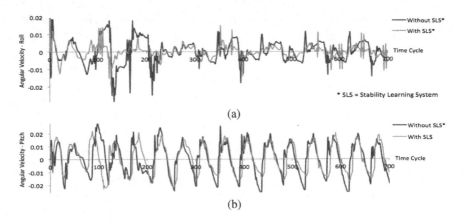

Fig. 10. Comparison angular velocity data between locomotion system without SLS and with SLS (a) Roll direction (b) Pitch direction

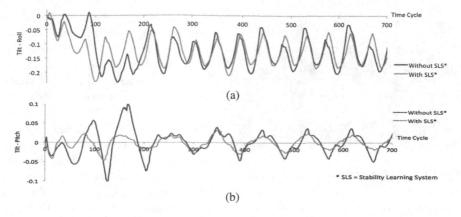

(a)

(b)

Fig. 11. Comparison angular velocity data between locomotion system without SLS and with SLS (a) Roll direction (b) Pitch direction

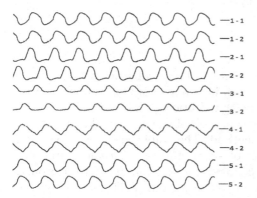

Fig. 12. Signal output of coupled neuron

In the second experiment, we continued to implement the stability system that used recurrent neural network model to support the balancing system. We defined $\eta_v = 0.02$, $\eta_u = 0.02$, and $\eta_w = 0.01$ as the learning rate. We analyzed the signal oscillation without stability learning system (SLS) and with SLS to compare the difference among them. Figures 10 and 11 shows the angular velocity oscillation and tilt oscillation comparison in both of locomotion system. The locomotion with SLS has smaller tilt and angular velocity oscillation, which imply the locomotion with SLS more stable than locomotion without SLS. Beside that, the SLS also increase the speed of walking in robot locomotion as shown in Fig. 9. The locomotion with SLS can reach the length of movement longer because the SLS can effectively reduce the disturbance that hamper the robot walking. The weight between sensoric neuron and joint neuron is changing every time sampling depending on the robot condition. SLS is effective for balancing system of humanoid robot locomotion. Beside that, the values of time sampling configuration are important. We have captured the running process simulation as seen in Fig. 13. This model is expected forming the human-like locomotion system.

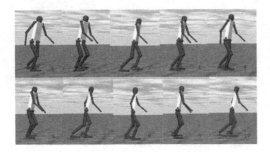

Fig. 13. Running process in ODE simulation

The output of coupled neuron representing the signal of joints in robot when walking on flat surface was recorded as shown in Fig. 12. The hip-x, knee, and ankle-x joints have phase difference 180 degrees, while hip-y and ankle-y joints have the same phase. These signals have been influence with ground reaction. Therefore, these signals have different pattern depending on the environment condition. In this paper, we only consider the disturbances in the flat surface. We proved that the stability system can reduce the oscillation of robot walking and also improve the speed of robot walking. This biological approach based robot locomotion are needed improvement, therefore the robot can walk in different direction, walk in uneven surface, and reduce the external disturbance.

6   Conclusion

In this research, we have designed the humanoid biped robot locomotion based on neural oscillator. The locomotion pattern in humanoid robot can be acquired by controlling the weight of synapse by using evolutionary algorithm. The design of neural structure between sensoric neuron and joint neuron could change the stabilization approach by using ZMP method. The stabilization of locomotion can be increased by learning the weight of synapse using recurrent neural network. Signal pattern resulted from the coupled neuron is suitable for the ground reaction. We have designed the synapse from sensoric to joint neuron and the walking movement without turning left or right. As the future research, we are designing to add a neuron systems adapted as brain neuron systems that process the information from sensoric neuron system and generate signal to motoric neuron system. Brain neuron systems also control the timing impulse for controlling the signal generated by motoric neuron.

References

1. Kajita, S., et al.: Biped walking pattern generation by using preview control of zero-moment point. In: Proceedings of the International Conference on Robotics and Automation, vol. 2, pp. 1620–1626 (2003)
2. Vukobratovic, M., et al.: Zero-moment point - thirty five years of its life. Int. J. Humanoid Rob. **1**(1), 157–173 (2004). World Scientific Publishing Company
3. Kim, J., et al.: Experimental realization of dynamic walking of biped humanoid robot KHR-2 using ZMP feedback and inertial measurement. Adv. Rob. **20**(6), 707–736 (2006)

4. Saputra, A.A., et al.: Combining pose control and angular velocity control for motion balance of humanoid robot soccer EROS. In: Proceedings of the IEEE-RiiSS, USA, pp. 16–21 (2014)
5. Saputra, A.A., et al.: Acceleration and deceleration optimization using inverted pendulum model on humanoid robot EROS-2. In: Proceedings of the 2nd ISRSC, Indonesia, pp. 17–22 (2014)
6. Missura, M., et al.: Self-stable omnidirectional walking with compliant joints. In: Proceedings of 8th Workshop on Humanoid Soccer Robots, Atlanta, USA (2013)
7. Taga, G., et al.: Self-organized control of bipedal locomotion by neural oscillators in unpredictable environment. Biol. Cybern. **65**(3), 147–159 (1991)
8. Victor, M., et al.: Towards goal-directed biped locomotion: combining CPGs and motion primitives. Rob. Auton. Syst. **62**, 1669–1690 (2014)
9. Baydin, A.G.: Evolution of central pattern generators for the control of a five-link bipedal walking mechanism. Paladyn J. Behav. Rob. **3**(1), 45–53 (2012)
10. Shan, J., et al.: Design of central pattern generator for humanoid robot walking based on multi-objective GA. In: Proceedings of the IEEE/RSJ International Conference on Intelligent Robots and Systems, vol. 3, pp. 1930–1935 (2000)
11. Mori, T., et al.: Reinforcement learning for CPG-driven biped robot. In: McGuinness, D.L., Ferguson, G. (eds.) pp. 623–630. AAAI Press/The MIT Press (2004)
12. Guanjiao, R., et al.: Multiple chaotic central pattern generators with learning for legged locomotion and malfunction compensation. J. Inf. Sci. **294**, 666–682 (2015). Elsevier
13. Nassour, J., et al.: Multi-layered multi-pattern CPG for adaptive locomotion of humanoid robot. Biol. Cybern. **108**(3), 291–303 (2014)
14. He, B., et al.: Real-time walking pattern generation for a biped robot with hybrid CPG-ZMP algorithm. Int. J. Adv. Rob. Syst. **11**(160), 1–10 (2014)
15. Matsuoka, K.: Sustained oscillations generated by mutually inhibiting neurons with adaptation. Biol. Cybern. **53**, 367–376 (1985)
16. Matsuoka, K.: Mechanisms of frequency and pattern control in the neural rhythm generators. Biol. Cybern. **56**(5), 345–353 (1987)
17. Saputra, A.A., et al.: Efficiency energy on humanoid robot walking using evolutionary algorithm. In: Proceedings of the IEEE Congress on Evolutionary Computation, Sendai, Japan, pp. 573–578 (2015)
18. RoboCup Soccer Humanoid League Rules and Setup Book (2015)

A Dynamic and Efficient Active Vision System for Humanoid Soccer Robots

Matías Mattamala[✉], Constanza Villegas, José Miguel Yáñez, Pablo Cano, and Javier Ruiz-del-Solar

Department of Electrical Enginering, Advanced Mining Technology Center (AMTC), Universidad de Chile, Av. Tupper 2007, Santiago, Chile
{mmattamala,jruizd}@ing.uchile.cl

Abstract. This paper presents an efficient active vision system which controls the head of a humanoid soccer robot. The system explicitly separates static information obtained offline from the map, and dynamic information from mobile objects, such as the ball and other players. Both types of information are mapped and handled in a simplified structure called *action space*, which assigns scores to each possible action of the robot's head. Scores also consider the movement constraints of the robot's head. Due to its simplicity and efficient information handling, the proposed active vision system is able to run in real-time in less than 1 ms. The performance of the system in a robot soccer environment is tested via simulation and real experiments.

Keywords: Active vision · Camera control · Robot soccer · Self-localization · RoboCup · Standard Platform League

1 Introduction

Achieving a goal and staying self-localized is a natural task for a human, but not for a robot. This task is even more complex if the robot is in a highly dynamic environment -as a soccer match-, where several landmarks must be considered in order to have a good performance in both self-localization and playing during the game. In this context, it is not recommended to use passive self-localization systems based on predefined routines or heuristics because these are not related with the real world; in fact, the number of possible configurations of the ball and players on the field is huge. Then, it is necessary to use active localization methodologies to cope with the environment dynamics. Active localization systems are frequently divided into two types: active navigation and active sensing. Active navigation involves robot displacement in order to reduce self-localization uncertainties; on the other hand, active sensing refers to the manipulation of the robot's sensor in order to get more information from the world. In particular, an active vision system uses a camera as main sensor.

This work describes an active vision system for a humanoid soccer robot which aims to choose a head pose that maximizes a score in the so-called *action*

© Springer International Publishing Switzerland 2015
L. Almeida et al. (Eds.): RoboCup 2015, LNAI 9513, pp. 316–327, 2015.
DOI: 10.1007/978-3-319-29339-4_26

space. The space is updated during the game using *a priori* information from the map, and dynamic information obtained from the ball and other players' positions. The proposed system also takes into account the movement constraints of the robot's head. This approach allows not only to control the robot's head but also to calculate the action to be performed using minimum computational resources in real-time.

The paper is organized as follows: relevant related work is presented in Sect. 2. The active vision system proposed is described in detail in Sect. 3. Finally, Sect. 4 presents the experimental results obtained, and Sect. 5 ends with the main conclusions.

2   Related Work

Active vision is a field mainly related to computer vision which origins date back to late 80's under the name of *active perception* [2], *animate vision* [3] and *active vision* [1]. These paradigms, despite having different backgrounds and motivations, are based on the same principle: in order to get better and most valuable information from the environment it is necessary to control the camera gaze, such as humans do with their vision system. This idea establishes a direct relation between perception and action, reason for which not only robotics have been involved in its development, but also neurosciences and cognitive sciences.

In this work we are concerned about applications of active vision systems in mobile robotics. Active vision has developed a strong relation with the problem of robot self-localization, currently one of the most important topics in robotics. For instance, Burgard [4] introduced an entropy minimization criteria to actively explore the environment. Davison [6] presented the first active vision approach to the SLAM problem using a stereo head, then Vidal-Calleja extended it to the monocular case [14]. Seara [10] introduced an intelligent gaze control system for a walking robot.

In the RoboCup context, several works have been published for the Aibo quadruped robots, such as Fukase [7], Mitsunaga [9], Stronger [12] and Guerrero [8]. Regarding NAO humanoid robots, currently used in the Standard Platform League, Seekircher et al. [11] presented an entropy minimization approach which handles both self-localization and ball uncertainty as well as the costs of choosing among them by using a camera control policy. Czarnetzki et al. [5] proposed another entropy minimization approach using a particle filter and testing the method in static and dynamic environments. However, neither the ball nor other mobiles objects are considered in that systems. Both approaches rely on an expensive modified particle filter to determine the target to gaze, requiring several optimizations that reduce the accuracy of the system. These modifications allow both systems to run in 5 ms in average. In contrast, our approach not only avoids online state estimations but it also considers the information of the ball and other players, achieving self-localization improvements and running in less than 1 ms.

3   Active Vision

3.1   General View

This work describes an efficient active vision system for robot soccer applications. It is based on an explicit division between the a *priori* information that can be obtained off-game from the known-map using its inherent landmarks, and the dynamic information provided by the moving ball and other robots in play. Our approach maps all information sources into an *action space* which modify certain scores associated to each possible head action. Then, the next target to be seen is selected as the action with highest score.

3.2   Action Space

Before describing the active vision system it is necessary to introduce a major concept for this work: the action space. The action space A of an end-effector is defined as follows:

$$A := \{(a, \omega) : a \in \mathbb{R}^N, \omega \in [0, 1]\} \tag{1}$$

where N is the number of *degrees of freedom* given by the effector's joints, a is an *action* which denotes a possible state of the effector, and ω is the *score* of the action a. Then, every possible state of the effector has a score assigned according to a criteria that depends on the application. The idea is to perform decision making over the next effector's actions just by selecting the highest score.

In this work we are concerned about controlling the robot's head, so the set of actions a will be reduced to reachable head poses $a = (\phi_{pan}, \phi_{tilt}) \in \Phi_{pan} \times \Phi_{tilt}$, where Φ_{pan} and Φ_{tilt} denotes the set of allowed angles for each joint. Since the robot will move its head in order to improve its self-localization, each action a is associated with real world observations that affect the robot's localization system. Hence, the criteria that defines ω will be to quantify the accuracy of the self-localization given a determined action a.

The next sections present a methodology to assign scores to each possible action, using a discrete representation of the action space. In a first formulation only prior information obtained from the known map is considered. Later, the static model is improved by adding some robot's physical constraints as well as dynamic information from the environment.

3.3   Obtaining a Priori Information

Most of the active vision systems presented in the Sect. 2 modify a state estimator using simulated perceptions in order to find the optimal action to be performed by the robot's head. Guerrero [8] proposed a probabilistic framework to infer which landmarks could be gazed depending on the current head pose. On the other hand, Seekircher [11] used a learned model for each perceptor (goal posts, lines, corners) to perform a Monte Carlo Exploration [13].

In order to avoid expensive online calculations using simulated perceptions, we rather propose to exploit the prior information of the map by sampling a measure of self-localization improvement while executing different actions, i.e. observing different landmarks. This process avoids calculating accurate sensor models because it fuses their localization improvements. Then, the collected data can be stored in a look-up-table, facilitating access during execution.

We develop an offline routine which uses the particle filter self-localization, and samples the particles' weighting sum for each possible robot pose (x, y, θ) and head pose $(\phi_{pan}, \phi_{tilt})$. In order to make the data acquisition tractable, we run the algorithm in a predefined discrete grid for the x, y, θ and ϕ_{tilt} dimensions, using the grid pose as input of the particle filter estimation. We did not consider the ϕ_{pan} dimension into the sampling process because this information can be approximated from poses with the same orientation θ while the active vision system runs.

Algorithm 1 presents the main procedure of the sampling process, which is run via simulation. We create a high dimensional table according to the sampling setup (the init_table function). Then, the system iterates moving the robot over the field, changing its pose and head tilt. The robot executes the particle filter localization system but we modified it to use the ground truth pose as the resulting motion update, resetting the previous frame estimation. Afterwards, the sensor update is executed using the exact observations expected for the current pose, which should provide the best possible response of the observation models. We also modified the output of the particle filter in order to extract the particles' weighting sum ω as a measure of self-localization accuracy; this is the score that will be assigned to the action performed.

Algorithm 1. Algorithm to sample the Particle Filter result over the field.

Generate_Field_Table()
  init_table(T)
  for all $\bar{x}$ in x_{grid} **do**
    for all $\bar{y}$ in y_{grid} **do**
      for all $\bar{\theta}$ in θ_{grid} **do**
        move_to_pose($\bar{x},\bar{y},\bar{\theta}$)
        for all $\bar{\phi}$ in ϕ_{pgrid} **do**
          set_head_tilt($\bar{\phi}$)
          $\omega \leftarrow$ sensor_update($\bar{x}, \bar{y}, \bar{\theta}$)
          save_in_table($T, \bar{x}, \bar{y}, \bar{\theta}, \bar{\phi}, \omega$)
        end for
      end for
    end for
  end for

The results obtained for the sampling process are shown in Fig. 1. Please notice that every position on field has a white ring assigned which represents the pan-tilt action space for each pose.

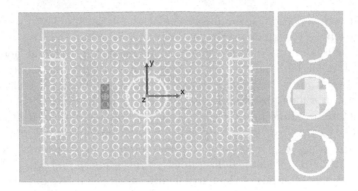

Fig. 1. Left: Example of field scores obtained for each grid position (x, y) after sampling. Each white ring denotes a set of pan-tilt actions. Right: A zoom of field area, marked in red. White rings are formed by small circles which denote the possible actions. The radius of each circle indicates the score assigned to the action (only the highest score is shown). In this case, we can notice that if we are close to the right penalty cross, the best action to perform in order to improve self-localization is to point at the center circle (Color figure online).

Using this approach, the action space for a determined pose in the field is obtained directly from the look-up-table. The action space changes while the robot navigates, allowing it to perform different head movements. Nevertheless, it works only in an empty field, because no other objects are being considered; that is why we call it a *static action space*. Next section will cover how to solve this problem by adjusting the scores dynamically.

3.4   Dynamic Information

In the previous section we sampled data from the field to determine a set of possible targets to gaze for each field position, each one having an assigned score. However, this information is not enough to cope with the environment dynamics because it only considers the map information. We have to improve the action space in order to deal with this issue, using dynamic models to modify the data. This section presents three adjustments to the *static action space* using the variable information of the field during a game. These are modeled by 2-dimensional functions that weight the scores ω.

Head Limits Constraints. A first improvement consist of adding the head limits information to the model. A humanoid robot has a limited range for the pan and tilt head angles, so this information must be considered. Denoting by Φ_{limit} the set of actions inside the constrained range, we can define a head-limiting function as:

$$f_{limits}(\phi_{pan}, \phi_{tilt}) = \mathbf{1}_{\Phi_{limit}}(\phi_{pan}, \phi_{tilt}) \tag{2}$$

Where $1_{\Phi_{limit}}$ denotes the *indicator* or *characteristic function*. Then, every score of actions out of the pan or tilt ranges defined by the robot head is set to zero.

Head Movement Penalization. In general, we are concerned about choosing the best action in the next frame of execution, so we are interested in choosing an action that provides the most information at the lower *angular cost* for the head. According to this fact, it is necessary to consider a penalization function which reduces the score of the actions further away from the actual pose of the head. Assuming independence between coordinates in the action space, we can model the penalization function as follows:

$$f_{head}(\phi_{pan}, \phi_{tilt}) = v^h_{\mu,\kappa}(\phi_{pan})g^h_{\mu,\sigma}(\phi_{tilt}) \tag{3}$$

where $v^h_{\mu,\kappa}(\phi)$ corresponds to the *von Mises function* or *circular gaussian function* (see (4)), whereas $g^h_{\mu,\sigma}(\phi)$ denotes a *gaussian function* (see (5)). The election of von Mises function is related with the periodicity of the action space with respect to pan angle as shown in Fig. 1.

$$v^h_{\mu,\kappa}(\phi) = V e^{\kappa \cos(\phi-\mu)} \tag{4}$$

$$g^h_{\mu,\sigma}(\phi) = G e^{-\frac{(\phi-\mu)^2}{2\sigma^2}} \tag{5}$$

In this case, the means of both functions are set at the current head pose. The dispersion κ and variance σ are parameters set by-hand. The constants V and G are chosen so as to normalize the peak of each function to 1.

Avoiding Obstacles in the Field of View. Obstacles occlude the information obtained from the map and affect the observations expected by the robot, so we would like to avoid gazing in those directions. This does not affect the obstacle detection, because the robot keeps moving the head during the active vision execution. In addition, obstacle detections are shared among the player of the same team. We model each obstacle using the same one-dimensional functions as before but with different parameters:

$$f_{obstacle}(\phi_{pan}, \phi_{tilt}, r) = k(r)(1 - v^o_{\mu,\kappa}(\phi_{pan}))g^o_{\mu,\sigma}(\phi_{tilt}) \tag{6}$$

The function shown in Eq. 6 further reduces the score of the actions closer to an obstacle on the action space. However, it also includes a correction function $k(r)$ which weights the score reduction depending on the obstacle distance, denoted by r. This function is estimated using real data. The means (μ_{pan}, μ_{tilt}) associated to the von Mises function $v^o_{\mu,\kappa}$ and the Gaussian function $g^o_{\mu,\sigma}$, respectively, are estimated by projecting the obstacle center onto the action space, using the inverse kinematics of the robot camera (Fig. 2 left). The von Mises' dispersion κ and the Gaussian variance σ are estimated in a similar way, by projecting the vertices of the obstacle bounding box onto the action space, and calculating the midpoint for each dimension p_{pan} and q_{tilt}. Using these points and the previously estimated mean, it is possible to estimate κ and σ as follows:

$$\hat{\kappa}_{pan} = \frac{1}{\kappa_0 |p_{pan} - \mu_{pan}|} \tag{7}$$

$$\hat{\sigma}_{tilt}^2 = \sigma_0^2 |q_{tilt} - \mu_{tilt}|^2 \tag{8}$$

where κ_0 and σ_0 are compensation factors to adjust the parameter estimation empirically.

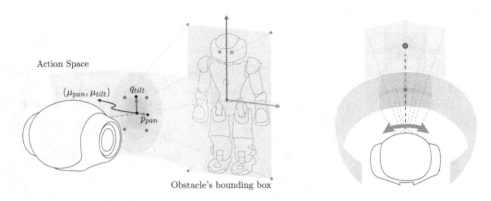

Fig. 2. Left: $f_{obstacle}$ parameter estimation using the bounding box vertices projected onto the action space. Right: Actions in which the ball appears in the Field of View define a region in the *ball action space* (Color figure online).

Updating Static Information. Finally, the *localization action space* can be updated using the information described anteriorly. The previous functions are evaluated in each action $a = (\phi_{pan}, \phi_{tilt})$ of the current static action space, and used to weight and, therefore, update its scores:

$$\omega_{loc}(a) = \omega(a) \cdot f_{limits}(a) \cdot f_{head}(a) \cdot f_{obstacle}(a) \tag{9}$$

3.5   Adding the Ball to the Model

The model presented shows a clear strategy to cope with both static and dynamic field information. This allows the robot to keep self-localized despite having several obstacles and to choose the best target under the principles previously exposed. This strategy, however, cannot handle the most important element in the game: the ball.

In order to have a good game performance, all players must maintain self-localized and track the ball frequently. This can be achieve in a simple form by using the same principles used to update the action space. Let Φ_{ball} be a set of actions where the ball appears inside the robot's field of view. This can be calculated by using inverse kinematics as we did for the obstacles. We define a complementary *ball action space* as an action space which handles the ball gazing scores. The scores ω_{ball} are set to 1 if the action is in Φ_{ball}, and 0 if not. This defines a subset of actions where the ball can be seen (Fig. 2 right). This information will be included into the active vision model.

3.6   Action Space-Based Decision Making

Finally, we merge the localization information handled in the *localization action space* with the ball information (Eq. 10). This is done by defining a cost function which provides a new score for each action. In order to cope with the localization and ball gazing *trade-off*, a factor α is introduced to determine ball importance. This factor is modified by the robot's running behaviors depending on the player's role or the game state. A graphical 3D visualization of the final action space is shown in Fig. 3.

$$\Omega(\phi_{pan}, \phi_{tilt}) = \omega_{loc}(\phi_{pan}, \phi_{tilt}) + \alpha\omega_{ball}(\phi_{pan}, \phi_{tilt}) \tag{10}$$

Afterwards, the best action is selected as the action with the maximum fused score associated, as is shown in (11).

$$\phi_{pan}^*, \phi_{tilt}^* = \arg\max_{\phi_{pan}, \phi_{tilt}} \Omega(\phi_{pan}, \phi_{tilt}) \tag{11}$$

Fig. 3. Resulting action space after information fusion. A red circle denotes the selected target. Orange circles show actions with highest score for each pan angle where the ball can be observed (Color figure online).

4   Experimental Results

In order to test the proposed methodology, we prepared two different experimental settings: the first one is a simulated setup considering ten robots on a RoboCup SPL field, whereas the second one is a real experiment using the robot in a reduced field.

4.1   Simulated Full Field Experiment

The use of simulations allow to carry out repeatable experiments and to explore the use of different sets of parameters. For this reason, the simulated setup was used to analyze the proposed active vision system. Besides the robot under analysis, i.e. the one that uses the robot's vision system, nine other robots were added on the field to represent a real game setup. The robots were located arbitrarily following a common team formation. The ball was located at position $(0, 1900)$ as shown in Fig. 4 left.

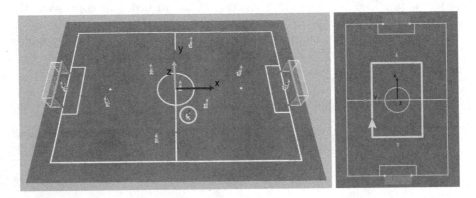

Fig. 4. Left: Simulated experiment configuration. 9 robots were added to represent a real game. The ball was located at position $(0, 1900)$. The robot using the active vision system, indicated with a yellow circle, had to walk 10 times over the shown in the right side. Right: Path followed by the robot in the experiment (Color figure online).

This experiment consisted of following a rectangular path over the full field (Fig. 4 right). The proposed active vision system was compared with a passive one. When using the passive vision system the robot followed a predefined head control routine. It pointed the camera at the estimated ball position for 3 s, and to the other landmarks such as goals and corners for 1 s.

The active vision system used the targets calculated using the scores. The look-up table that stores the results of the particle filters used 40 and 24 bins for the x and y dimensions, whereas 64 and 6 bins where used for the pan and tilt angles respectively. Head pan penalization parameter κ was set to 2 [rads], whereas tilt penalization parameter σ was fixed to 5 [rads]. Obstacle compensation factors κ_0 and σ_0 were set to 1 and 0.05 [rads] respectively, according to experimental observations. The ball importance factor α was varied in several tests.

Translational and rotational errors as well as ball seen percentage were measured for both systems. Main results obtained for several ball importance factor values are shown in Table 1. A graphical comparison of the paths walked using both approaches is shown in Fig. 5.

Table 1. Self-localization errors and ball seen percentage for the simulated setup

Head control system	Self-localization			Ball
	Error (mean, mm)	Error (std, mm)	Error (rad)	Seen (%)
Passive	96.8	139.2	0.18	36.8
Active, $\alpha = 0.5$	44.927	23.926	0.08	22.9
Active, $\alpha = 1$	57.2	35.5	0.11	31.9
Active, $\alpha = 2$	59.4	40.3	0.11	32.8

This quantitative results show that the active vision systems outperforms the passive one with respect to translational and rotational error, in both mean and standard deviation. The proposed active vision method reduced the pose estimation errors in 40 % in average, with respect to the passive system's errors. However, the ball seen percentage depends strongly of the ball importance factor α chosen, being slightly reduced in a 10 % from 36.8 % in the passive case, to 32.8 % for the active system with $\alpha = 2$. Translational and rotational errors are also affected by the ball importance as is expected.

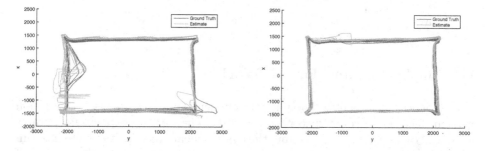

Fig. 5. Left: Comparison of the trajectory followed by the robot while using the passive vision system (in red) and the estimated ground truth position (shown in blue). Erratic behavior is mainly caused by field side ambiguities as well as by obstacle occlusion. Right: Same path using the active system with $\alpha = 2$ (Color figure online).

Most of the errors before changing the robot path direction. Passive vision systems assume full landmark observation and do not consider neither occlusions nor current robot position. Several errors are caused by wrong observations that affect the state estimator, inducing systematic errors in the system. The proposed active vision system can cope with the environment dynamics, choosing a target strongly related with the current robot position. Nevertheless, ball gaze frequency is reduced at the same time, according to the ball importance factor selected.

4.2   Real Robot Experiment

A second test was performed in a half field using a real robot, by utilizing the ground truth system of the RoboCup Small Size League, just as in [11]. The robot was configured with the same parameters as the simulated experiment, using a ball importance factor of 2, and had to follow a straight path between the points $(1000, -1000)$ and $(1000, 1000)$. The ball was positioned at the center of the field. The aim of this experiment was to test the system in both self-localization and ball gazing, as well as to measure the time required for its execution. Main results are shown in Table 2.

Time execution measurements show that the system uses minimal computational resources to run, requiring less than 1 ms. These results were obtained while executing other modules required for a soccer robot, such as perception, self-localization and decision making.

Table 2. Self-localization errors and ball seen percentage for the real setup

Head control system	Self-localization			Ball
	Error (mean, mm)	Error (std, mm)	Error (rad)	Seen (%)
Passive	151.4	86.9	0.58	52.2
Active, $\alpha = 2$	86.6	56.1	0.32	51.9

5   Conclusions and Future Work

In this paper we present a dynamic and efficient approach to the active vision problem focused on robot soccer applications. The method explicitly separates static and dynamic information from the environment, allowing the system to run in real-time by calculating offline expensive procedures, such as simulations of the resulting state estimation for each action candidate. In addition, information representation by the *action space* allows us to choose a head pose considering different information sources such as the position of other robots and the ball, as well as movement constraints of the robot's head. The proposed system could considerably improve the accuracy of the self-localization estimation process by reducing both translation and rotational errors, using low computational resources.

Further developments include an extension of the action space to the real case, which would require a replacement of the particle filter discrete look-up-table used for the *static action space* by an interpolated model or learned model. Since we are coping with the particle filter results, this should not increment the computational cost considerably during the game. In addition, parameter estimation could also be improved through learning algorithms, such as reinforcement learning.

Acknowledgments. The authors thank Pablo Saavedra for his assistance to prepare the ground truth system and the UChile Robotics Team for their general support. We also thank the BHuman SPL Team for sharing their coderelease, contributing the development of the Standard Platform League. This work was partially funded by FONDECYT Project 1130153.

References

1. Aloimonos, J.: Purposive and qualitative active vision. In: Proceedings of the 10th International Conference on Pattern Recognition, 1990, vol. 1, pp. 346–360. IEEE (1990)
2. Bajcsy, R.: Active perception. Proc. IEEE **76**(8), 966–1005 (1988)
3. Ballard, D.H.: Animate vision. Artif. Intell. **48**(1), 57–86 (1991)
4. Burgard, W., Fox, D., Thrun, S.: Active mobile robot localization by entropy minimization. In: Proceedings of the 2nd EUROMICRO Workshop on Advanced Mobile Robots, 1997, pp. 155–162, October 1997
5. Czarnetzki, S., Kerner, S., Kruse, M.: Real-time active vision by entropy minimization applied to localization. In: Ruiz-del-Solar, J. (ed.) RoboCup 2010. LNCS, vol. 6556, pp. 266–277. Springer, Heidelberg (2010)
6. Davison, A.J., Murray, D.W.: Simultaneous localization and map-building using active vision. IEEE Trans. Pattern Anal. Mach. Intell. **24**(7), 865–880 (2002)
7. Fukase, T., Yokoi, M., Kobayashi, Y., Ueda, R., Yuasa, H., Arai, T.: Quadruped robot navigation considering the observational cost. In: Birk, A., Coradeschi, S., Tadokoro, S. (eds.) RoboCup 2001. LNCS (LNAI), vol. 2377, pp. 350–355. Springer, Heidelberg (2002)
8. Guerrero, P., Ruiz-del-Solar, J., Romero, M.: Explicitly task oriented probabilistic active vision for a mobile robot. In: Iocchi, L., Matsubara, H., Weitzenfeld, A., Zhou, C. (eds.) RoboCup 2008. LNCS, vol. 5399, pp. 85–96. Springer, Heidelberg (2009)
9. Mitsunaga, N., Asada, M.: Visual Attention Control by Sensor Space Segmentation for a Small Quadruped Robot Based on Information Criterion. In: Birk, A., Coradeschi, S., Tadokoro, S. (eds.) RoboCup 2001. LNCS (LNAI), vol. 2377, pp. 154–163. Springer, Heidelberg (2002)
10. Seara, J.F., Schmidt, G.: Intelligent gaze control for vision-guided humanoid walking. Rob. Auton. Syst. **48**(4), 231–248 (2004)
11. Seekircher, A., Laue, T., Röfer, T.: Entropy-based active vision for a humanoid soccer robot. In: Ruiz-del-Solar, J., Chown, E., Plöger, P.G. (eds.) Robocup 2010. LNCS, vol. 6556, pp. 1–12. Springer, Heidelberg (2011)
12. Stronger, D., Stone, P.: Selective visual attention for object detection on a legged robot. In: Lakemeyer, G., Sklar, E., Sorrenti, D.G., Takahashi, T. (eds.) RoboCup 2006. LNCS (LNAI), vol. 4434, pp. 158–170. Springer, Heidelberg (2007)
13. Thrun, S., Burgard, W., Fox, D.: Probabilistic Robotics (Intelligent Robotics and Autonomous Agents). The MIT Press, Cambridge (2005)
14. Vidal-Calleja, T., Davison, A., Andrade-Cetto, J., Murray, D.: Active control for single camera SLAM. In: IEEE International Conference on Robotics and Automation, Orlando, May 2006

Development Track

A Realistic RoboCup Rescue Simulation Based on Gazebo

Masaru Shimizu[1]($\boxtimes$), Nate Koenig[2], Arnoud Visser[3], and Tomoichi Takahashi[4]

[1] Chukyo University, Nagoya, Japan
shimizu@sist.chukyo-u.ac.jp
[2] Open Source Robotics Foundation, San Francisco, USA
[3] Universiteit van Amsterdam, Amsterdam, The Netherlands
[4] Meijo University, Nagoya, Japan

Abstract. Since the first demonstration of the Virtual Robot Competition, USARSim has been used as the simulation interface and environment. The underlying simulation platform, Unreal Engine, has seen three major upgrades (UT2004, UT3 and UDK). These upgrades required a whole new USARSim simulator to be built from scratch. Yet, between those versions the USARSim interface has not been modified, which made USARSim a stable platform for more than 10 years. This stability allowed developers to concentrate on their control and perception algorithms. This paper describes a new prototype of the USARSim interface; implemented as plugin to Gazebo, the simulation environment native to ROS. This plugin would facilitate a shift of the maintenance of the simulation environment to the Open Source Robotics foundation and attract new teams to the Virtual Robot Competition.

1 Introduction

The challenge provided by the RoboCup Rescue Simulation League is to have a team of robots cooperate inside a devastated area. Most research institutes have access to only a few robots with a limited sensor suite and do not have access to all the robotic hardware necessary to build a complete rescue team. Simulators allow teams to experiment with algorithms for cooperation between robots in a safe, low-cost environment. However, to be useful, the simulator should provide realistic noise models for sensors and actuators; noise models which should be validated [1,2,4,5,8,13,14,17].

The Robot Operating System (ROS) has been steadily gaining popularity among robotics researchers as an open source framework for robot control [15]. Gazebo is the simulation environment used by ROS, although it was originally developed for the Player-Stage environment [11]. Gazebo is based on the Open Dynamics Engine (ODE), although it has the flexibility to switch between physics engines. The Pioneer robot is validated based on the default ODE physics engine [7].

© Springer International Publishing Switzerland 2015
L. Almeida et al. (Eds.): RoboCup 2015, LNAI 9513, pp. 331–338, 2015.
DOI: 10.1007/978-3-319-29339-4_27

2   Related Research

The Unified System for Automation and Robot Simulation (USARSim) environment has been used for many years by robotics researchers and developers as a validated framework for simulation [3,6]. The original version was developed in 2003, based on the concept of GameBots [10].

The validation approach applied to USARSim is to perform the same experiment in simulation and with a real world system, and to quantitatively compare the results. This effort may sometimes be costly, because it entails developing accurate models of the robotic systems at hand, but it has proved to be a formidable advantage which makes it possible to extrapolated from simulation to reality quickly and to identify early which algorithms are not generally applicable. Part of the USARSim success [3] draws from this extensive validation efforts. This validation has been performed for the Pioneer robot [4], the Kurt3D robot [1], the Kenaf robot [14], the Nao robot [13], the AR.Drone robot [17], the camera sensor [5], the laser sensor [8] and the GPS sensor [2]. It would be nice if not only the interface, but also part of this validation effort could be ported from USARSim to Gazebo.

Note that there exists an interface between ROS and USARSim [12], but this interface works precisely the other way around, making it possible for ROS-nodes to connect to simulation of USARSim. USARSim has many benefits (for instance, the realism of the lighting from the Unreal Engine), but is difficult to maintain with the current small developers community. Gazebo, the simulation environment native to ROS, is a much better choice for the future. To demonstrate the benefits of the change from USARSim based on Unreal to a USARSim based on Gazebo, this paper describes a prototype of such simulation environment. Table 1 shows functional comparison between USARSim and Gazebo. Table 2 shows merits of using Gazebo compared to USARSim. Those tables clearly show the possibilities of USARSim based on Gazebo for the RoboCup Virtual Robot League.

3   Benefits

There are several additional benefits of making this choice. Firstly the progress made by the Open Source Robotics foundation in improving Gazebo would be directly available to the RoboCup Rescue Simulation League community. The Open Source Robotics foundation recently extended the open source Gazebo robot simulator extensively on request of the Defense Advanced Research Projects Agency (DARPA). The new interface described in this paper would allow the teams active in the Rescue Simulation and the research institutes active in the DARPA Robotics Challenge to use the same simulation environment, allowing for cross development. Thirdly the maintenance of the simulation environment of the Virtual Robot competition would come in professional hands, now that USARSim is no longer actively supported by the National Institute of Standards and Technology (NIST). Last, but not least, this would allow to attract new teams to the RoboCup Rescue Simulation League.

Table 1. Functional comparison between UDK and Gazebo

	USARSim with UDK	USARSim with Gazebo
Simulator	UDK	Gazebo
Changeability	Modification packages on UDK(except for UDK)	All
Physics Engine	Unreal Engine	ODE(Default), Bullet, DART, Simbody
3D Simulation	Possible	Possible
Performance for	Real Time	Accuracy
Kinds of robot included by simulator	AirRobot, ATRVJr, Cooper, ERS, HMMWV, Kenaf, Kurt, Lisa, P2AT, P2DX, Pssarola, QRIO, Rugbot, Sedan, SnowStorm, Soryu, Submarine, Talon, Tarantula, TeleMax, Zerg	Atlas, Kuka, Pioneer 2DX, Pioneer 3AT, PR2, Robo Naut, Quad Rotor, Kuka, youbot
Flying robots	Possible	Possible
Multi robots	Possible	Possible
Capability of adding objects and fields by users	Possible (Not so easy in scaling)	Possible
Capability of adding robot by users	Possible (Not so easy in scaling)	Possible
Realistic rendering	Impressive	Possible
Lighting and shadowing	Impressive	Possible
Simulation of water, mud, sand	Limited	Possible by change physics engine
Capability of connection with ROS	Possible	Possible
Arbitrary viewpoint	1	Any numbers of cameras which you want
Capability of connection with each camera video stream	Impossible (solved by tiled approach)	Possible
Getting Ground Truth data from map	Possible (done for competition visualization)	Possible
Active environment	Possible	Possible
Disaster environment	Possible	Possible
Movable obstacles	Possible	Possible
Fog effect	Possible	Possible

4   Design

The interface between Gazebo and USARSim is designed as a WorldPlugin; the preferred method to modify the simulation environment. The plugin starts a server-routine, which listens to port 3000. Multiple clients (currently limited to teams of 16 robot controllers, as in the original UT2004 version) can connect to this port and spawn a robot into the Gazebo world.

At the moment, the robot is spawned with a specific sensor-suite. In principal, an user can modify this configuration in the Gazebo GUI. The USARSim interface has commands to query the current configuration (GETCONF and GETGEO commands), but the format of the STATUS message should be updated to notify the robot controller that the configuration has been changed (and should be queried again).

The location of the spawn-position is important, because with a team of robots a designer wants each robot to have an unique start position. In an Unreal

Table 2. Merits of using Gazebo instead of UDK

	USARSim with UDK	USARSim with Gazebo
Commands and sensor data transferring protocol	GameBot Protocol	Topic or Any protocols which you need
Transferring data protocol with ROS	GameBot Protocol (Need protocol converter at ROS side)	Topic or Any protocols which you need
Changeability inside of simulator	Not yet (Unreal Engine 4 is Open Source)	Possible (Open Source)
Programming language	Unreal Script	C, C++, Python

world, start-positions are specified by inserting PlayerStart positions with their corresponding coordinate system to the world. This is a native feature of Unreal, because Unreal Tournament was a multi-player game where each player also needed an unique start position. Gazebo has a comparable way to specify start positions, once a PlayerStart model is created.

Once a robot is spawned and configured, the regular sense-plan-act cycle starts. Sensor messages are received via SEN messages, the actuators are controlled by DRIVE, SET and MISPKG messages.

The implementation of the interface is based on the efficient Boost-library [16], the same library which is used inside Gazebo.

5   Requirements

The goal of this study is to create a fully functional prototype, which will allow to control robots inside Gazebo via the USARSim interface. This would mean that

- New robots could be dynamically spawned in the world by an INIT command.
- The robots could be configured with SET commands.
- The robot's sensor suite could be queried with GETCONF and GETGEO commands.
- The robots could be steered by sending DRIVE commands.
- Sensor updates would be send via SEN messages.
- Camera images would be published via a separate high-speed socket (typically port 5003)
- A private socket (typically port 50000) should be available for the Wireless Server Simulation, which needs Ground Truth information to calculate distances between robots and the number of walls in the line of sight between the robots.

6   Architecture

This USARSim prototype is built by a diagram which indicates how the USARSim interface should be incorporated in the Gazebo architecture. This diagram (Fig. 1) shows connections between an USARSim user client and Gazebo simulator via the new plugin.

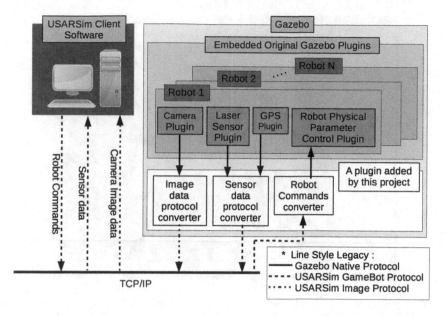

Fig. 1. Diagram of our prototype simulator with Gazebo. At right middle of the diagram, white blocks are our added plugin software in Gazebo. The plugin translate bidirectionally USARSim commands and Camera images and Sensor data between USARSim protocol (GameBot) and Gazebo Native protocol (Topic).

7 Results

In the current implementation[1] it is possible to query for start poses, to spawn a Pioneer 3AT robot, drive this robot through the environment and follow its progress by watching the camera images published over a high speed binary channel. The later accomplishment was the most critical, because this was estimated as the most difficult feature to replicate in an equivalent way in Gazebo.

We were also able to import a world generated in Unreal Editor into Gazebo, as can be seen in Fig. 2. This is quite attractive scenario, because the Unreal Editor is really very professional. The latest version of Unreal, the Open Source version 4.7, is able to create very large maps, which is essential in rescue situations. In addition, a lot of effort is spent to create the realistic worlds for the RoboCup Virtual Robot competition. With this method the existing maps can be ported to Gazebo.

8 Future Work

The USARSim has a rich set of sensors, such as laserscans, sonar, encoders and inertial sensors. Each of this sensors is publishing their measurements in a

---

[1] Can be downloaded from https://github.com/m-shimizu/RoboCupRescuePackage.

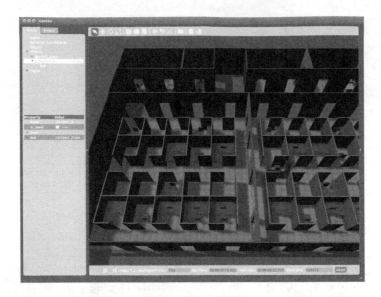

Fig. 2. A screenshot of a world in Gazebo, converted from a USARSim world created in Unreal Editor: RoboCup 2013 Virtual Robot competition Preliminary 1.

slightly different format. In addition, walking, flying and diving robots have a slightly different drive commands compared to wheeled robots. To finalize the plugin, the interface of each of the sensors and actuators have to be replicated. No fundamental problems are foreseen here.

One of the assets of USARSim is the focus on validation. It would be an advantage if all validation effort [1,2,4,5,8,13,14,17] could be repeated in the Gazebo environment. This is not only beneficial for the RoboCup community, but for all users of Gazebo. The first validation performed which could be performed is to perform driving experiments with the Pioneer 3AT; accelerating along a straight trajectory and turning circles. A comparison could be made between the real system, USARSim based on the Unreal Engine and USARSim based on Gazebo.

This could be continued with experiments on slopes and trajectories over obstacles. The NIST institute has made a very useful test-document for Rescue Robots, with a wide variety of experiments which could be performed [9].

Because the effort of many RoboCup Rescue teams concentrate on mapping and object recognition, a comparable set of experiments could be performed to estimate the level of realism for each sensor.

9   Conclusion

The new prototype will be presented at the RoboCup 2015 in China. In addition, the teams will be allowed to stress-test the solution in their laboratory. When tested by the experienced teams, the new USARSim interface to Gazebo will be presented on the Leagues website and in relevant newsgroups and social media. It is the intention to accompany this announcement with an invitation for a workshop, at an international robotics conference, to introduce the new design to a larger audience. It will be used as showcase for both the RoboCup and the robotics rescue community as a whole.

Acknowledgement. This project was supported by RoboCup Foundation.

References

1. Albrecht, S., Hertzberg, J., Lingemann, K., Nüchter, A., Sprickerhof, J., Stiene, S.: Device level simulation of kurt3d rescue robots. In: Proceedings of the 3rd International Workshop on Synthetic Simulation and Robotics to Mitigate Earthquake Disasters (SRMED 2006). Citeseer (2006)
2. Balaguer, B., Balakirsky, S., Carpin, S., Lewis, M., Scrapper, C.: USARSim: a validated simulator for research in robotics and automation. In: Workshop on Robot Simulators: Available Software, Scientific Applications, and Future Trends at IEEE/RSJ (2008)
3. Balakirsky, S., Carpin, S., Lewis, M.: Robots, games, and research: success stories in USARSim. In: Proceedings of the 2009 IEEE/RSJ International Conference on Intelligent Robots and Systems. IEEE Press (2009)
4. Carpin, S., Lewis, M., Wang, J., Balakirsky, S., Scrapper, C.: Bridging the gap between simulation and reality in urban search and rescue. In: Lakemeyer, G., Sklar, E., Sorrenti, D.G., Takahashi, T. (eds.) RoboCup 2006: Robot Soccer World Cup X. LNCS (LNAI), vol. 4434, pp. 1–12. Springer, Heidelberg (2007)
5. Carpin, S., Stoyanov, T., Nevatia, Y., Lewis, M., Wang, J.: Quantitative assessments of usarsim accuracy. In: Proceedings of PerMIS, vol. 2006 (2006)
6. Jacoff, A., Lewis, M., Birk, A., Carpin, S., Wang, J.: High fidelity tools for rescue robotics: results and perspectives. In: Bredenfeld, A., Jacoff, A., Noda, I., Takahashi, Y. (eds.) RoboCup 2005. LNCS (LNAI), vol. 4020, pp. 301–311. Springer, Heidelberg (2006)
7. Drumwright, E., Shell, D., Koenig, N., Hsu, J.: Extending open dynamics engine for robotics simulation. In: Ando, N., Balakirsky, S., Hemker, T., Reggiani, M., von Stryk, O. (eds.) SIMPAR 2010. LNCS, vol. 6472, pp. 38–50. Springer, Heidelberg (2010)
8. Formsma, O., van Noort, S., Visser, A., Dijkshoorn, N.: Realistic simulation of laser range finder behavior in a smoky environment. In: Ruiz-del-Solar, J. (ed.) RoboCup 2010. LNCS, vol. 6556, pp. 336–349. Springer, Heidelberg (2011)
9. Jacoff, A., Messina, E., Huang, H.M., Virts, A., Norcross, A.D.R., Sheh, R.: Guide for evaluating, purchasing, and training with response robots using DHS-NIST-ASTM International Standard Test Methods. Technical report, Intelligent Systems Division, Engineering Laboratory, National Institute of Standards and Technology (2009)
10. Kaminka, G.A., Veloso, M.M., Schaffer, S., Sollitto, C., Adobbati, R., Marshall, A.N., Scholer, A., Tejada, S.: Gamebots: a flexible test bed for multiagent team research. Commun. ACM **45**(1), 43–45 (2002)
11. Koenig, N., Howard, A.: Design and use paradigms for gazebo, an open-source multi-robot simulator. In: Proceedings of the 2004 IEEE/RSJ International Conference on Intelligent Robots and Systems (IROS 2004), vol. 3, pp. 2149–2154. IEEE (2004)

12. Visser, A., Balakirsky, S., Kootbally, Z.: Enabling codesharing in rescue simulation with USARSim/ROS. In: Behnke, S., Veloso, M., Visser, A., Xiong, R. (eds.) RoboCup 2013. LNCS, vol. 8371, pp. 592–599. Springer, Heidelberg (2014)
13. van Noort, S., Visser, A.: Validation of the dynamics of an humanoid robot in USARSim. In: Proceedings of the Workshop on Performance Metrics for Intelligent Systems, pp. 190–197. ACM (2012)
14. Okamoto, S., Kurose, K., Saga, S., Ohno, K., Tadokoro, S.: Validation of simulated robots with realistically modeled dimensions and mass in USARSim. In: 2008 IEEE International Workshop on Safety, Security and Rescue Robotics, SSRR 2008, pp. 77–82. IEEE (2008)
15. Quigley, M., Conley, K., Gerkey, B., Faust, J., Foote, T., Leibs, J., Wheeler, R., Ng, A.Y.: Ros: an open-source robot operating system. In: ICRA Workshop on Open Source Software, vol. 3, p. 5 (2009)
16. Schaeling, B.: The Boost C++ Libraries. XML Press, Laguna Hills (2014)
17. Visser, A., Dijkshoorn, N., van der Veen, M., Jurriaans, R.: Closing the gap between simulation and reality in the sensor and motion models of an autonomous AR. Drone. In: Proceedings of the International Micro Air Vehicle Conference and Flight Competition (IMAV11), pp. 40–47 (2011)

Hambot: An Open Source Robot for RoboCup Soccer

Marc Bestmann, Bente Reichardt, and Florens Wasserfall[✉]

Hamburg Bit-Bots, Fachbereich Informatik, Universität Hamburg,
Vogt-Kölln-Straße 30, 22527 Hamburg, Germany
{0bestman,9reichar,wasserfall}@informatik.uni-hamburg.de
http://robocup.informatik.uni-hamburg.de

Abstract. In this paper a new robot is presented which was designed especially for RoboCup soccer. It is an approach to evolve from the standard Darwin based skeleton towards a robot with more human motion capabilities. Many new features were added to the robot to adapt it for the special requirements of RoboCup Soccer. Therefore, the interaction possibilities with the robot were improved and it has now more degrees of freedom to easier grip a ball and balance itself while walking. The design is open source, thus allowing other teams to easily use it and to encourage further development. Furthermore, nearly all parts can be produced with a standard 3D printer.

Keywords: RoboCup · Humanoid · Open source · Robot · Design · 3D printing · Rapid prototyping

1   Introduction

Since the release of the Darwin-OP robot [7] in 2010 there was not much development of hardware in the humanoid league. Nearly all currently used platforms use the same structure as the Darwin-OP and are still expensive. Therefore, we started developing our own robot platform, designed to be cheap, open source and usable in kid- and teen-size league. The first prototype was developed in 2014 and was presented at the RoboCup world championship in Brazil. An improved second prototype was build in early 2015, almost completely from 3D printed parts. It will be used by us, the Hamburg Bit-Bots, in the 2015 season of RoboCup tournaments. Throughout this paper we frequently use the Darwin-OP as reference platform, due to its influence in the league.

2   Current Problems in the RoboCup Humanoid League

The existing specification limits for robots in the Humanoid League are quite open. However, nearly all teams in the kid- and teen-size league use a Darwin-OP or a robot with the same skeleton and DOF layout. It contains only the

L. Almeida et al. (Eds.): RoboCup 2015, LNAI 9513, pp. 339–346, 2015.
DOI: 10.1007/978-3-319-29339-4_28

major joints of a human, e.g. shoulder, hip or knee. Some teams made small
modifications to the Darwin, e.g. changing the camera or mainboard. Other
teams built a robot on their own, for example CIT Brains [5] and Hanuman
KMUTT [13], but used the Darwin motor layout as well. The only group which
did a major change is the FUmanoids team [12]. They added an additional joint
between the hip and the upper body, similar to the human lumbar spine, and
used parallel kinematics in the robot's legs. Even the newer platforms Nimbro-
OP [11] and the robot of team Baset [3] use nearly the same layout. Most of the
robots in the competition in 2014 were not especial designed for RoboCup. The
battery layout is a good illustration for this problem. Batteries must be changed
during the game due to their limited capacity. Although this is time critical the
batteries are located inside the robot and connected to the electronics by an
extra cable. Further the robot is often difficult to handle during development,
because most interaction is done via a connected laptop and not on the robot
itself. Therefore, the attention of the developer is divided between the laptop
and robot.

The existing platforms are quite expensive, approx. 10,000 € [8] for the too
small Darwin-OP and approx. 22,000 € [14] for the newer Nimbro-OP. These
costs make it difficult for a new team to start in the league, because they need
at least 4 robots. Even the existing teams need to buy parts and new robots as
well when the number of players increases, which is planned in the Humanoid
League proposed roadmap [2].

3   Goals of the New Robot

After analyzing the current robots and their limitations, we extracted the following goals for the new robot platform. Achieving these goals should improve
the performance and the usability of the robot during competitions.

Costs: Reducing the hardware cost lowers the barrier for new teams and enables
established teams to upgrade their robots. The motors are a major part of
the price, therefore the costs can be reduced by reusing servos which are
currently used. The Dynamixel servos, especially the MX-28 [9], are very
common in the league. For the mechanical parts, simple aluminum sheet
metal and 3D printed plastic parts are low cost alternatives to carbon parts,
used in the Nimbro, which are expensive and harder to obtain.

Interaction: The RoboCup competitions usually start with some set-up days
for the teams to prepare their robots. This is required because many algorithms, e.g. walking or vision, need to be adapted for the new environment.
To simplify these tasks, the robot should be equipped with a direct human
robot interface. This includes simple buttons as well as higher level controls
to do recurrent tasks, such as parameter adjustments, without the need of a
laptop. Good debug information is crucial to find bugs quickly. In addition
to the wireless network, audio and visual output is desired to simplify this
task. A human understandable voice is very helpful for debugging purposes
and is a step towards robot to robot communication with natural language.

Open Source: Established platforms like the Darwin or Nimbro are not entirely open source due to the restrictions on third party standard parts (e.g. CM-730 motor controller board and the Dynamixel servos). This leads to difficulties concerning replacement, repairs and changes in the firmware. It also limits development of the hardware. This is one reason why there are so few modifications of the Darwin platform. Most of the extensions are limited to the replacement of cameras, batteries or motor controller boards. Although a more open platform is clearly desirable, we currently stick with the existing Dynamixel motors due to the low upgrading cost from the Darwin and the lack of an adequate alternative. Another difficulty is changing the plastic parts of the Darwin-OP because these are made by injection molding. Thus the production method has to be simple to enable other teams to produce their own parts.

Designed for RoboCup Soccer Competition: There are no robots available which are designed exclusively for RoboCup soccer, because the market is too small for it. Even the NAO robot, which has a league on its own, is used in RoboCup because it is common in other research fields. Standard robots are missing helpful features and have functions which are not needed. For example a fast battery change as well as a handle to pick up the robot is required during a RoboCup game but not necessary for many other research activities. Therefore, we wanted to design the robot from scratch for RoboCup. Features such as fast repair and an anatomy made especially for soccer playing are major concerns.

Progress in Relation to the Darwin-OP: As mentioned in Sect. 2, most of the currently used robots in humanoid soccer have a very similar body composition to the Darwin-OP. With more degrees of freedom (DOF) in the torso of the robot, it is possible to bend the upper body in pitch and roll direction independently from the legs. This is useful for kicking, walking stabilization and for standing up. A third DOF in the shoulder would enable the robot to move the arms more freely. This is important for the throw-ins.

4 "GOAL"

The first prototype GOAL was made out of aluminum sheets in 2014. Only three additional MX-64 [9] and two additional MX-28 motors were used to upgrade one Darwin to a 87 cm tall robot with 24 DOF. The robot was able to stand and walk, but was unable to get up, because the motors were not able to lift the weight of the upper body.

Besides the problem of getting up, we experienced high lead times for the production of the sheet metal parts, significantly delaying the development. The production method considerably constrains the design of complex parts, which are required especially in the torso, where parts have to be connected in all three dimensions. This particularly affects the cable routing. Changing existing parts later on is complicated due to these constraints and the high production time. Another problem which is already known from the Darwin-OP is loosening nuts and screws, resulting in instable part connections (Fig. 1).

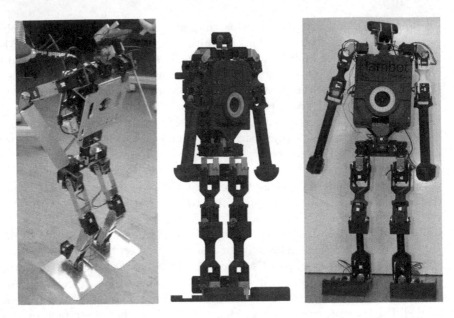

Fig. 1. Prototypes of GOAL (**left**), Hambot (**right**) and CAD model (**center**).

5   "Hambot"

The second and current prototype version is designed to be fabricated using 3D printers. Almost all parts have been designed again from scratch and refined by several steps of evolution in a rapid prototyping and testing process. Upgrading from a Darwin only five MX-106 and four MX-64 are needed. This minimize the costs of production.

5.1   Body Composition

Feet: The capability to control the toes extends the feet by another DOF, potentially improving both walking stability and standing up movements. This first design is simple but it can be replaced by one which is more useful for high kick. The toes can be bend up to 90° (Fig. 2).

Legs: The legs are designed in a pipe-like shape. This has advantages compared to the U-profile composition, which is often used with sheet metals. Cables are routed directly through these pipes, preventing abrasion of the wires. Besides, it looks more human and is more comfortable to touch and hold (Fig. 2).

Waist: At the waist an additional joint with two DOF has been added at the waist, which allows the robot to move its upper body independently from the legs. The joint should be similar to the lumbar spine and the lower thoracic spine of humans, but only in two directions. This flexibility is necessary for human walking because the upper body moves and rotates during every step

[6]. By doing so, human walking is very energy efficient [10]. With the roll axis of the new waist joint this movement is possible. Furthermore it allows to move the center of mass over the supporting leg during a kick. The pitch axis improves standing up and picking up the ball. With this joint we can move the upper body about 45° to left and right, 35° to the front and 12° to the back.

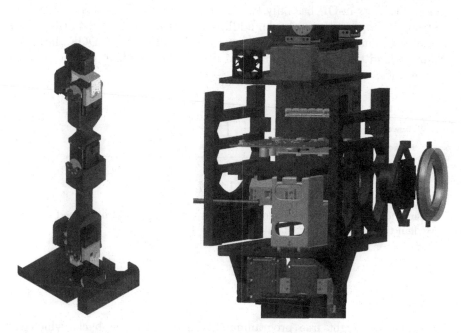

Fig. 2. CAD model of the leg (**left**) and explosion view of the torso (**right**). Note the additional shoulder servos (blue) and the new waist joint (purple). The two batteries (yellow, only one shown) are inserted from the left and locked by a bolt (brown) which fits into a bayonet socket. The powerboard (red) and motor controller board (not shown) are directly plugged into the backbone board (green) and locked by a side plate (not shown) (Color figure online).

Torso: The torso consists of a cage-like structure with detachable side plates for easy access. 3D printing allows the production of complex parts that exactly fit into each other or are slidable to one side. The electronics consist of the main computer board (Odroid XU3 lite [4]), the powerboard and a subboard. The powerboard manages the power sources and the voltage conversions. The subboard controls the servos and other peripheral electronics, such as the LEDs and the audio output. These two boards are directly plugged into a backbone board, which is located at the side of the robot and replaces the cables, which would normally run through the torso. Therefore, no cables are necessary inside the torso, which simplifies maintenance. All boards except

the Odroid XU3 lite are open source and developed by ourself. The two
batteries in the robot can be hot swapped. Each has an own printed case
which can be slid into the robot and locked with a bolt, thus enabling a
fast change, approximately 13 s for both batteries (Fig. 3). LEDs on the side
are showing the current battery charge levels and which battery is currently
used.

Shoulders: The human shoulder is a joint with three DOF, whereas the shoulder of the Darwin-OP has only two DOF. For many tasks this is sufficient,
but not for a good throw-in. An additional motor was added to the robots
shoulder to enable a movement in yaw direction. This third DOF allows the
robot to hold the ball behind his head like a human, while his elbows point
to the sides.

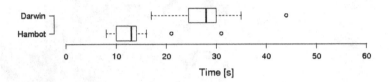

Fig. 3. Time for battery change, tested by a group of trained and untrained test persons. Median time: Darwin 28 s, Hambot 13 s.

5.2   Interaction

The Hambot has eight free programmable buttons on the back, which are
equipped with LEDs to indicate their state. There is a LCD touch display in
the back for laptop free interaction (Sect. 3). A ring of RGB LEDs is embedded into the front. Every LED can be individually controlled. This is handy to
express the robots current beliefs, e.g. the position of the ball. The audio output
is used for debugging. Therefore, a dedicated text-to-speech chip and a speaker
with a human resonance frequency is installed to ensure good speech quality.

5.3   3D Printing

All parts were designed to be smaller than 20 × 20× 10 cm and therefore printable
with a low cost fused deposition modeling (FDM) consumer printer. The two
most used plastic print materials are acrylonitrile butadiene styrene (ABS) and
polylactic acid (PLA) which can be printed by almost all consumer printers.
While it is possible to build the robot with both materials, ABS is preferred due
to its better strength and heat resistance [1]. The printing direction is important
for the stability of prints. Therefore, all parts are printed in a direction that
maximizes the plane between two layers and improves adhesion. Standard socket
cap screws (ISO 4762 12.9) and nuts (ISO 4032) were used to connect the parts.

Screwing directly into the plastic would be possible, but threads in plastic tend to wear off very fast. Multiple disassembles due to repairs would destroy the parts. Therefore, steel nuts were used for tightening. They are inserted into prepared holes, which clamp them into their position. Thus all screws can be tightened without a wrench and the parts can be assembled multiple times. The estimated print time for a complete Hambot is two weeks with a standard FDM printer. Parallel printing reduces this time.

5.4 Costs

The estimated hardware costs for the parts to build a whole Hambot are listed in Table 1. It is possible to reduce them significantly by reusing the parts of the Darwin, because only four additional MX-64 and five MX-106 servo motors are needed. Therefore, the cost for upgrading a Darwin reduces to approximately 4350 €. Expenses for maintaining the 3D printers are not included in

Table 1. Estimated hardware costs for one Hambot (**left**) and for an upgrade from a Darwin (**right**).

Filament	100 €	100 €
Odroid XU3 lite	120 €	120 €
Logitech C910	60 €	60 €
Other electronics	550 €	550 €
Dynamixel servos	6800 €	3520 €
Total	7630 €	4350 €

this calculation. It is possible to use the electronics from the Darwin to save more money, but the additional interaction possibilities would not be usable. It is also possible to reuse some metal connectors of the Darwin, but these can be replaced by 3D printed parts.

6 Conclusion and Further Work

This work introduces an open source humanoid robot. The costs are significantly lower than buying new robots that are currently available on the market. The costs for switching from Darwin-like robots with approximately 45 cm height over to a 87 cm robot are even lower. This becomes even more important, with increasing robot size and number of players in the next years [2]. Due to its size, Hambot is currently allowed to play in the Kid- and Teen-size league. The increased interaction possibilities enable a faster development as well as a better handling during the game. First tests in real environment were done at the IranOpen and GermanOpen 2015 and showed that the 3D printed structure is sufficiently stable. The size of the motors was increased to enable more stable getting up and walking. The new version will be used by us, the Hamburg Bit-Bots, at the world championship in China during July 2015. Next steps include the development of more human feet and a better camera system.

We encourage other teams to use the projects source code which is available at: https://github.com/bit-bots.

Acknowledgments. Thanks to the RoboCup team Hamburg Bit-Bots. Thanks for help building this robot to Marcel Hellwig, Dennis Reher and special thanks to Nils Rokita.

References

1. Plastic Properties of Acrylonitrile Butadiene Styrene (ABS) (2015). http://www.dynalabcorp.com/technical_info_abs.asp
2. Baltes, J., Missoura, M., Seifert, D., Sadeghnejad, S.: Robocup soccer humanoid league. Technical report (2013)
3. Farazi, H., Hosseini, M., Mohammadi, V., Jafari, F., Rahmati, D., Bamdad, D.E.: Baset humanoid team description paper. Technical report, Humanoid Robotic Laboratory, Robotic Center, Baset Pazhuh Tehran cooperation, No. 383 (2014)
4. Hardkernel co., Ltd. ODROID-XU3 Lite product specification (2015). http://www.hardkernel.com/main/products/prdt_info.php?g_code=G141351880955
5. Hayashibara, Y., Minakata, H., Irie, K., Fukuda, T., Loong, V.T.S., Maekawa, D., Ito, Y., Akiyama, T., Mashiko, T., Izumi, K., Yamano, Y., Ando, M., Kato, Y., Yamamoto, R., Kida, T., Takemura, S., Suzuki, Y., Yun, N.D., Miki, S., Nishizaki, Y., Kanemasu, K., Sakamoto, H.: CIT brains (kid size league). Technical report, UnChiba Institute of Technology (2015)
6. Mochon, S., McMahon, T.A.: Ballistic walking. J. Biomech. **13**(1), 49–57 (1980)
7. Robotis. Darwin OP Project Information (2015). http://darwinop.sourceforge.net
8. Robotis. Robotis international shop (2015). http://www.robotis-shop-en.com/?act=shop_en.goods_list&GC=GD070001
9. Robotis. Robotis international shopp (2015). http://www.robotis.com/xe/dynamixel_en
10. Romeo, F.: A simple model of energy expenditure in human locomotion. Rev. Bras. Ensino Fis. **31**(4), 4306–4310 (2009)
11. Schwarz, M., Schreiber, M., Schueller, S., Missura, M., Behnke, S.: Nimbro-op humanoid teensize open platform. In: Proceedings of the 7th Workshop on Humanoid Soccer Robots. IEEE-RAS International Conference on Humanoid Robots (2012)
12. Seifert, D., Freitag, L., Draegert, J., Gottlieb, S.G., Schulte-Sasse, R., Barth, G., Detlefsen, M., Rughöft, N., Pluhatsch, M., Wichner, M., Rojas, R.: Berlin united - FUmanoids team description paper. Technical report, Freie Universität Berlin, Institut für Informatik (2015)
13. Suppakun, N., Wanitchaikit, S., Jutharee, W., Sanprueksin, C., Phummapooti, A., Tirasuntarakul, N., Maneewarn, T.: Hanuman KMUTT: team description paper. Technical report, King Mongkut's University of Technology Thonburi (2014)
14. Universität Bonn, Institute for Computer Science. Nimbro Homepage (2015). http://www.nimbro.net/OP/

Polyurethane-Based Modular Series Elastic Upgrade to a Robotics Actuator

Leandro Tomé Martins[1], Christopher Tatsch[1], Eduardo Henrique Maciel[2],
Renato Ventura Bayan Henriques[2], Reinhard Gerndt[3],
and Rodrigo Silva da Guerra[1(✉)]

[1] Centro de Tecnologia, Universidade Federal de Santa Maria,
Av. Roraima, 1000 Santa Maria, RS, Brazil
rodrigo.guerra@ufsm.br

[2] Programa de Pós-Graduação em Engenharia Elétrica, Universidade Federal do Rio
Grande do Sul, Avenida Osvaldo Aranha, 103, Porto Alegre, RS, Brazil

[3] Department of Computer Sciences, Ostfalia University of Applied Sciences,
Am Exer 2, 38302 Wolfenbüttel, Germany

Abstract. This article extends previous work, presenting a novel
polyurethane based compliant spring system designed to be attached
to a conventional robotics servo motor, turning it into a series elastic
actuator (SEA). The new system is composed by only two mechanical
parts: a torsional polyurethane spring and a round aluminum support for
link attachment. The polyurethane spring, had its design derived from
a iterative FEM-based optimization process. We present also some system identification and practical results using a PID controller for robust
position holding.

Keywords: Series elastic actuator · Passive compliance

1 Introduction

Traditional robots usually operate at a low speed and with high torque, demanding large peak power output for short periods, accurate feedback sensing, and
suitability in shape, size and mass [11]. With the advances on fast and powerful
controllers and precise sensors, the demand for such decoupling between a manipulator and its load can be relaxed without compromising the performance. Moreover, the demands of the field of human-robot interaction rise concern on the
safety of the actuation mechanism and on its behaviour towards uncertainties in
the environment. A low impedance torque control scheme is usually required for
stable and robust human-robot dynamic interaction [3]. Low impedance means
that the actuator source force (torque) to the load, rather than commanding the
load's position.

The design of compliant robot joints can be divided in two main groups:
(1) active (or simulated) compliance and (2) passive (or real) compliance. Simulated compliance is achieved through software, by continuously controlling the

© Springer International Publishing Switzerland 2015
L. Almeida et al. (Eds.): RoboCup 2015, LNAI 9513, pp. 347–355, 2015.
DOI: 10.1007/978-3-319-29339-4_29

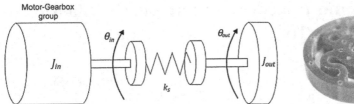

Fig. 1. Series elastic actuator topology [8].

Fig. 2. Polyurethane spring proposed in our system (Color figure online).

impedance of back-drivable electric motors (see for instance [6]). Real or passive compliance is achieved by inserting an elastic element between motor and load. This is typically done through the use of mechanical springs in the design of the joints (see for instance [4]). For a while there has been some debate on the advantages and disadvantages of choosing active versus passive compliance [10]. However, with regard to human/robot interactions, there is consensus that passive compliance ensures higher levels of safety.

A Series Elastic Actuator [8] basically consists of traditional stiff servo actuator in series with a spring connected to the load, as shown in Fig. 1. This topology allows the load to be partially decoupled from the motor, and the force exerted on the output of the compliant element can be evaluated by simply measuring the deflection of the spring.

The device that we propose in this work is an evolution on the previously design proposed by the same authors [9]. Recently, Ates et al. have also independently published a similar work [2]. The upgraded design presented here consists of a two-part component, using a modular polyurethane-based spring. More specifically, we designed this spring using a thermoplastic polyurethane (TPU) elastomer. The material is cheap, tough, easy to mill and presents rubber-like elasticity [1]. This is an extremely low-cost design since it can be easily manufactured using a 3-axis CNC router. Our aim is toward applications on lower budget humanoid robots, such as the ones seen in robot soccer competitions, specially trying to provide a better support for impact on the knees during walking and protecting shoulder joints during a fall. Our designed device consists of software, firmware, electronics and a mechanical accessory that can be easily attached to the popular Dynamixel MX series servo actuators, manufactured by Robotis, transforming it into a SEA. This servo actuator was chosen due to its wide popularity within the RoboCup community, however the general idea could be easily adapted to fit most servo actuators of similar "RC-servo-style" design (Fig. 2).

The remainder of this work is organized as follows: Sect. 2 explains the main details regarding the design as well as the modelling of the SEA. Section 3 shows some data regarding the actual construction of the device and a robot upgrade case. Section 4 presents the closing remarks and future work.

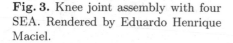

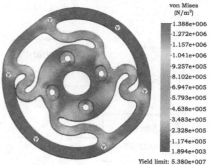

Fig. 3. Knee joint assembly with four SEA. Rendered by Eduardo Henrique Maciel.

Fig. 4. Finite Element Analysis showing the von Mises stress well below the maximum yield of the material (Color figure online).

2 Methodology

This section is divided in 3 subsections: Subsection 2.1 presents the mechanical design of the SEA. Subsection 2.3 talks about the manufacturing, the electronics and the firmware. Subsection 2.2 presents the system identification and control methods.

2.1 Design Requirements

The elastic element presented in this paper was designed aiming the application on the knees of a 1.2 m tall humanoid robot which uses a parallel leg mechanism. The robot is being developed by the joint RoboCup team WF Wolves (Germany) and Taura Bots (Brazil) [5]. This robot employs the Dynamixel MX-106R servo actuators manufactured by Robotis in a redundant arrangement, allowing the springs to be compressed against each other for leg rigidity modulation. The assembly is shown in Fig. 3, and the components are: (1) circuit board, (2) leg link frame, (3) attachment cover, (4) polyurethane torsional spring, (5) Dynamixel MX-106R servo actuator.

With the humanoid robot knee application in mind and based on the choice of the servo-motor, the SEA design was elaborated so as to ensure a symmetrical response, without saturation when exposed to the maximum torque supported by the motor. The spring consists on four "s" shapes, with the width of 3 mm. This dimension was decided after a Finite Element Analysis study (see Fig. 4).

2.2 System Identification and Control

The open-loop SEA system shown in Fig. 5 is composed by an input signal, an output signal, a disturbance signal and two transfer functions. One transfer

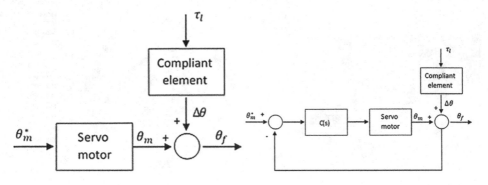

Fig. 5. Opened-loop system.   **Fig. 6.** Closed-loop system.

function corresponds to the dynamics of the servo motor, which combines the behaviors of its internal PID controller, its DC motor driver and the DC motor inside its case. From the command of a desired position $\theta_m{}^*$, an error signal is intrinsically compensated by a PID controller and then converted into voltage level to the motor armature generating θ_m. The other transfer function corresponds to the compliant element behaviour, which has, as input, an external load τ_L, and produces a angular deflection $\Delta\theta$. The output of the SEA system, is the final position θ_f, given by the sum $\theta_m + \Delta\theta$. The external load can be seen as a disturbance to the system.

The problem we are faced with is to identify these two transfer functions. In order to find a theoretical SEA model, we present a system identification method based on *Matlab System Identification Toolbox*, and then a control law is presented. The controller has the goal of providing the final position to track the desired position, even under the effect of an external load.

Here we assume the system can be reasonably approximated by a general linear polynomial model. In this paper, an Auto-Regressive with External Input (ARX) model structure is chosen to represent the SEA system. The algorithm involved in the ARX model estimation is fast and efficient when the number of data points is very large.

The controller is designed in order to let the final position θ_f track the set-point position θ_m^* of the servo motor. The closed-loop system is shown in Fig. 6.

Another goal of the compensated system is to reject disturbances. A discrete PID controller was used, since this is a simple method to match the specifications of project.

2.3   Manufacture and Electronics

The manufacturing of the two mechanical parts was all done on an ordinary 3-axis CNC router, using a 2mm cutter. Both the polyurethane and the aluminum parts can be milled in a single operation each, without the need for fixing the parts in different orientations. Refrigeration fluid was not needed.

In order to read the spring's angular displacement a magnet/magnetometer based circuit was designed (see Fig. 7). A radially polarized cilindrical rare earth magnet is placed on the center of the polyurethane part, and the circuit board is placed on top of the assembly so that the magnetometer chip is aligned with it. For educational purposes the electronics was designed to be Arduino compatible [7]. The firmware mimics Dynamixel's protocol: an id is assigned to each SEA, as if these were additional torque-disabled servo-motors, answering queries about their angular positions on the same RS485 bus.

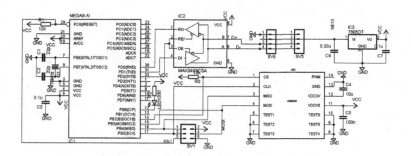

Fig. 7. Schematic of the instrumentation electronics on the SEA.

The interface's firmware was programmed to communicate using Dynamixel's RS485 protocol. Each device can be programmed to receive a distinct id thus allowing them to communicate through the same bus as the original servo actuators, using the same protocol.

3 Results

3.1 Obtaining the Stiffness

For a linear spring, the stiffness can be described by Hook's law, given by $\tau = -k.\Delta\theta$. In order to assess the stiffness of the spring an experiment was performed, applying a known mass on the tip of the frame attached to the compliant element output, and measuring the resulting angle deflection $\Delta\theta$. The rotational torque derived from the known mass can be determined by $\tau = Fl = mgl\cos(\Delta\theta)$.

The experiment was repeated for 20 different values of mass and the results were plotted, as shown in Fig. 8. The line which fits the data was found by linear regression, where its slope represents the inverse of the stiffness. The estimated stiffness was $k = \frac{1}{11.33} = 0.088\,Nm/deg$.

3.2 Obtaining the Transfer Functions

In order to find the transfer function of the compliant element system, the step response to an input torque applied to the system was measured, as shown in

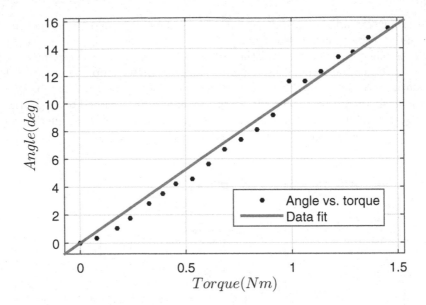

Fig. 8. Linear regression of data set plot

Fig. 9. We can realize the output behaves as a spring-damper combination. In order to identify the transfer function, we applied an optimization process to identify which ARX model better fits the data set, and the transfer function found, for a sample time of $T_s = 0.02449$ was

$$G(z) = \frac{8.431z}{z^2 - 0.743z + 0.4229}$$

Comparing the identified model with the experimental data response, we can see that the identified model fits very well, with a confidence of 88.97 %.

Similarly to the method previously described, to obtain the transfer function of the servo motor, an input signal (desired position) was applied, and the output signal (servo motor position) was measured, performing a step response. The ARX model transfer function found for the servo motor and the same sample time used before was

$$G(z) = \frac{0.03044z^2}{z^3 - 1.774z^2 + 0.865z - 0.06091}$$

When compared, the step responses for the identified and experimental systems, we found a confidence of 81.87 %.

3.3 Controller

A PID controller was implemented in a ROS package for real-world validation in a single joint experiment setup, the PID gains used were: $k_p = 0.8$, $k_i = 3$

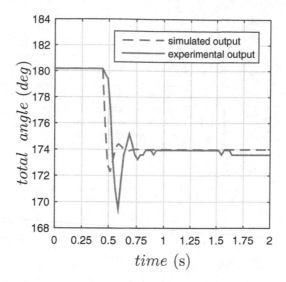

Fig. 9. Open loop response releasing weight at specific time.

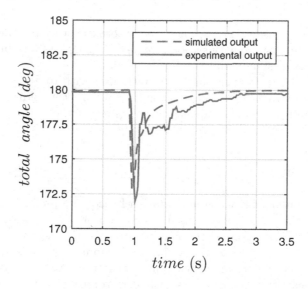

Fig. 10. Closed loop response releasing weight at specific time.

and $k_d = 0.025$ and the sample time was $T_s = 0.02449$. A MX-28R motor was fixed on a wrench and a lever arm of 88 mm was attached to the SEA module. We tested the closed loop response when a disturbance is applied to the system at a specific time and a step reference input response of the system holding a 633.58 g mass.

Figure 10 shows the simulated and experimental results for the first case, where an external disturbance is applied at 1 s to the system. In comparison to

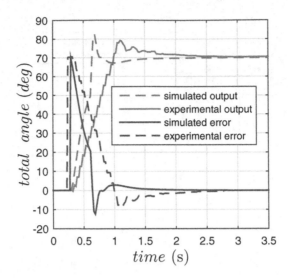

Fig. 11. Step closed loop response.

the open loop response we can see now the external disturbance being rejected by the closed loop system, with setpoint angle of 180 deg. As we can see, it takes about 1.5 s to the system stabilize after disturbance.

In the second case The setpoint was initialized with 0 deg, and after 0.25 s the setpoint was switched to 70 deg, while the system was holding a mass. The results are shown in Fig. 11, and the curves represent the output and the PID error of the simulated and experimental results.

4   Discussion

This work presented a SEA upgrade solution based on an affordable module to be mounted to the output of an existing servo-motor. The two-part mechanical design was shown to be simple to manufacture, and the electronics circuit was designed around the popular Arduino platform, communicating angular displacements through the bus using the same infrastructure. We have also performed system identification and we have shown how robust position control can be achieved.

For future work the authors also want to explore the use of torque mode control, available in the Dynamixel models MX-64 and MX-106.

References

1. Ashby, M., Johnson, K.: Materials and Desing: The Art and Science of Material Selection in Product Design. Elsevier Editora Ltda, Rio de Janeiro (2011)

2. Ates, S., Sluiter, V.I., Lammertse, P., Stienen, A.H.A.: Servosea concept: cheap, miniature series-elastic actuators for orthotic, prosthetic and robotic hands. In: Proceedings of 5th IEEE RAS & EMBS International Conference on Biomedical Robotics and Biomechatronics (BioRob) (2014)
3. Carpino, G., Accoto, D., Sergi, F., Tagliamonte, N.L., Guglielmelli, E.: A novel compact torsional spring for series elastic actuators for assistive wearable robots. J. Mech. Des. **134**(121002), 1–10 (2012)
4. Guizzo, E., Ackerman, E.: The rise of the robot worker. IEEE Spectr. **49**(10), 34–41 (2012)
5. Hannemann, A.K., Stiddien, F., Xia, M., Krebs, O., Gerndt, R., Krupop, S., Bolze, T., Lorenz, T.: WF Wolves - Humanoid kid size team description for RoboCup 2014. RoboCup Tournament 2014 (2014). http://www.wf-wolves.de
6. Jain, A., Kemp, C.C.: Pulling open doors and drawers: coordinating an omni-directional base and a compliant arm with equilibrium point control. In: IEEE International Conference on Robotics and Automation (ICRA), pp. 1807–1814 (2010)
7. Kushner, D.: The making of arduino. IEEE Spectrum 26 (2011)
8. Laffranchi, M., Sumioka, H., Sproewitz, A., Gan, D., Tsagarakis, N.: Compliant actuators. In: Adaptive Modular Architectures for Rich Motor Skills (2011)
9. Gerndt, R., de Mendonça Pretto, R., da Silva Guerra, R., Martins, L.T.: Design of a modular series elastic upgrade to a robotics actuator. In: Bianchi, R.A.C., Akin, H.L., Ramamoorthy, S., Sugiura, K. (eds.) RoboCup 2014. LNCS, vol. 8992, pp. 701–708. Springer, Heidelberg (2015)
10. Wang, W., Loh, R.N.K., Gu, E.Y.: Passive compliance versus active compliance in robot-based automated assembly systems. Industr. Rob. **25**(1), 48–57 (1998)
11. Wyeth, G.: Control issues for velocity sourced series elastic actuators. In: Australasian Conference on Robotics and Automation (2006)

RLLib: C++ Library to Predict, Control, and Represent Learnable Knowledge Using On/Off Policy Reinforcement Learning

Saminda Abeyruwan[✉] and Ubbo Visser

Department of Computer Science, University of Miami, 1365 Memorial Drive,
Coral Gables, FL 33146, USA
{saminda,visser}@cs.miami.edu

Abstract. RLLib is a lightweight C++ template library that implements incremental, standard, and gradient temporal-difference learning algorithms in reinforcement learning. It is an optimized library for robotic applications and embedded devices that operates under fast duty cycles (e.g., $\leq 30\,\mathrm{ms}$). RLLib has been tested and evaluated on RoboCup 3D soccer simulation agents, NAO V4 humanoid robots, and Tiva C series launchpad microcontrollers to predict, control, learn behavior, and represent learnable knowledge.

Keywords: RLLib · Reinforcement learning · Gradient temporal-difference

1   Overview

The RoboCup initiative presents a real-time, dynamic, complex, adversarial, and stochastic multi-agent environments for agents to learn and reason about different problems [10]. Reinforcement Learning (RL) is a framework that provides a set of tools to design sophisticated and hard-to-engineer behaviors [11]. RL agents naturally formalize their learning goals in two layers: (1) physical layers – where controlling or predicting of functions relate to sensorimotor interactions such as walking, kicking, so forth; and (2) decision layers – with dynamically emerging behaviors through actions, options, or knowledge representations. Therefore, the RL paradigm includes controlling and prediction; it also provides means to represent highly expressive knowledge using General Value Functions (GVFs) from sensorimotor streams [14].

RLLib is a lightweight C++ template library that implements incremental, standard, and gradient temporal-difference learning algorithms in RL to effectively solve problems defined in physical and decision layers. The library is designed and written specifically for robotic applications and embedded devices such as RoboCup 3D soccer simulation agents, physical NAO V4 humanoid robots, and Tiva C series launchpad microcontrollers that operate under fast duty cycles. The library is first released on May 17, 2013 under the open source license "Apache License, Version 2.0", and the latest release version is v2.2.

© Springer International Publishing Switzerland 2015
L. Almeida et al. (Eds.): RoboCup 2015, LNAI 9513, pp. 356–364, 2015.
DOI: 10.1007/978-3-319-29339-4_30

The main objective of this application paper is to describe the design, correctness, implementation, and efficacy of RLLib across multiple hardware platforms. Henceforth, we have organized the paper as follows. Section 2 describes the existing applications. A brief description on the implemented algorithms and features are described in Sect. 3, while, the interpretations related to the algorithms are summarized in Sect. 4. The main concepts of the framework is described in Sect. 5. The experiments and evaluations are provides in the penultimate Sect. 6. Finally, we conclude the paper with a summary and extensions in Sect. 7.

2   Related Work

There exist RL development platforms published by researchers for specific RL problems and for general use. Notably, the most common RL development platform is RL-GLUE[1] with RL-LIBRARY[2] packages [29]. It provides a language-independent platform over text based massage passing among agents and environments. RL-GLUE is being used in RL competitions in ICML and NIPS workshops. PyBrain[3] [19] and RLPy[4] are libraries written in Python to formulate and learn from RL problems. RL toolbox[5], libpgrl[6], YORLL[7], and rllib[8] [8] are C++ based platforms to develop RL algorithms in different scenarios, while CLSquare[9] [9] is a standardized platform for testing RL problems with on-policy batch controllers. BURLAP[10] [7], PIQLE[11] [6], MMF[12], QCON[13], and RLPark[14] are Java platforms that model and learn from RL problems. MDP Toolbox[15] is an Octave based RL development platform. dotRL[16] [18] is a .NET platform for rapid RL method development and validation.

RLLib is significantly differs from the existing platforms because of the following: (1) the library is written and designed specifically for applications and devices where limited computational resources are available. Therefore, the memory footprint as well as the computational requirements that are needed by the library have been optimized; (2) a configurable C++ template functions to

---

[1] http://glue.rl-community.org.
[2] http://library.rl-community.org.
[3] http://pyrain.org/.
[4] http://acl.mit.edu/RLPy/.
[5] http://www.igi.tu-graz.ac.at/gerhard/ril-toolbox.
[6] https://code.google.com/p/libpgrl/.
[7] http://www.cs.york.ac.uk/rl/software.php.
[8] http://malis.metz.supelec.fr/spip.php?article122.
[9] http://ml.informatik.uni-freiburg.de/research/clsquare.
[10] http://burlap.cs.brown.edu/.
[11] http://piqle.sourceforge.net/.
[12] http://mmlf.sourceforge.net/.
[13] http://sourceforge.et/p/elsy/wiki/Home/.
[14] http://rlpark.github.io/.
[15] http://www7.inra.fr/mia/T/MDPtoolbox/.
[16] http://sourceforge.net/projects/dotrl/.

synchronize with application or device hardware requirements; (3) the library emphasizes more on learnable knowledge representation and reasoning from sensorimotor streams; (4) a clean and transparent API exists that enables users to model their RL problems easily; and (5) a self-contained C++ template library covers plethora of incremental, standard, and gradient temporal-difference learning algorithms in RL that is published to-date (e.g., [22,28]). Our library has been successfully used in [1] to learn role assignment in RoboCup 3D soccer simulation agents.

3 Features

RLLib implements and features: (1) off-policy prediction algorithms: (GTD(λ) and GQ(λ)) [14]; (2) off-policy control algorithms: Greedy-GQ(λ) [14] and (Softmax GQ(λ) and Off-PAC) [5]; (3) on-policy algorithms: (TD(λ), SARSA(λ), Expected SARSA(λ), and Actor-Critic (continuous and discrete actions, discounted, averaged reward settings, so forth)) [25], (Alpha Bound TD(λ) and SARSA(λ)) [3], and (True TD(λ) and SARSA(λ)) [21]; (4) incremental supervised learning algorithms: Adaline [2], (IDBD and Semi-Linear IDBD) [26], and Auto-Step [17]; (5) discrete and continuous policies: (Random, Random X percent bias, Greedy, ϵ-greedy, Boltzmann, Normal, and Softmax); (6) sparse feature extractors (e.g., Tile Coding) with pluggable hash functions [25]; (7) an efficient implementation of the dot product for sparse coder based feature representations; (8) benchmark environments: (Mountain Car, Mountain Car 3D, Swinging Pendulum, Helicopter, and Continuous Grid World) [25]; (9) optimization for very fast duty cycles (e.g., using culling traces, RLLib is tested on the RoboCup 3D simulator agents, physical NAO V4 humanoid robots, and Tiva C series launchpad microcontrollers); (10) a framework to design complex behaviors, predictors, controllers, and represent highly expressive learnable knowledge representations in RL using GVFs; (11) a framework to visualize benchmark problems; and (12) a plethora of examples demonstrating on-policy and off-policy control experiments.

4 Prelude

The standard RL models an AI agent and its environment interactions in discrete time steps $t = 1, 2, 3, \ldots$. The agent senses the state of the world at each time step $S_t \in \mathcal{S}$ and selects an action $A_t \in \mathcal{A}$. One time step later the agent receives a scalar reward $R_{t+1} \in \mathbb{R}$, and senses the state $S_{t+1} \in \mathcal{S}$. The rewards are generated according to a reward function $r : S_{t+1} \to \mathbb{R}$. The objective of the RL is to learn a stochastic action-selection policy $\pi : \mathcal{S} \times \mathcal{A} \to [0, 1]$, that gives the probability of selecting each action in each state, such that the agent maximizes rewards summed over the time steps [25].

An interpretation assigns semantics to a set of entities in a domain of discourse. In RL, a set of arbitrary value functions contains the entities over which the interpretation is defined [14–16,27]. In this context, knowledge is represented as a large number of approximate value functions each with its: (1) target policy (π); (2) pseudo-reward function ($r : \mathcal{S} \rightarrow \mathbb{R}$); (3) pseudo-termination function ($\gamma : \mathcal{S} \rightarrow [0, 1]$); and (4) pseudo-terminal-reward function ($z : \mathcal{S} \rightarrow \mathbb{R}$).

The GVFs are defined over the four functions: π, γ, r, and z. However, γ function is more substantive than reward functions as the termination interrupts the normal flow of state transitions. In pseudo mode, the standard termination is omitted. As an example, in robotic soccer, a base problem can be defined as duration in which a goal is scored by either teams. We can consider a pseudo-termination has occurred when a striker is changed. A GVF with respect to a state-action function is defined as $q^{\pi,\gamma,r,z}(s, a)$. Therefore, these four functions provide the meaning to the GVF in the form of a question.

In continuous state and action spaces, a value function is represented using a function approximator, $\hat{q}$, which amounts to a majority of problems encountered in practice. As an example, in linear function approximation, there exists a weight vector, $\boldsymbol{\theta} \in \mathbb{R}^N$, to be learned. Hence, an approximate GVF is defined as: $\hat{q}(s, a, \theta) = \boldsymbol{\theta}^{\mathrm{T}} \boldsymbol{\phi}(s, a)$, such that, $\hat{q} : \mathcal{S} \times \mathcal{A} \times \mathbb{R}^N \rightarrow \mathbb{R}$. The weights are learned using on/off-policy algorithms implemented in Sect. 3. The approximate value function, $\hat{q}$, is the answer to the question formed by its value function, q. Therefore, the truth value of q is measured by $\hat{q}$, which in return completes the intended interpretation.

5 Platform

RLLib closely follows the design principles and recommendations presented in [13, 24]. The development of the library has taken significant efforts to minimize memory footprint as well as computational requirements that are requested by RL problems. Since RLLib specifically emphasizes on learnable knowledge representation and reasoning, it has been modularized based on on-policy and off-policy RL methods. In addition, RLLib provides implementation of incremental supervised learning algorithms that can be used simultaneously with RL problems.

Listings 1 and 2 provide the minimal pseudo-code to setup on/off-policy RL agents. The controlling, behavior learning, and learnable knowledge representation problems should use "ControlAlgorithm.h" header file. The algorithms related to predictions and supervised learning problems are implemented in "PredictorAlgorithm.h" and "SupervisedAlgorithm.h" header files. All C++ templates implemented in the library are under the namespace RLLib, and use a single parameter T. This parameter could be a C++ primitive type as shown in Listings 1 and 2 or a complex object defined by the user. It is our experience that majority of RL problems can be defined using a primitive type. Devices such as Tiva C launchpad microcontrollers supports single precision floating point representations. Therefore, the templates should be initialized with float primitive type.

```
// --------------------------------------------
 1 include "ControlAlgorithm.h"
 2 include "RL.h"
 3 using namespace RLLib;
// RL Problem ----------------------------------
 4 RLProblem<double>* problem = ...;
// Projector -----------------------------------
 5 Hashing<double>* hashing = ...;
 6 Projector<double>* projector = ...;
 7 StateToStateAction<double>* toStateAction = ...;
// Predictor -----------------------------------
 8 Trace<double>* e = ...;
 9 GQ<double>* gq = ...;
// Policies π and π_b ---------------------------
10 Policy<double>* target = ...;
11 Policy<double>* behavior =...;
// ---------------------------------------------
12
13
// Controller ----------------------------------
14 OffPolicyControlLearner<double>* control = ...;
// Runner --------------------------------------
15 RLAgent<double>* agent = ...;
16 Simulator<double>* sim = ...;
17 sim->run(); // OR sim->step();
// ---------------------------------------------
```

```
// --------------------------------------------
 1 include "ControlAlgorithm.h"
 2 include "RL.h"
 3 using namespace RLLib;
// RL Problem ----------------------------------
 4 RLProblem<double>* problem = ...;
// Projectors ----------------------------------
 5 Hashing<double>* hashing = ...;
 6 Projector<double>* projector = ...;
 7 StateToStateAction<double>* toStateAction = ...;
// Critic --------------------------------------
 8 Trace<double>* critice = ...;
 9 GTDLambda<double>* critic = ...;
// Policies π and π_b ---------------------------
10 PolicyDistribution<double>* target = ...;
11 Policy<double>* behavior =...;
// Actor ---------------------------------------
12 Traces<double>* actoreTraces = ...;
13 ActorOffPolicy<double>* actor = ...;
// Controller ----------------------------------
14 OffPolicyControlLearner<double>* control = ...;
// Runner --------------------------------------
15 RLAgent<double>* agent = ...;
16 Simulator<double>* sim = ...;
17 sim->run(); // OR sim->step();
// ---------------------------------------------
```

Listing 1: Pseudo-code for action-value methods.   Listing 2: Pseudo-code for policy gradient methods .

In line 4, we define the RL problem using an instance of the template `RLProblem<T>`. This template as well as `RLAgent<T>` and `Simulator<T>` (lines 15-16) templates are defined in `"RL.h"` header file. An instance of the template `Projector<T>` (line 6) extracts features from the state variables. These features are part of a function approximation architecture (linear or non-linear), e.g., tile based sparse features [25, Sect. 8.3.2] or compact features [12], suitable for the problem. Some feature extractors require a hashing function, that is defined in line 5. For action-value functions, the features could also include actions (or options), that is defined in `StateToStateAction<T>` (line 7). In Listing 1, lines 8-9 define the predictor, which is used in off-policy controller (line 14), while Listing 2 defines the critic that is used in the actor-critic controller. Lines 12-13 in Listing 2 define the actor for the prior controller. Lines 10-11 define the target policy (the smooth policy distributions in Listing 2) and the behavior policy. Line 15 defines the RL agent that is used in the simulator (line 16) simultaneously with the RL problem.

In simulations (e.g., [23]), specifically when the simulator has the control over the perception-actuation cycles, the runner is executed with `run()` in line 17. In practical problems (e.g., [20]), where the agents and the environments run on disjoint processes, the runner will wait for the percepts, then updates the agent, which in return transmits the actions to the environment. In such situations, the runner is executed with `step()` in line 17. It is to be noted that either in simulations or in practical problems the runner will execute the same update steps, such that, the user will experience the same set of execution steps. A practitioner can construct the C++ objects in lines 4-16 in an initialization subroutine, and execute line 17 in a subroutine that calls in every duty cycle.

There are complex combination of RL algorithms used in practice. RLLib allows many combination of these algorithms by changing a few lines of code in Listings 1 and 2. For example, in oder to implement on-policy action-value methods, a practitioner can change the predictor in line 9 to `Sarsa<T>` and

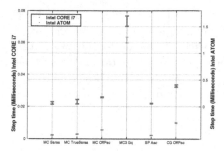

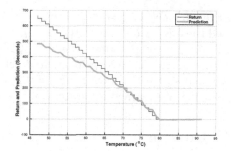

Fig. 1. Step update times in milliseconds. The thick error bars (blue) show the step time for Intel ATOM, while the thin error bars (red) show the step update time for Intel CORE–i7 (Color figure online).

Fig. 2. Predicting the time to shutdown in seconds. The bold line (red) shows the prediction to shutdown from a given temperature of NAO left knee. The thin line (blue) shows the actual return (Color figure online).

the controller in line 14 to OnPolicyControlLearner<T> in Listing 1. Similarly, different combination of RL algorithms can be included in Listing 2 for actor-critic, and parameterized policies.

6   Experiments

This section provides experiments: (1) to validate the effectiveness of the library across multiple hardware platforms; and (2) the ability to answer a subjectively posed predictive question using the conceptualization of GVFs.

Figure 1 shows the step time in milliseconds, i.e., the time an algorithm requires to update its parameters from the observations, for a set of benchmark problems popular in RL literature. We have considered two hardware platforms: (1) Intel CORE–i7 2.2GHz laptop; and (2) Intel ATOM 1.6GHz CPU available on NAO humanoid robot. MC Sarsa, MC TrueSarsa, and MC OffPac represent the two dimensional mountain car benchmark problem solved using sarsa, true sarsa, and off-policy actor-critic algorithms. The reader is referred to [5, 21, 23] that describe the problems, feature extractions, and parameter settings respectively. MC3 Gq describes the three dimensional mountain car [30] solved using greedy-GQ algorithm [14]. SP Aac and CG OffPac represent swinging pendulum and continuous grid-world problems solved using average [4] and off-policy [5] actor-critic algorithms. Even though the step update on NAO platform is on average 30.64 times slower than the laptop, the step update is suitable for real-time operations (worst case time is approximately 1.5 ms).

Our second experiment poses a question of the form "How much time remaining on NAO before the left knee temperature reaches above 80 °C?". We have configured a NAO robot to walk in the standard platform league soccer field. We have started the robot in resting temperature, and stopped the experiment when at least one of the joints reached the critical temperature (>80 °C). We have used GQ(λ) [5] with the GVF question encoding: $\pi(s,a) = 1$, $r(s) = 0.01$, $z(s) = 0$ seconds (we have queried the robot sensor values at 100 Hz), $\forall s \in \mathcal{S}$, and $\gamma(s) = 0$

if the left knee temperature is greater than $80\,°C$, otherwise $\gamma(s) = 1$. We have used the answer encoding: $\lambda(s) = 0.4$, $\forall s \in \mathcal{S}$ and single tiling with 28 regions of the effective temperature range ($47\,°C$, $91\,°C$) with 512 memory (the size of the feature, eligibility, and primary and secondary parameter vectors). We set the two step size parameters to $\alpha_v = 0.2$ and $\alpha_w = 0.000001$. As shown in Fig. 2, the agent has learned accurate predictions to shutdown from returns.

7 Conclusion

RLLib framework and its features, as presented in Sects. 1 and 5, have been fully implemented, empirically tested, and released to public from project website at: http://web.cs.miami.edu/home/saminda/rllib.html. In addition, RLLib provides testbeds, intuitive visualization tools, and extension points for complex combination of RL algorithms, agents, and environments. Compared to existing platforms, RLLib has been successfully deployed in different hardware configurations due to its low memory footprint and computational efficiency. The library is also compatible with devices support by Energia, an open-source electronics prototyping platform.

References

1. Abeyruwan, S., Seekircher, A., Visser, U.: Dynamic role assignment using general value functions. In: AAMAS 2013, Adaptive Learning Agents Workshop (2013)
2. Bishop, C.M.: Pattern Recognition and Machine Learning. Information Science and Statistics, 1 edn. Springer, Heidelberg (2007)
3. Dabney, W., Barto, A.G.: Adaptive step-size for online temporal difference learning. In: AAAI Conference on Artificial Intelligence (2012)
4. Degris, T., Pilarski, P.M., Sutton, R.S.: Model-free reinforcement learning with continuous action in practice. In: American Control Conference (ACC), pp. 2177–2182. IEEE (2012)
5. Degris, T., White, M., Sutton, R.S.: Off-policy actor-critic. In: Proceedings of the 29th International Conference on Machine Learning (ICML), pp. 457–464 (2012)
6. Delepoulle, F.D.C.S.: PIQLE: a platform for implementation of q-learning experiments. In: Neural Information Processing Systems (NIPS), Workshop on Reinforcement Learning Benchmarks and Bake-off II (2005)
7. Diuk, C., Cohen, A., Littman, M.L.: An object-oriented representation for efficient reinforcement learning. In: Proceedings of the 25th International Conference on Machine Learning (ICML), pp. 240–247 (2008)
8. Frezza-Buet, H., Geist, M.: A C++ template-based reinforcement learning library: fitting the code to the mathematics. J. Mach. Learn. Res. (JMLR) **14**(1), 625–628 (2013)
9. Hafner, R., Riedmiller, M.: Case study: control of a real world system in CLSquare. In: Proceedings of the NIPS Workshop on Reinforcement Learning Comparisons, Whistler, British Columbia, Canada (2005)
10. Kitano, H., Asada, M., Kuniyoshi, Y., Noda, I., Osawai, E., Matsubara, H.: RoboCup: a challenge problem for AI and robotics. In: Kitano, H. (ed.) RoboCup 1997. LNCS, vol. 1395, pp. 1–19. Springer, Heidelberg (1998)

11. Kober, J., Peters, J.: Reinforcement learning in robotics: a survey. In: Wiering, M., van Otterlo, M. (eds.) Reinforcement Learning. ALO, vol. 12, pp. 579–610. Springer, Heidelberg (2012)
12. Konidaris, G., Osentoski, S., Thomas, P.: Value function approximation in reinforcement learning using the Fourier basis. In: Proceedings of the 25th Conference on Artificial Intelligence, pp. 380–385 (2011)
13. Kovacs, T., Egginton, R.: On the analysis and design of software for reinforcement learning, with a survey of existing systems. Mach. Learn. **84**(1–2), 7–49 (2011)
14. Maei, H.R.: Gradient temporal-difference learning algorithms. Ph.D. thesis, University of Alberta (2011)
15. Maei, H.R., Sutton, R.S.: GQ(λ): a general gradient algorithm for temporal-difference prediction learning with eligibility traces. In: Proceedings of the 3rd Conference on Artificial General Intelligence (AGI), pp. 1–6. Atlantis Press (2010)
16. Maei, H.R., Szepesvári, C., Bhatnagar, S., Sutton, R.S.: Toward off-policy learning control with function approximation. In: Proceedings of the 27th International Conference on Machine Learning (ICML), pp. 719–726 (2010)
17. Mahmood, A.R., Sutton, R.S., Degris, T., Pilarski, P.M.: Tuning-free step-size adaptation. In: Acoustics, Speech and Signal Processing (ICASSP), pp. 2121–2124. IEEE (2012)
18. Papis, B., Wawrzynski, P.: dotRL: a platform for rapid reinforcement learning methods development and validation. In: 2013 Federated Conference on Computer Science and Information Systems (FedCSIS), pp. 129–136 (2013)
19. Schaul, T., Bayer, J., Wierstra, D., Sun, Y., Felder, M., Sehnke, F., Rückstieß, T., Schmidhuber, J.: PyBrain. J. Mach. Learn. Res. **11**, 743–746 (2010)
20. Seekircher, A., Abeyruwan, S., Visser, U.: Accurate ball tracking with extended Kalman filters as a prerequisite for a high-level behavior with reinforcement learning. In: The 6th Workshop on Humanoid Soccer Robots at Humanoid Conference, Bled (Slovenia) (2011)
21. Seijen, H.V., Sutton, R.: True online TD(λ). In: Jebara, T., Xing, E.P. (eds.) Proceedings of the 31st International Conference on Machine Learning (ICML). JMLR Workshop and Conference Proceedings, pp. 692–700 (2014)
22. Sigaud, O., Buffet, O.: Markov Decision Processes in Artificial Intelligence. Wiley, New York (2013)
23. Sutton, R.S.: Generalization in reinforcement learning: successful examples using sparse coarse coding. In: Advances in Neural Information Processing Systems 8, pp. 1038–1044. MIT Press (1996)
24. Sutton, R.S.: a standard interface for reinforcement learning software in C++. http://webdocs.cs.ualberta.ca/sutton/RLinterface/RLI-Cplusplus.html. Accessed 12 July 2015
25. Sutton, R.S., Barto, A.G.: Reinforcement Learning: An Introduction. MIT Press, Cambridge (1998)
26. Sutton, R.S., Koop, A., Silver, D.: On the role of tracking in stationary environments. In: Proceedings of the 24th International Conference on Machine Learning, pp. 871–878. ACM (2007)
27. Sutton, R.S., Modayil, J., Delp, M., Degris, T., Pilarski, P.M., White, A., Precup, D.: Horde: a scalable real-time architecture for learning knowledge from unsupervised sensorimotor interaction. In: Proceedings of the 10th International Conference on Autonomous Agents and Multiagent Systems (AAMAS), pp. 761–768 (2011)

28. Szepesvári, C.: Algorithms for Reinforcement Learning. Synthesis Lectures on Artificial Intelligence and Machine Learning. Morgan & Claypool Publishers, San Rafael (2010)
29. Tanner, B., White, A.: RL-Glue: language-independent software for reinforcement-learning experiments. J. Mach. Learn. Res. **10**, 2133–2136 (2009)
30. Taylor, M.E., Kuhlmann, G., Stone, P.: Autonomous transfer for reinforcement learning. In: Proceedings of the 7th International Joint Conference on Autonomous Agents and Multiagent Systems (AAMAS), vol. 1, pp. 283–290 (2008)

Fawkes for the RoboCup Logistics League

Tim Niemueller[1](✉), Sebastian Reuter[2], and Alexander Ferrein[3]

[1] Knowledge-based Systems Group, RWTH Aachen University, Aachen, Germany
niemueller@kbsg.rwth-aachen.de
[2] Institute Cluster IMA/ZLW & IfU, RWTH Aachen University, Aachen, Germany
sebstian.reuter@ima-zlw-ifu.rwth-aachen.de
[3] MASCOR Institute, Aachen University of Applied Sciences, Aachen, Germany
ferrein@fh-aachen.de

Abstract. Autonomous mobile robots comprise a great deal of complexity. They require a plethora of software components for perception, actuation, task-level reasoning, and communication. These components have to be integrated into a coherent and robust system in time for the next RoboCup event. Then, during the competition, the system has to perform stable and reliably. Providing a software framework for teams to use tremendously eases that effort. Even more so when providing a fully integrated system specific for a particular domain.

We have recently released our full software stack for the *RoboCup Logistics League* (RCLL) based on the Open Source *Fawkes Robot Software Framework*. This release includes all software components of the RoboCup 2014 winning team Carologistics (in cooperation with the AllemaniACs RoboCup@Home team). It specifically also includes the parts of the software which are domain or platform specific or which we consider our competitive edge and were kept private until now. We think that this will make the league much more accessible to new teams and might help existing teams to improve their performance.

1 Introduction

Over the past eight years, we have developed the *Fawkes Robot Software Framework* [1] as a robust foundation to deal with the challenges of robotics applications in general, and in the context of RoboCup in particular. It has been developed and used in the Middle-Size [2] and Standard Platform [3] soccer leagues, the RoboCup@Home [4,5] service robot league, and now in the *RoboCup Logistics League* [6]. The frameworks or parts of it have also been used in other contexts [7,8].

The Carologistics are the first team in the RCLL to publicly release their software stack. Teams in other leagues have made similar releases before. What makes ours unique is that it provides a complete and *ready-to-run package with the full software* (and some additions and fixes) that we used in the competition in 2014 – which we won. This in particular *includes* the complete *task-level executive* component, that is the strategic decision making and behavior generating

© Springer International Publishing Switzerland 2015
L. Almeida et al. (Eds.): RoboCup 2015, LNAI 9513, pp. 365–373, 2015.
DOI: 10.1007/978-3-319-29339-4_31

software. This component was typically held back or only released in small parts in previous software releases by other teams (for any league).

In the RCLL all teams use the same hardware platform "Robotino" by Festo Didactic. This means that there is *no hardware barrier* that prevents teams from using the software effectively and quickly. Even more so, with the *3D simulation environment* based on Gazebo which we have developed [9] and provide, teams can immediately start using our software system for their own development. We provide extensive documentation and are expanding it continuously.

The public release, documentation and videos are available at https://www.fawkesrobotics.org/p/llsf2014-release.

In the following we will briefly describe the framework, some major components[2], and our simulation environment in Section 2 with a high-light on the task-level executive in Sect. 3. We conclude in Sect. 4.

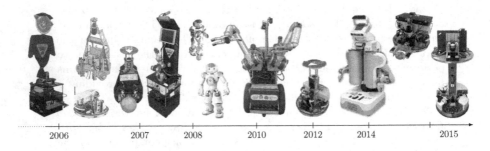

Fig. 1. Robots running (parts of) Fawkes which were or are used for the development of the framework and its components.

2   Fawkes Robot Software Framework

At the core of the public release for the RCLL is the *Fawkes Robot Software Framework*[1] which is Open Source software.

Fawkes was initially started in 2006 as an effort to build a capable and faster software platform for a new generation of Mid-Size league robots of the *AllemaniACs*[2] *RoboCup Team* (cf. Fig. 1). It was used for the first time at RoboCup 2007 in Atlanta. Since then it was also used on our domestic service robot Caesar [5] in the RoboCup@Home league winning the RoboCup in 2006 and 2007, placing second in 2008, and winning the German Open 2007 and 2008 [4]. From 2008 to 2010 we participated as team ZaDeAt [3], a joint team from University of Cape Town (ZA), RWTH Aachen University (DE) and Technical University of Graz (AT), in the Standard Platform League. During this time we developed the Lua-based Behavior Engine [10], a component which was ported to ROS in 2010 and used, for example, on HERB at CMU [7]. Since 2012 we participate in the RoboCup Logistics League as the *Carologistics*[3] *joint team* consisting of the

---

[1] Fawkes website at http://www.fawkesrobotics.org.

[2] Website of the AllemaniACs at https://robocup.rwth-aachen.de.

[3] Website of the Carologistics at https://www.carologistics.org.

Knowledge-Based Systems Group, the Institute Cluster IMA/ZLW & IfU (both RWTH Aachen University), and the Institute for Mobile Autonomous Systems and Cognitive Robotics (Aachen University of Applied Sciences). In 2014 we won both, the World Champion title and the German Open, including the technical challenges. Fawkes is also used in combination with ROS on a PR2 in a project on hybrid reasoning [8]. The Carologistics have been driving the development of the RCLL with active members in the Technical and Organization Committees. We developed the referee box of the competition [11] and proposed the use of physical processing machines [12].

The overall software structure is designed as a three-layer architecture [13] and follows a component-based paradigm [14–16]. It consists of a deliberative layer for high-level reasoning, a reactive execution layer for breaking down high-level commands and monitoring their execution, and a feedback control layer for hardware access and functional components. The communication between single components – implemented as *plugins* – is realized by a hybrid blackboard and messaging approach [1]. Fawkes has adapter plugins to integrate it with the ROS [17] ecosystem. On the Carologistics and AllemaniACs robots, we use it for visualization purposes.

Software Components. Fawkes already contains a wide variety of more than 125 software components and more than two dozen software libraries, many of which are used in the RCLL. These cover a wide range of functionalities, from plugins providing infrastructure, over functional components for self-localization and navigation, and perception modules via point clouds, laser range finders, or computer vision, to behavior generating components. In the following we describe some examples with a particular focus on the RCLL. The behavior components are explained in more detail in Sect. 3.

Fig. 2. Machine signal detection used in the RCLL 2014. The markings denote the detected lights.

Navigation. Fawkes comes with an implementation of Adaptive Monte Carlo Localization which is an extended port from ROS. In the RCLL, we use a pre-specified map and a laser range finder to determine and track the position of the robot on the field. For locomotion path planning we use a layered structure. A component called *navgraph* has a graph overlay over the playing field, where nodes specify travel points or points of interest like machines, and edges are passages free from static obstacles. When moving to a specific point the navgraph plugin determines a path on this graph to reach the goal. It then instructs the *colli* [18], a local path planner and collision avoidance module we have developed. Based on the next (intermediate) goal on the path it plans a collision free path on the grid. It may divert as much as necessary from the global path, sometimes resulting in bypassing whole parts of the planned path, for example if it is blocked by an obstacle like an opponent robot. The navgraph recognizes this and skips the bypassed intermediate nodes.

Perception. Another important component is detection and recognition of the light signal of a machine as shown in Fig. 2. While it might seem like a routine task for computer vision, it is complicated by several factors. Since the lights can be on and off, the brightness of the image varies significantly. Additionally, the field contains green walls in the area of the delivery gates (upper left corner in Fig. 3(a) and (b)). A full search for the light signal in an image therefore results in many false positives and negatives. Thus we use a laser cluster detection of the light signal to reduce the search space.

Simulation. The RCLL game emphasizes research and application of methods for efficient planning, scheduling, and reasoning on the optimal work order of production processes handled by a group of robots. An aspect that distinctly separates this league from others is that the environment itself acts as an agent by posting orders and controlling the machines (*environment agency*).

Therefore, we have created an *open simulation environment* [9] to support research and development. There are three core aspects in this context: (1) The simulation should be a turn-key solution with simple interfaces, (2) the world must react as close to the real world as possible, including in particular the machine responses and signals, and (3) various levels of abstraction are desirable depending on the focus of the user, e.g. whether to simulate laser data to run a self-localization component or to simply provide the position.

In recent work [9], we provide such an environment. It is based on the well-known Gazebo simulator addressing these issues: (1) its wide-spread use and open interfaces already adapted to several software frameworks in combination with our models and adapters provide an easy to use solution; (2) we have connected the simulation directly to the referee box, the semi-autonomous game controller of the RCLL, so that it provides precisely the reactions and *environment agency* of a real-world game; (3) we have implemented *multi-level abstraction* that allows to run full-system tests including self-localization and perception or to focus on high-level control reducing uncertainties by replacing some lower-level components using simulator ground truth data. This allows to develop an idealized strategy first, and only then increase uncertainty and enforce robustness by failure detection and recovery.[4]

(a) Carologistics (three Robotino 2 with laptops on top) and BBUnits (two Robotino 3) during the RCLL 2014 finals.

(b) The simulation of the RCLL 2014 in Gazebo. The circles above the robots indicate their localization and robot number.

Fig. 3. The RCLL 2014 in real and simulation.

---

[4] Simulation is available at https://www.fawkesrobotics.org/p/llsf-sim/.

We have recently joined effort with the BBUnits team of TU Munich and DLR to generalize the simulation further and establish a simulation sub-league.[5]

The released software package contains all of these components which will reduce the initial development and integration effort especially for new teams.

Framework Interoperability. The ability to interact with other robot software frameworks is essential, as it allows to build on already available packages, and to integrate into an existing system. Fawkes has been used in both contexts, for example with ROS [17]. Several generic integration plugins allow for the exchange of data like point clouds or coordinate transformations. Additionally, there is a ROS node that embeds a Fawkes instance into the ROS ecosystem.

Fawkes has also been integrated with, e.g., reasoning systems by means of a remote middleware access library. This is useful for software components that are not based on a specific framework but which provide independent run-times with, for example, a plugin mechanism to integrate with execution systems.

3   Task-Level Coordination and Execution

Task coordination is performed using an incremental reasoning approach [19]. In the following we describe the behavior components, and the reasoning process from the rules in 2014. For computational and energy efficiency, the behavior components also coordinate activation of the lower level components.

Behavior Components for the RCLL. Tasks that the reasoning component of a robot in RCLL LLSF must fulfill are:

Exploration. Gather information about the machine types by visiting them and reporting on the recognized light signal.

Production. Complete the production chains as often as possible dealing with incomplete knowledge.

Execution Monitoring. Instruct and monitor the reactive mid-level Lua-based Behavior Engine.

Three robots perform these steps cooperatively as a group, that is, they communicate information about their current intentions, acquire exclusive control over resources like machines, and share their beliefs about the current state of the environment. This continuous updating of information suggests an incremental reasoning approach. As facts become known, the robot needs to adjust its plan.

---

[5] Code is available at https://github.com/robocup-logistics.

Lua-Based Behavior Engine.
In previous work we have developed the Lua-based Behavior
Engine (BE) [10]. It mandates
a separation of the behavior
in three layers, as depicted in
Fig. 4: the low-level processing
for perception and actuation, a
mid-level reactive layer, and a
high-level reasoning layer. The

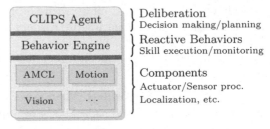

Fig. 4. Behavior layer separation

layers are combined following an adapted hybrid deliberative-reactive coordination paradigm with the BE serving as the reactive layer to interface between
the low- and high-level systems.

The BE implements individual behaviors – called skills – as hybrid state
machines (HSM). They can be depicted as a directed graph with nodes representing states for action execution and monitoring. Edges denote jump conditions
implemented as Boolean functions. For the active state of a state machine, all
outgoing conditions are evaluated, typically at about 15 Hz. If a condition fires,
the active state is changed to the target node of the edge. A table of variables
holds information like the world model, for example storing numeric values for
object positions. It remedies typical problems of state machines like fast growing
number of states or variable data passing from one state to another. Skills are
implemented using the light-weight, extensible scripting language Lua.

Incremental Reasoning Agent. The problem at hand with its intertwined
world model updating and execution naturally lends itself to a representation
as a fact base with update rules for triggering behavior for certain beliefs. We
have chosen the CLIPS rules engine [20]. Incremental reasoning means that the
robot does not create a full-fledged plan at a certain point in time and then
executes it. Rather, when idle it commits to the 'then-best' action. This avoids
costly re-planning (as with approaches using planners), it allows to cope with
incomplete knowledge, and it is computationally inexpensive. As soon as the
action is completed and based on its knowledge, the next best action is chosen.
High priority tasks can also interrupt running tasks. In Fig. 5 we show a simplified
rule for the production process. The game is in the production phase, the robot is
currently idle and holds a raw material puck S_0 or no puck: (phase PRODUCTION)
(state IDLE)(holding NONE|S0). Furthermore there is a $T5$-machine, with
matching team-color, has no produced puck, is not already loaded with an S_0,
and no other robot is currently bringing an S_0. If these conditions are satisfied
and the rule has the highest priority of currently active rules, the rule fires
proposing to load the machine appropriately.

There is a set of such production rules with their conditions and priorities
determining what the robot does in a certain situation, or – in other terms –
based on a certain belief about the world in the fact base. This simplifies adding
new decision rules. The decisions can be made more granular by adding rules
with more restrictive conditions and a higher priority.

```
(defrule load-T5-with-S0
  (declare (salience ?*PRIORITY-LOAD-T5-WITH-S0*))
  (phase PRODUCTION)
  ?s <- (state IDLE)
  (holding NONE|S0)
  (team-color ?team-color)
  (machine (mtype T5) (loaded-with $?l&~:(member$ S0 ?l))
    (incoming $?i&~:(member$ BRING_S0 ?i)) (name ?name)
    (produced-puck NONE) (team ?team-color))
  =>
  (printout t "PROD: Loading T5 " ?name " with S0" crlf)
  (assert (proposed-task (name load-with-S0)
          (args (create$ ?name))))
  (retract ?s)
  (assert (state TASK-PROPOSED))
)
```

Fig. 5. CLIPS production process rule

After a proposed task was chosen, the coordination rules of the agent cause communication with the other robots to announce the intention and ensure that there are no conflicts. If the coordination rules accept the proposed task, process execution rules perform the steps of the task (e.g. getting an S_0 from the input storage and bringing it to the machine). Here, the agent calls the Behavior Engine to execute the actual skills like driving to the input storage and loading a puck.

4 Conclusion

The integration of a complete robot system even for medium-complex domains such as the RCLL can be tedious and time consuming. We had made the decision early in 2012 when joining the RCLL to go for a more complex, but then also more robust and flexible system. In 2014 this was finally rewarded by winning the RoboCup 2014 RCLL competition.

The public release of a fully working and thoroughly tested integrated software stack lowers the barrier of entry for new teams to the league and fosters research and exchange among members of the RoboCup community in general, and in the RoboCup Logistics League in particular.

Acknowledgments. T. Niemueller was supported by the German National Science Foundation (DFG) research unit *FOR 1513* on Hybrid Reasoning for Intelligent Systems (http://www.hybrid-reasoning.org).

The Carologistics team members in 2014/2015 are: D. Ewert, A. Ferrein, S. Jeschke, N. Limpert, G. Lakemeyer, K. Leonardic, M. Löbach, R. Maaßen, V. Mataré, T. Neumann, T. Niemueller, F. Nolden, S. Reuter, J. Rothe, and F. Zwilling.

References

1. Niemueller, T., Beck, D., Lakemeyer, G., Ferrein, A.: Design principles of the component-based robot software framework fawkes. In: Ando, N., Balakirsky, S., Hemker, T., Reggiani, M., von Stryk, O. (eds.) SIMPAR 2010. LNCS, vol. 6472, pp. 300–311. Springer, Heidelberg (2010)

2. Beck, D., Niemueller, T.: AllemaniACs 2009 team description. Technical report, KBSG, RWTH Aachen University (2009)
3. Ferrein, A., Steinbauer, G., McPhillips, G., Niemueller, T., Potgieter, A.: Team ZaDeAt 2009 - team report. Technical report, RWTH Aachen University, Graz University of Technology, and University of Cape Town (2009)
4. Schiffer, S., Lakemeyer, G.: AllemaniACs team description RoboCup@Home. Technical report, KBSG, RWTH Aachen University (2011)
5. Ferrein, A., Niemueller, T., Schiffer, S., G.L.: Lessons learnt from developing the embodied AI platform caesar for domestic service robotics. In: Proceedings of AAAI Spring Symposium, 2013 - Designing Intelligent Robots: Reintegrating AI (2013)
6. Niemueller, T., Ferrein, A., Jeschke, S., Ewert, D., Reuter, S., Lakemeyer, G.: Decisive factors for the success of the carologistics RoboCup team in the RoboCup logistics league 2014. In: Bianchi, R.A.C., Akin, H.L., Ramamoorthy, S., Sugiura, K. (eds.) RoboCup 2014. LNCS, vol. 8992, pp. 155–167. Springer, Heidelberg (2015)
7. Srinivasa, S.S., Berenson, D., Cakmak, M., Collet, A., Dogar, M.R., Dragan, A.D., Knepper, R.A., Niemueller, T., Strabala, K., Vande Weghe, M., Ziegler, J.: HERB 2.0: lessons learned from developing a mobile manipulator for the home. Proc. IEEE 100(8), 2410–2428 (2012)
8. Niemueller, T., Abdo, N., Hertle, A., Lakemeyer, G., Burgard, W., Nebel, B.: Towards deliberative active perception using persistent memory. In: Proceedings of the Workshop on AI-based Robotics at the International Conference on Intelligent Robots and Systems (IROS) (2013)
9. Zwilling, F., Niemueller, T., Lakemeyer, G.: Simulation for the RoboCup logistics league with real-world environment agency and multi-level abstraction. In: Bianchi, R.A.C., Akin, H.L., Ramamoorthy, S., Sugiura, K. (eds.) RoboCup 2014. LNCS, vol. 8992, pp. 220–232. Springer, Heidelberg (2015)
10. Ferrein, A., Lakemeyer, G., Niemüller, T.: A lua-based behavior engine for controlling the humanoid robot nao. In: Baltes, J., Lagoudakis, M.G., Naruse, T., Ghidary, S.S. (eds.) RoboCup 2009. LNCS, vol. 5949, pp. 240–251. Springer, Heidelberg (2010)
11. Niemueller, T., Ewert, D., Reuter, S., Ferrein, A., Jeschke, S., Lakemeyer, G.: RoboCup logistics league sponsored by festo: a competitive factory automation benchmark. In: RoboCup Symposium 2013 (2013)
12. Niemueller, T., Lakemeyer, G., Ferrein, A., Reuter, S., Ewert, D., Jeschke, S., Pensky, D., Karras, U.: Proposal for advancements to the LLSF in 2014 and beyond. In: ICAR - 1st Workshop on Developments in RoboCup Leagues (2013)
13. Gat, E.: Three-layer architectures. In: Artificial Intelligence and Mobile Robots (1998)
14. McIlroy, M.D.: 'Mass Produced' Software Components. Software Engineering: Report on a Conference Sponsored by the NATO Science Committee (1968)
15. Brugali, D., Scandurra, P.: Component-based robotic engineering (Part I). IEEE Rob. Autom. Mag. 16(4), 84–96 (2009)
16. Brugali, D., Shakhimardanov, A.: Component-based robotic engineering (Part II). IEEE Rob. Autom. Mag. 17(1), 100–112 (2012)
17. Quigley, M., Conley, K., Gerkey, B.P., Faust, J., Foote, T., Leibs, J., Wheeler, R., Ng, A.Y.: ROS: an open-source robot operating system. In: ICRA Workshop on Open Source Software (2009)

18. Ferrein, A., Schiffer, S., Lakemeyer, G., Jacobs, S., Beck, D.: Robust collision avoidance in unknown domestic environments. In: Baltes, J., Lagoudakis, M.G., Naruse, T., Ghidary, S.S. (eds.) RoboCup 2009. LNCS, vol. 5949, pp. 116–127. Springer, Heidelberg (2010)
19. Niemueller, T., Lakemeyer, G., Ferrein, A.: Incremental task-level reasoning in a competitive factory automation scenario. In: Proceedings of AAAI Spring Symposium 2013 - Designing Intelligent Robots: Reintegrating AI (2013)
20. Wygant, R.M.: CLIPS: a powerful development and delivery expert system tool. Comput. Ind. Eng. **17**(1–4), 546–549 (1989)

Benchmarking Workshop

Synthetical Benchmarking of Service Robots: A First Effort on Domestic Mobile Platforms

Min Cheng[1(✉)], Xiaoping Chen[1], Keke Tang[1], Feng Wu[1], Andras Kupcsik[3], Luca Iocchi[2], Yingfeng Chen[1], and David Hsu[3]

[1] University of Science and Technology of China, Hefei, China
ustccm@mail.ustc.edu.cn
[2] Sapienza University of Rome, Rome, Italy
[3] National University of Singapore, Singapore, Singapore

Abstract. Most of existing benchmarking tools for service robots are basically qualitative, in which a robot's performance on a task is evaluated based on completion/incompletion of actions contained in the task. In the effort reported in this paper, we tried to implement a synthetical benchmarking system on domestic mobile platforms. Synthetical benchmarking consists of both qualitative and quantitative aspects, such as task completion, accuracy of task completions and efficiency of task completions, about performance of a robot. The system includes a set of algorithms for collecting, recording and analyzing measurement data from a MoCap system. It was used as the evaluator in a competition called the BSR challenge, in which 10 teams participated, at RoboCup 2015. The paper presents our motivations behind synthetical benchmarking, the design considerations on the synthetical benchmarking system, the realization of the competition as a comparative study on performance evaluation of domestic mobile platforms, and an analysis of the teams' performance.

1 Introduction

Benchmarking robotic systems is challenging [1–3]. Robotic competitions are believed to be a feasible way to overcome the difficulty by appealing research groups to take their experimental results to be compared under the same test conditions [4]. Lots of well recognized competitions are held every year around the world. The focus of AAAI [5–7] and IJCAI [8] Robot Competitions is putted on benchmarking AI and robotic technology with relevance to real-life applications and changes yearly. DARPA Robotics Challenge [9] aims to develop semi-autonomous ground robots that can do complex tasks in dangerous, degraded, human-engineered environments. RoboCup[1], an initiative to promote research in AI, robotics, and related fields, currently is the largest robotics competition, with a number of leagues such as RoboCup Soccer, RoboCup Rescue, RoboCup@Work, RoboCup@Home [10–12]. RoboCup@Home aims to drive

---

[1] http://www.robocup.org/.

© Springer International Publishing Switzerland 2015
L. Almeida et al. (Eds.): RoboCup 2015, LNAI 9513, pp. 377–388, 2015.
DOI: 10.1007/978-3-319-29339-4_32

research on domestic robotics towards robust techniques and useful applications and to stimulate teams to compare their approaches on a set of common tests, and has resulted in improvement of capabilities of domestic service robots (DSRs) such as mobile manipulation [13], human-robot interaction, object recognition. RoCKIn [14,15], a project of FP7, broadens the scope of RoboCup@Home and RoboCup@Work in terms of scientific validity by being organized as a scientific benchmarking competition.

Most of existing competitions focus on qualitative evaluation on the performance of a robot, do not provide quantitative evaluation on what degree of performance a robot achieves. The objective of this effort is to advance and extend benchmarking competition by introducing quantitative evaluation. We share the same objective with RoCKIn, while taking a different approach. RoCKIn is a top-down endeavor by starting from a global framework for its long-term goals. Our effort is bottom-up in the sense that we started our endeavor from a much smaller case study—synthetical benchmarking of domestic mobile plarforms (DMPs).

We describe our motivations of introducing synthetical benchmarking in Sect. 2. A set of prescribed features for benchmarking DSRs are given in Sect. 3. Based on these features, the BSR challenge was organized at RoboCup 2015. The implementation of the BSR challenge is presented in Sect. 4. We provide an analysis on performance of participating teams to the BSR challenge in Sect. 5. A brief discussion and future work are given in Sect. 6. We draw conclusions in Sect. 7.

2   Why Synthetical Benchmarking

In this paper, by synthetical benchmarking we mean benchmarking that includes both qualitative and quantitative benchmarking. A qualitative benchmarking evaluates robot performance based on completion/incompletion of the actions contained in a task, where only two outcomes, i.e., completion or incompletion of each of these actions, are considered. Then some statistics on the qualitative outcomes of the actions may be made as an evaluation of the task. As an example, consider a task consisting of only one action pick up a can. In current competition of @Home league, one can only observe whether the action is completed or not by a robot, as an evaluation of the robot's performance on this task.

A quantitative benchmarking provides quantitative evaluation of robot performance on tasks. For example, when a robot completes a task/action, accuracy (such as errors) of the task/action completion can be acquired in quantitative evaluation. For instance, consider action *move to a waypoint*. In quantitative benchmarking, one can acquire quantitative measurement, the errors, of the robot's moving performance on the task. Without this quantitative evaluation, it is very hard to acquire any objective and accurate evaluation on the moving performance.

There are strong reasons why synthetical benchmarking of service robots is needed by introducing quantitative evaluation. First, quantitative benchmarking can generates finer evaluation than qualitative benchmarking can. Suppose either

Robot-1 or Robot-2 complete a task (say, pick up a can) with 80 % success. Then one cannot distinguish between the two robots performance on the task. However, there may be significant difference between accuracies of the two robots completions of the task. In some scenarios (as for the task of move to a waypoint) and applications, accuracy is a necessary factor in performance evaluation of robots.

Second, inclusion of both quantitative and qualitative aspects in performance evaluation of service robots supports better trade-offs among these aspects. Tasks in the @Home and @Work competitions are complicated and thus should be evaluated based on trade-offs among multiple performance factors. Generally, a comprehensive evaluation of such a task should include the following performance factors: completion of the task, accuracy of completions, and efficiency of completions (which can be measured simply with the time a robot spends for its completion of a task). A more reasonable overall evaluation should reflect some trade-offs among these factors, with accuracy being included in.

Third, accuracy data enable new solutions to some of costly work in development of service robots. For many functionalities of a service robot, even if an algorithm is correct, there are still a lot of parameters in the algorithm need to be tuned. Currently, manual tuning is the only solution, which costs a lot of time and is very low efficient. However, auto-tuning based on Machine Learning technology becomes possible if a sufficient amount of relevant accuracy data can be acquired. In this case, benchmarking supports research and development of service robots in a more direct and efficient way.

Based on these considerations, we have launched this long-term effort on synthetical benchmarking of service robots. At the first phase of this effort, we have done the following work. First, we have implemented a (semi-)automatic real-time evaluation system (ARES) for benchmarking DMP performance. The system includes a set of algorithms for collecting, recording and analyzing measurement data from a MoCap system [16,17], OptiTrack[2]. Second, we organized the Demo Challenge on Service Robots, a competition at RoboCup 2015. 25 teams applied and 11 of them were qualified for the competition. 10 qualified teams actually participated in the competition, in which our ARES was used for evaluating performance of the competing robots. Third, we organized a workshop on the same subject during the competition. About 100 participants from more than 10 countries attended the workshop.

The MoCap system (showed in Fig. 1) we used is an optical detection system with passive markers [18], which uses several fixed high speed cameras around the measurement area to triangulate a precise marker position. A set of markers (showed in Fig. 2) are attached on a robot, so that the robots behaviors can be captured by the MoCap system and then evaluated by our ARES real-time.

---

[2] http://www.optitrack.com/.

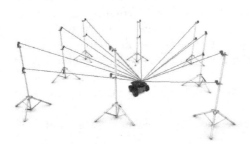

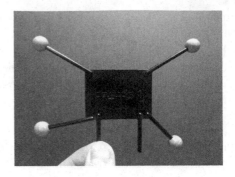

Fig. 1. The MoCap system   **Fig. 2.** The marker

3   Benchmarking on DMPs

In order to make benchmarks of DMP specific, an initial set of key features (hardware properties and functionalities) was derived from an analysis of DMPs and from experiences and observations of the common working scenes of DSRs. These features are evaluation criteria for the performance of DMPs. Furthermore, these features not only help design the benchmarks and the score system for the competition, but also allow for a later analysis of a team's performance. These features are divided into two groups: *hardware properties* and *functionalities*.

Hardware Properties. Taking into account DSRs' working environments and application demands, we propose *hardware properties* that must be implemented in each DMP in order to perform properly in the tests. To achieve these hardware properties, many technical details should be considered appropriately during mechanical designing and component selection progress. An appropriate trade-off is also needed between cost and the DMP's performance. The proposed hardware properties are characterized below.

Cost Limitation. Unlike pure theoretical research, one of goals of robotics research is to improve human life by bringing robust robotic technology to industry to create robotic applications. However, there is big gap between robotics research results and robotic products. Frequently robotics research pay more attention to verifying hypotheses and increasing knowledge, paying little attention to the cost and marketability of research outcomes. Thus, we insist that cost should be a important benchmarking condition.

Motor Feedback. Motor feedback captures the rotation angle of each wheel per control cycle, which is the source data to compute the odometry that is usually used as the input data in localization module. Besides, in the case of robot precise relative pose adjustment (e.g. mobile manipulation), odometry is the basis for adjusting robot pose, since global poses, generated by global localization techniques, are generally not as precise as odometry.

Payload Capacity. DMPs are expected to be extensible and customizable. Additional accessories, e.g. robotic arms and manipulators are expected to be integrated with DMPs to implement specific functionalities. According to the weight of common accessories, we believe that the payload capacity of DMPs should not be less than 20 kg.

Traversable Ability. Despite the fact that the floors of every-day environments are even, minor unevenness such as carpets, transitions in floor covering between different areas, and minor gaps (e.g. gaps between the floor and the elevator) are inevitable and also reflected in the RoboCup@Home competition. DMPs should be designed to adapt to these environment diversities.

Functionalities. The overall robotic system performance depends on the performance of integrated functional modules, which can be described as functional abilities or *functionalities*. As to DMPs, localization, navigation, and obstacle avoidance are the main functionalities.

Localization. The ability to estimate the real poses of a robot in the working environment.

Navigation. The task of path-planning and safely navigating to a specific target position in the environment.

Obstacle Avoidance. The ability to avoid collisions during a robot travel in the environment. Robots should be able to avoid not only static obstacles but also dynamic ones.

4   Implementation of the BSR Challenge

A robot qualified for the BSR challenge is expected to have a basic mobile platform (i.e., a robot base) and extended sensors such as camera, Laser Range Finder (LRF). The hardware cost of the basic mobile platform (including the costs of materials and components) or the market retail price (not discounted or second-hand price) should be less than 1,600 USD (about 10,000 RMB). The hardware cost or the market retail price (not discounted or second-hand price) of extended sensors should be less than 50 % of the basis mobile platform.

In order to enable the BSR challenge a synthetical benchmarking, we introduced a MoCap system to measure and, at the same time, record the movement of a DMP, with high accuracy in real time. The recorded data can not only enable quantitative analysis of the performance of DMPs in the competition, but also help make the DMP performance reproducible, which are taken as being utmost important to scientific experiments. After competitions, teams have free access to the record data.

The BSR challenge was organized in three tests and a presentation session. The mentioned key features are evaluated either as functional abilities, or as an

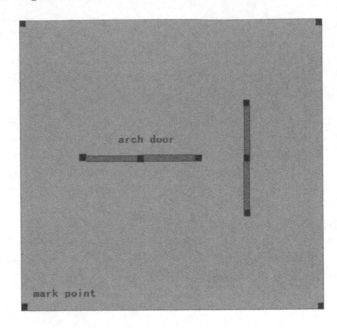

Fig. 3. The competition area (Color figure online)

integrated test. These tests and the score system are designed carefully ensuring each feature be contained in a test and be reflected in the final score. In the presentation session, each team was required to report its technical approach and share their experience with other teams.

4.1   Competition Area Layout

DMPs are tested in an indoor competition area (about 7 m × 7 m) where part of the ground may be uneven (within 3 cm of ups and downs) and there may be some obstacles on the floor. Obstacles include, but are not limited to: hollow obstacles (such as arches), furniture, small common objects, or even moving persons. Large obstacles such as arch and furniture are part of the field.

Figure 3 illustrates the setup of the competition area, where there are two sets of double arches (the greed blocks). The 10 landmarks (the red points in Fig. 3) are given as shown in Fig. 3. Among these landmarks, six are located on the arches and the other four are located in the corners. The coordinates of the landmarks in the MoCap system are provided for the participants to map the local coordinates of their robot to the coordinates of the MoCap system.

The double arch is shown in Fig. 4. The door width is 100 cm. There is a slider for each door, the height of which is adjusted randomly by the referee before a test in competition. A robot must decide autonomously whether it is able to go through a door according to its own height and the height of the slider on the door. In this case, the robot has to provide the capability of perception of 3D environment and reaction to dynamic environment. Besides, there is a plastic

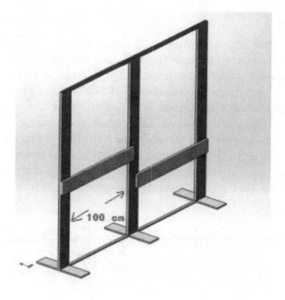

Fig. 4. The double arch.

bar (1.5 cm heights) at the bottom of each door. Robots go through a door may take a risk of being blocked by the bar, which is a trick to test their traversable ability.

(a)   (b)

Fig. 5. Competition area

The MoCap system and four HD video cameras were installed for the competition, covering the whole competition area (showed in Fig. 5), by which robots' movement data and videos were recorded, in real time, from beginning to end.

4.2   Stage I Test

In stage I, robots are allowed to use only odometry as sensor. The robots are required to do two separate actions (moving in a straight line and turning at a given spot) under each payload condition: empty, 10 kg, 20 kg (showed in Fig. 1). Based on the feedback from the odometry of the robot and the measurement data collected by the MoCap system, the accuracy of the robot's movement for performing the tasks is computed. Each team is encouraged to try an extra payload once, which must exceed the maximum routine at least 20 kg, by given a bonus score. According to the movement errors measured by the MoCap system, each robot performance can be evaluated by being compared to the minimal error among all the teams under the same payload condition (Fig. 6).

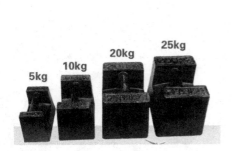

Fig. 6. Loads

Fig. 7. Obstacles

A final score for each team is computed by normalizing the scores of their performance under different load conditions with different score weights (showed in Table 1).

Table 1. The score weight under different load conditions.

Load	0 kg	10 kg	20 kg	≥40 kg
Score weight	0.2	0.3	0.5	0.2 (bonus)

4.3   Stage II Test

In stage II, a robot is allowed to use sensors besides the odometry to build a global map of the field before test. The map will also be used for evaluating the robot's performance. In the competition area, the robot is required to reach 7 way-points in the correct order (specified by the referee) under each payload condition: empty, 10 kg, 20 kg. The robot trajectory is recorded, and the distance between each way-point and robot stop point is measured by the MoCap system automatically. Before each team test, obstacles (showed in Fig. 7) in the field and sliders on the arches are rearranged. A team gets punished when the robot colliding with obstacles or facilities in the field, and rewarded when the robot successfully passing through an arch.

The task in the final test is similar to that of stage II, but is more difficult, by adding more obstacles into the field and decreasing the maximum acceptable distance error.

5   Analysis of Team Performance

There were 10 qualified teams (partly showed in Fig. 8) participated in the BSR challenge. All the teams completed stage I test. Moreover, 7 teams could bear the payload of 40 kg. Table 2 shows the average and minimal motion errors. According to Table 2, the average motion errors (both distance and direction error) increases with the weight of the payload, which indicates that the payload has effect on the motion accuracy. Additionally, from the Table 2, we can see that some teams could achieve quite small motion errors under different payload conditions.

Fig. 8. Participating teams and robots

Table 2. Average and the minimal motion error in stage I

Load		0 kg	10 kg	20 kg
Distances error (mm per 1 m)	Average	3.7	4.96	9.88
	Minimal	0.5	0.25	0.25
	Best team	0.5	0.32	0.25
Direction error (degree per a round)	Average	7.18	7.97	8.31
	Minimal	0.35	0.47	0.42
	Best team	0.35	0.47	0.42

Since tests in stage II and final involved the localization and navigation abilities, perception sensors had great influence on the robot performance. Being limited by the cost restriction (1, 600 USD), robots can't be equipped with high price LRFs (e.g. Sick LMS100, HOKUYO URG-04LX, etc.). RPLIDARs[3], a kind of low-cost 360 degree LRF, and kinects were commonly used among teams. According to the sensor configurations, teams can be divided into three categories: teams with a low-cost LRF, teams with a kinect, and teams with a low-cost LRF and a kinect. As tasks in stage II and the final test were the same, their results are combined and analyzed according to the sensor configurations.

Table 3 presents the statistical results of stage II and the final test. From this table, it is evident that teams with a low-cost LRF got smaller motion errors and fewer collisions than teams only equipped with a kinect. This is because that the kinect provide depth data only in a limited distance interval (typically from 0.3 m to 5 m), however, the low-cost LRF can offer better observations in 2d point cloud. The Passing Door column of Table 3 shows that only teams equipped with both a low-cost LRF and a kinect could successfully make passing door actions.

Table 3. Statistical result in stage II and the final test

Sensor configuration	Number of teams	Average error (m)	Minimale error (m)	Collision (number of times)	Passing door (number of times)
Low-cost LRF	4	0.26	0.08	4	0
Kinect	3	0.47	0.29	7	0
Low-cost LRF Kinect	3	0.24	0.08	3	4

6   Discussion and Future Work

Our goal is to establish a set of synthetical benchmarks for DMPs, as a matter of fact, the BSR challenge had some limitations. Although, we proposed a set of key features of DMPs, these features can't cover every aspect of service robot benchmarks. In the future, we are going to broaden the scope of key features of DMPs, allowing more features (e.g. moving velocity, battery capacity) being evaluated. Moreover, the BSR challenge only evaluated the motion accuracy of a robot. More aspects such as time consumption will be included in the benchmarking scope, impelling teams to make trade-offs in these performance factors.

A comprehensive service robot benchmarking system contains three different levels: feature/ability benchmarking, subsystem benchmarking and system benchmarking. As an integrated system, the overall performance of a service robot not only depends on the performance of each single feature/ability, but also depends on the integration of single feathers/abilities and subsystems. But, only the feature/ability benchmarking was involved in the BSR challenge. More effort will be devoted to subsystem and overall system benchmarking.

---

[3] http://www.robopeak.com/.

The BSR challenge was a combination of benchmarking test and competition. However, ranking-oriented property of competition is a significant disadvantage for benchmarking. Attracted by the ranking, teams may develop solutions that converge to "local optimum" performance, by exploiting the vulnerability of rules. In the future, efforts need to be made both on organization and rules changes to overcome this drawback.

7   Conclusion

Robotic competitions play an important role in benchmarking robot systems, and hence provide a basis for this effort. However, most of existing bench-marking tools are qualitative, while in many cases quantitative evaluation is needed. Synthetical benchmarking consists of both qualitative and quantitative aspects, such as task completion, accuracy of task completions and efficiency of task completions, about performance of a robot. This paper presents our idea of introducing synthetical benchmarking into evaluation of service robots and a first realization of our synthetical benchmarking system on domestic mobile platforms. The system includes a set of algorithms for collecting, recording and analyzing measurement data from a MoCap system. We used the system as the evaluator in the BSR challenge, in which 10 teams participated. The competition was organized mainly as a comparative study on performance evaluation of domestic mobile platforms. An analysis of teams' performance is also given in the paper. Observations and future directions are made from the analysis.

Acknowledgments. A special thank is given to Intel China for its sponsorship of the BSR challenge and workshop. The authors from USTC are supported by the Natural Science Foundation of China under grants 60745002 and 61175057, as well as the USTC Key Direction Project. All authors are thankful for contributions from all collaborators who are not an author of the paper and all participants in the event.

References

1. del Pobil, A.P.: Why do we need benchmarks in robotics research? In: Proceedings of the Workshop on Benchmarks in Robotics Research, IEEE/RSJ International Conference on Intelligent Robots and Systems (2006)
2. Nardi, L., Bodin, B., Zia, M.Z., Mawer, J., Nisbet, A., Kelly, P.H., Davison, A.J., Luján, M., O'Boyle, M.F., Riley, G., et al.: Introducing slambench, a performance and accuracy benchmarking methodology for slam. arXiv preprint arXiv:1410.2167 (2014)
3. Fontana, G., Matteucci, M., Sorrenti, D.G.: Rawseeds: building a benchmarking toolkit for autonomous robotics. In: Amigoni, F., Schiaffonati, V. (eds.) Methods and Experimental Techniques in Computer Engineering, pp. 55–68. Springer, New York (2014)
4. Behnke, S.: Robot competitions-ideal benchmarks for robotics research. In: Proceedings of IROS-2006 Workshop on Benchmarks in Robotics Research (2006)

5. Schultz, A.C.: The 2000 AAAI mobile robot competition and exhibition. AI Mag. **22**(1), 67 (2001)
6. Maxwell, B.A., Smart, W., Jacoff, A., Casper, J., Weiss, B., Scholtz, J., Yanco, H., Micire, M., Stroupe, A., Stormont, D., et al.: 2003 AAAI robot competition and exhibition. AI Mag. **25**(2), 68 (2004)
7. Balch, T., Yanco, H.: Ten years of the AAAI mobile robot competition and exhibition. AI Mag. **23**(1), 13 (2002)
8. Firby, R.J., Prokopowicz, P.N., Swain, M.J., Kahn, R.E., Franklin, D.: Programming CHIP for the IJCAI-95 robot competition. AI Mag. **17**(1), 71 (1996)
9. Pratt, G., Manzo, J.: The DARPA robotics challenge [competitions]. IEEE Rob. Autom. Mag. **20**(2), 10–12 (2013)
10. Wisspeintner, T., Van Der Zant, T., Iocchi, L., Schiffer, S.: Robocup@ home: scientific competition and benchmarking for domestic service robots. Interact. Stud. **10**(3), 392–426 (2009)
11. Wisspeintner, T., van der Zan, T., Iocchi, L., Schiffer, S.: RoboCup@Home: results in benchmarking domestic service robots. In: Baltes, J., Lagoudakis, M.G., Naruse, T., Ghidary, S.S. (eds.) RoboCup 2009. LNCS, vol. 5949, pp. 390–401. Springer, Heidelberg (2010)
12. Holz, D., Iocchi, L., van der Zant, T.: Benchmarking intelligent service robots through scientific competitions: the Robocup@ home approach. In: AAAI Spring Symposium: Designing Intelligent Robots (2013)
13. Stuckler, J., Holz, D., Behnke, S.: Demonstrating everyday manipulation skills in Robocup@ home. IEEE Rob. Autom. Mag. **19**(2), 34–42 (2012)
14. Amigoni, F., Bonarini, A., Fontana, G., Matteucci, M., Schiaffonati, V.: Benchmarking through competitions. In: European Robotics Forum-Workshop on Robot Competitions: Benchmarking, Technology Transfer, and Education (2013)
15. Ahmad, A., Awaad, I., Amigoni, F., Berghofer, J., Bischoff, R., Bonarini, A., Dwiputra, R., Fontana, G., Hegger, F., Hochgeschwender, N., et al.: Specification of general features of scenarios and robots for benchmarking through competitions. RoCKIn Deliverable D 1 (2013)
16. Corazza, S., Muendermann, L., Chaudhari, A., Demattio, T., Cobelli, C., Andriacchi, T.P.: A markerless motion capture system to study musculoskeletal biomechanics: visual hull and simulated annealing approach. Ann. Biomed. Eng. **34**(6), 1019–1029 (2006)
17. Kurihara, K., Hoshino, S., Yamane, K., Nakamura, Y.: Optical motion capture system with pan-tilt camera tracking and realtime data processing. In: ICRA, pp. 1241–1248 (2002)
18. Field, M., Stirling, D., Naghdy, F., Pan, Z.: Motion capture in robotics review. In: 2009 IEEE International Conference on Control and Automation. ICCA 2009, pp. 1697–1702, December 2009

Author Index

Printed in the United States
by Baker & Taylor Publisher Services